Protein Engineering

The Practical Approach Series

SERIES EDITORS

D. RICKWOOD
Department of Biology, University of Essex
Wivenhoe Park, Colchester, Essex CO4 3SQ, UK

B. D. HAMES
Department of Biochemistry and Molecular Biology,
University of Leeds, Leeds LS2 9JT, UK

Affinity Chromatography
Anaerobic Microbiology
Animal Cell Culture (2nd Edition)
Animal Virus Pathogenesis
Antibodies I and II
Biochemical Toxicology
Biological Membranes
Biomechanics—Materials
Biomechanics—Structures and Systems
Biosensors
Carbohydrate Analysis
Cell–Cell Interactions
Cell Growth and Division
Cellular Calcium
Cellular Neurobiology
Centrifugation (2nd Edition)
Clinical Immunology
Computers in Microbiology
Crystallization of Nucleic Acids and Proteins
Cytokines
The Cytoskeleton
Diagnostic Molecular Pathology I and II
Directed Mutagenesis
DNA Cloning I, II, and III
Drosophila
Electron Microscopy in Biology
Electron Microscopy in Molecular Biology
Enzyme Assays
Essential Molecular Biology I and II
Experimental Neuroanatomy
Fermentation
Flow Cytometry
Gel Electrophoresis of Nucleic Acids (2nd Edition)
Gel Electrophoresis of Proteins (2nd Edition)
Genome Analysis
Haemopoiesis
HPLC of Macromolecules
HPLC of Small Molecules
Human Cytogenetics I and II (2nd Edition)
Human Genetic Diseases
Immobilised Cells and Enzymes
In Situ Hybridization

Protein Engineering

A Practical Approach

Edited by

ANTHONY R. REES

Department of Biochemistry
University of Bath, UK

MICHAEL J. E. STERNBERG

Imperial Cancer Research Fund
Lincoln's Inn Fields, London, UK

and

RONALD WETZEL

Department of Macromolecular Sciences
SmithKline Beecham, King of Prussia, Pennsylvania, USA

OXFORD UNIVERSITY PRESS
Oxford New York Tokyo

Oxford University Press, Walton Street, Oxford OX2 6DP

Oxford   New York   Toronto
Delhi   Bombay   Calcutta   Madras   Karachi
Kuala Lumpur   Singapore   Hong Kong   Tokyo
Nairobi   Dar es Salaam   Cape Town
Melbourne   Auckland   Madrid
and associated companies in
Berlin   Ibadan

Oxford is a trade mark of Oxford University Press

A Practical Approach 🛈 is a registered trade mark
of the Chancellor, Masters, and Scholars of the University of Oxford
trading as Oxford University Press

Published in the United States
by Oxford University Press Inc., New York

A catalogue record for this book is available from the British Library

Library of Congress Cataloging in Publication Data
(Data available on request)
ISBN 0–19–963139–5 (hbk)
ISBN 0–19–963138–7 (pbk)

Typeset by Footnote Graphics, Warminster, Wilts
Printed in Great Britain by
Information Press, Eynsham, Oxon

Preface

Protein engineering is a curious yet fascinating mixture of molecular biology, protein structural studies, computation, and biochemistry. Straddling so many disciplines, it nevertheless has a form and style of its own. This uniqueness has its origins in the development of mutagenesis methods in the late 1970s that allow site-specific mutations to be incorporated into the gene for any protein. Thus, in principle, the role of every amino acid in a protein can be examined by mutating it in a conservative or even non-conservative fashion, provided the mutant protein can be produced in a suitable expression system.

While the methods for introduction of mutations have now become routine (see the companion volume *Directed Mutagenesis: A Practical Approach*), the interpretation of the effects of these mutations on the structure or activity of the target protein is less trivial. Thus, a large number of the more recent developments have addressed this aspect of the field.

In *Protein Engineering: A Practical Approach* the emphasis in Part I is heavily on methods that establish a sound structural basis for any protein engineering experiment and that allow the experimenter to both design the optimum strategy for mutation of a protein and computationally analyse the likely perturbation that mutation will impose on the structure. Stability measurements, described in Part II, are one means of assessing how well such designed alterations are accommodated by a protein. Modification of protein stability is also a major practical goal in the field. In addition, protein stability *in vivo* plays a major role in determining expression efficiency.

One of the most critical aspects of protein engineering is the problem of expression of mutant proteins in a heterologous host. In Part III expression systems encompassing mammalian, yeast, and bacterial hosts are described but should be seen more as a guide than as an exhaustive description of all the available expertise in the field.

Protein Engineering: A Practical Approach is the first practical text of its kind that attempts to lay a thorough theoretical foundation for the rational modification of proteins and at the same time addresses some of the more problematical experimental areas. We hope it will prove valuable to all research workers in the area, from graduate students to senior investigators.

Bath   A.R.R.
London   M.J.E.S.
King of Prussia   R.W.
June 1992

Contents

Contents

Contents

Contents

PART II   PROTEIN STABILITY AND
PHYSICO-CHEMICAL ANALYSIS OF MUTANTS

7. Principles of protein stability. Part 1— reversible unfolding of proteins: kinetic and thermodynamic analysis   167

*Bryan E. Finn, Xiaowu Chen, Patricia A. Jennings,
Susanne M. Saalau-Bethell, and C. Robert Matthews*

8. Principles of protein stability. Part 2— enhanced folding and stabilization of proteins by suppression of aggregation in vitro and in vivo   191

Ronald Wetzel

Contents

11. Protein engineering of antibody combining sites   253

Kate L. Hilyard, David Staunton, Alison E. Jones, and Anthony R. Rees

12. Selection of variants of antibodies and other protein molecules using display on the surface of bacteriophage fd   277

Ronald H. Jackson, John McCafferty, Kevin S. Johnson, Anthony R. Pope, Andrew J. Roberts, David J. Chiswell, Timothy P. Clackson, Andrew D. Griffiths, Hennie R. Hoogenboom, and Greg Winter

Contents

13. Expression of proteins in prokaryotic systems—principles and case studies

Geoffrey T. Yarranton and Andrew Mountain

Contents

Contributors

ANDREAS BACHMAIR
Institut für Botanik, Universität Wien, Rennweg 14, A-1030 Wien, Austria.

PAUL A. BATES
Biomolecular Modelling Laboratory, Imperial Cancer Research Fund, PO Box 123, 44, Lincoln's Inn Fields, London WC2A 3PX, UK.

RAINER BISCHOFF
Transgene S.A., 11, rue de Molsheim, 67000 Strasbourg, France.

MARK S. BOGUSKI
National Centre for Biotechnology Information, National Library of Medicine, National Institutes of Health, Building 38A, Room 8N-805, Bethesda, MD 20894, USA.

XIAOWU CHEN
Department of Chemistry, Penn State University, 152 Davey Laboratory, University Park, PA 16802, USA.

DAVID J. CHISWELL
Cambridge Antibody Technology Ltd, Science Park, Melbourn, Cambridgeshire SG8 6EJ, UK.

TIMOTHY P. CLACKSON
MRC Laboratory of Molecular Biology, Cambridge Centre for Protein Engineering, Hills Road, Cambridge CB2 2QH, UK.
Present address: Genentech Inc., 460, Point San Bruno Blvd, South San Francisco, CA 94080, USA.

G. MARIUS CLORE
Laboratory of Chemical Physics, Building 2, Room 123, National Institute of Diabetes and Digestive and Kidney Diseases, National Institutes of Health, Bethesda, MD 20892, USA.

MICHAEL COURTNEY
Transgene S.A., 11, rue de Molsheim, 67000 Strasbourg, France.

VALERIE DAGGETT
Department of Cell Biology, Stanford University Medical School, Stanford, CA 94305, USA.

BRYAN E. FINN
Department of Physical Chemistry II, Chemical Center, University of Lund, S-221 00 Lund, Sweden.

ANDREW D. GRIFFITHS
MRC Laboratory of Molecular Biology, Cambridge Centre for Protein Engineering, Hills Road, Cambridge CB2 2QH, UK.

ANGELA M. GRONENBORN
Laboratory of Chemical Physics, Building 2, Room 123, National Institute of Diabetes and Digestive and Kidney Diseases, National Institutes of Health, Bethesda, MD 20892, USA.

KATE L. HILYARD
Department of Biochemistry, University of Bath, Claverton Down, Bath BA2 7AY, UK.
Present address: Department of Biochemistry and Molecular Biology, Harvard University, 7, Divinity Avenue, Cambridge, MA 02138, USA.

HENNIE R. HOOGENBOOM
MRC Laboratory of Molecular Biology, Cambridge Centre for Protein Engineering, Hills Road, Cambridge CB2 2QH, UK.

MORIO IKEHARA
Protein Engineering Research Institute, 6-2-3, Furuedai, Suita, Osaka 565, Japan.

RONALD H. JACKSON
Cambridge Antibody Technology Ltd, Science Park, Melbourn, Cambridge-shire SG8 6EJ, UK.

PATRICIA A. JENNINGS
Department of Chemistry, Penn State University, 152 Davey Laboratory, University Park, PA 16802, USA.

STEFAN JENTSCH
Friedrich-Miescher-Laboratorium der Max-Planck-Gesellschaft, Biologische Arbeitsgruppen, Spemannstr. 37–39, Postfach 2109, D-7400 Tübingen 1, Germany.

KEVIN S. JOHNSON
Cambridge Antibody Technology Ltd, Science Park, Melbourn, Cambridge-shire SG8 6EJ, UK.

ALISON E. JONES
Department of Biochemistry, University of Bath, Claverton Down, Bath BA2 7AY, UK.

E. YVONNE JONES
Laboratory of Molecular Biophysics, The Rex Richards Building, South Parks Road, Oxford OX1 3QU, UK.

MASAKAZU KIKUCHI
Protein Engineering Research Institute, 6-2-3, Furuedai, Suita, Osaka 565, Japan.

PETER A. KOLLMAN
Department of Pharmaceutical Chemistry, University of California at San Francisco, San Francisco, California 94143, USA.

MICHAEL KRIEGLER
Department of Cell and Molecular Biology, Cytel Corporation, 3525 John Hopkins Court, San Diego, CA 92121, USA.

JOHN McCAFFERTY
Cambridge Antibody Technology Ltd, Science Park, Melbourn, Cambridgeshire SG8 6EJ, UK.

C. ROBERT MATTHEWS
Department of Chemistry, Penn State University, 152 Davey Laboratory, University Park, PA 16802, USA.

ANDREW MOUNTAIN
Celltech plc, 216 Bath Road, Slough SL1 4EN, UK.

ROBIN E. OFFORD
Department of Medical Biochemistry, Centre Medical Universitaire, 1 Michel Servet, 1211 Geneva 4, Switzerland.

JAMES OSTELL
National Center for Biotechnology Information, National Library of Medicine, National Institutes of Health, Building 38A, Room 8N-805, Bethesda, MD 20894, USA.

ANTHONY R. POPE
Cambridge Antibody Technology Ltd, Science Park, Melbourn, Cambridgeshire SG8 6EJ, UK.

ANTHONY R. REES
Department of Biochemistry, University of Bath, Claverton Down, Bath BA2 7AY, UK.

ANDREW J. ROBERTS
Cambridge Antibody Technology Ltd, Science Park, Melbourn, Cambridgeshire SG8 6EJ, UK.

SUSANNE M. SAALAU-BETHELL
Department of Chemistry, Penn State University, 152 Davey Laboratory, University Park, PA 16802, USA.

CHRIS SANDER
European Molecular Biology Laboratory, Postfach 10.2209, Meyerhofstrasse, 1, 6900 Heidelberg, Germany.

MICHAEL SCHARF
European Molecular Biology Laboratory, Postfach 10.2209, Meyerhofstrasse, 1, 6900 Heidelberg, Germany.

REINHARD SCHNEIDER
European Molecular Biology Laboratory, Postfach 10.2209, Meyerhofstrasse, 1, 6900 Heidelberg, Germany.

DENIS SPECK
Transgene S.A., 11, rue de Molsheim, 67000 Strasbourg, France.

DAVID J. STATES
National Center for Biotechnology Information, National Library of Medicine, National Institutes of Health, Building 38A, Room 8N-805, Bethesda, MD 20894, USA.

DAVID STAUNTON
Department of Biochemistry, University of Bath, Claverton Down, Bath BA2 7AY, UK.

MICHAEL J. E. STERNBERG
Biomolecular Modelling Laboratory, Imperial Cancer Research Fund, PO Box 123, 44, Lincoln's Inn Fields, London WC24 3PX, UK.

DAVID I. STUART
Laboratory of Molecular Biophysics, The Rex Richards Building, South Parks Road, Oxford OX1 3QU, UK.

RONALD WETZEL
Department of Macromolecular Sciences, SmithKline Beecham, King of Prussia, PA 19406-0939, USA.

GREG WINTER
MRC Laboratory of Molecular Biology, Cambridge Centre for Protein Engineering, Hills Road, Cambridge CB2 2QH, UK.

GEOFFREY T. YARRANTON
Celltech plc, 216 Bath Road, Slough SL1 4EN, UK.

Abbreviations

AAT	α1-antitrypsin
ACS	antibody combining site
apo-D	human apolipoprotein D
APS	ammonium persulphate
ASN.1	Abstract Syntax Notation 1
ATP	adenosine triphosphate
a.u.	absorbance unit
bGH	bovine growth hormone
BLAST	Basic Local Alignment Search Tool
Boc	t-butyloxycarbonyl
BSA	bovine serum albumin
CDR	complementarity determining region
COSY	correlated spectroscopy
c.p.m.	counts per minute
1D	one-dimensional
2D	two-dimensional
3D	three-dimensional
4D	four-dimensional
DANTE	delays alternating with nutation for tailored excitation
DEPC	diethylpyrocarbonate
DHFR	dihydrofolate reductase
DNase	deoxyribonuclease
DO	dissolved oxygen
DQF-COSY	double quantum filtered correlated spectroscopy
DTT	dithiothreitol
DW	dry weight
ECL	enhanced chemiluminescence
E.COSY	enhanced correlated spectroscopy
EDTA	ethylenediaminetetraacetate
ELISA	enzyme-linked immunosorbent assay
EMP	ethylmercury phosphate
ER	endoplasmic reticulum
EtOH	ethanol
FCS	fetal calf serum
FID	free induction decay
FMDV	foot-and-mouth disease virus
GAM	goat anti-mouse polyclonal antibodies
GIBB	GenInfo Backbone
HBM	high biomass medium

HBS	HEPES-buffered saline
HEL	hen egg lysozyme
HEPES	N-2-hydroxyethylpiperazine-N′-2-ethane sulphonic acid
HIV	human immunodeficiency virus
HMQC	heteronuclear multiple quantum coherence spectroscopy
HOHAHA	homonuclear Hartmann–Hahn spectroscopy
HPLC	high performance liquid chromatography
HRP	horseradish peroxidase
hsp	heat-shock protein
IB	inclusion body
ICAM-1	human intracellular adhesion molecule 1
IgG	immunoglobulin G
IPTG	isopropyl-β-D-thiogalactoside
KNTase	kanamycin nucleotidyltransferase (KNTase)
LB	Luria broth
LFA-1	lymphocyte function associated antigen
LTR	long terminal repeat
MACAW	multiple alignment construction and analysis workbench
MAD	multiple wavelength anomalous dispersion
MBS	Barth's saline
MIR	multiple isomorphous replacement
mol. wt	molecular weight
MPD	2-methyl-2,4-pentanediol
MR	molecular replacement
Msc	methanesulphonylethyloxycarbonyl
NCBI	National Center for Biotechnology Information
NCS	non-crystallographic symmetry
NLM	US National Library of Medicine
NMR	nuclear magnetic resonance
NMW	nominal molecular weight
NOE	nuclear Overhauser effect
NOESY	nuclear Overhauser enhancement spectroscopy
OD	optical density
OSI	open systems interconnection
PAGE	polyacrylamide gel electrophoresis
PAM	point acceptable mutations
PBS	phosphate-buffered saline
PCMBS	*p*-chloromercuri benzene sulphonate
P.COSY	primitive correlated spectroscopy
PCR	polymerase chain reaction
PDI	protein disulphide isomerase
PE.COSY	primitive exclusive correlated spectroscopy
PEG	polyethylene glycol
p.f.u.	plaque-forming units

Abbreviations

phOx	2-phenyl-5-oxazolone
phOx-CAP	4-ε-aminocaproic acid methylene-2-phenyl-5-oxazolone
PMSF	phenylmethylsulphonyl fluoride
PPb	glycogen phosphorylase b
PPIase	peptidyl-prolyl *cis-trans* isomerase
p.p.m.	parts per million
PVDF	polyvinyldifluoridene
rf	radiofrequency
RF DNA	replicating form DNA
RID	radial immunodiffusion
r.m.s.	root-mean-square
r.p.m.	revolutions per minute
SCR	structurally conserved region
SD	Shine–Dalgarno sequence
SDM	site-directed mutagenesis
SDS	sodium dodecyl sulphate
SIR	single isomorphous replacement
SL	spin lock pulse
SQL	Structured Query Language
STS	sequence tagged sites
TAMM	tetrakis(acetoxymercuri)methane
TBS	Tris-buffered saline
TCP/IP	transmission control protocol/internet protocol
TE	Tris–EDTA
TEMED	N,N,N′,N′-tetramethylethylenediamine
TGS	Tris–glycine SDS electrophoresis buffer
TIM	triosephosphate isomerase
TLC	thin layer chromatography
TNF	tumour necrosis factor
TOCSY	total correlated spectroscopy
tPA	tissue plasminogen activator
TPCK	1-chloro-4-phenyl-3-tosylamidobutane-2-one
TPPI	time proportional phase incrementation
Tris	tris(hydroxymethyl)-aminomethane
UAS	upstream activator sequence
UV	ultraviolet
UV/VIS	ultraviolet/visible
VR	variable region

PART I

Protein structure determination and prediction

1

Methods of structural analysis of proteins. Part 1—protein crystallography

E. YVONNE JONES and DAVID I. STUART

1. Introduction

X-ray crystallography provides a method by which the detailed three-dimensional structure of a protein may be revealed at or near to atomic resolution. A well-refined protein structure, at a resolution of 3 Å or better, provides a wealth of information: on specific atom positions, their interactions (hydrogen bonds, etc.), solvent accessibility, and variations in flexibility/mobility within the molecule. The constraints imposed on the protein molecule by its packing within the crystal may produce some slight perturbations in the conformation of certain flexible regions, but comparison of crystal (solved by X-ray crystallography) and solution (solved by nuclear magnetic resonance) structures of the same protein, as well as studies of the same protein in different crystal environments, clearly indicate that this is usually a minor effect. Many enzymes remain active when crystallized, and time-resolved structural studies of enzymic mechanisms will be touched on in this chapter.

Over the past few years there has been rapid development in many aspects of protein crystallography, often paralleling improvements in X-ray sources, X-ray detectors, and computers. These methodological advances have been spurred on by the wealth of systems now available for structural studies through the auspices of molecular biology. The technique has been employed over a spectrum of sizes ranging from peptides of about a dozen amino acids and short oligonucleotides to entire viruses with mass of the order of 10^7 daltons; the one essential prerequisite is the ability to produce highly ordered crystals of the chosen system.

Since we cannot hope to provide a comprehensive set of recipes for solving protein structures by X-ray crystallography, this chapter contains a personal and didactic set of ideas and prescriptions, all of which we would not follow all of the time. For fuller information, ref. 1 is probably the single most substantial, but rather disparate, collection of information available.

2. Overview

The fundamental principles of protein crystallography are covered in refs 2–4. X-ray crystallography involves the illumination of a crystal with electro-magnetic radiation of a wavelength comparable to the detail to be resolved (about 1 Å). A crystal comprises a well-ordered array of many billions of copies of a basic object (e.g. a protein molecule); the pattern of X-ray scattering by the electrons of the individual atoms in the repeated unit is additive over all the copies of the unit which form the crystal. The scattered radiation may be recombined to form an image of the electron density in the repeat unit (i.e. the protein). However, this requires a knowledge of both the amplitudes of the scattered beams (directly measurable) and their relative phases (not directly measurable, hence the phase problem). The key stages of a protein structure determination may be defined as follows.

(a) Crystallization. Precise conditions must be established under which the weak intermolecular forces between protein molecules produce highly ordered packing as crystals rather than random aggregation as a precipitate. The procedure depends on achieving

   i. supersaturation of the protein in solution
   ii. nucleation at a few sites for crystal formation
   iii. sustained growth of the crystals

(b) Data collection. In general, data are collected by rotating the protein crystal within the (single wavelength) X-ray beam while the intensities of the diffracted beams are recorded. These intensities yield the amplitudes of the structure factors. Alternatively, in the Laue method, the crystal usually remains stationary but data are collected using a continuous spectrum of X-ray wavelengths.

(c) Phasing. The amplitude and phase angle of a structure factor are physically independent quantities. Since all information on the relative phases of the structure factors is lost on collection of the data, an estimate of their values must be obtained via some form of additional information. This stage, with the exception of crystallization, remains potentially the most problematic part of a structure determination.

(d) Phase improvement. The quality, and hence interpretability, of the electron density map is critically dependent on the accuracy of the phases. Under certain favourable circumstances poor initial estimates of the phases may be greatly improved by making use of oversampling inherent in the data.

(e) Map interpretation. Successful phasing of the diffraction data allows an image of the electron density in the repeat unit of the crystal to be calculated. If this image is sufficiently detailed (i.e. based on high enough resolution data, which usually means diffraction data to a resolution of at least 3.5 Å) then the three-dimensional fold of the polypeptide chain may be followed. A model of the protein structure in terms of atomic coordinates may thus be constructed, given a knowledge of the primary sequence.

(f) Refinement. The initial molecular model is refined against the X-ray observations under the restraint of established stereochemical data (bond lengths, angles, etc.).

We will now highlight some of the key practical aspects inherent in each of these stages. Obviously, protein crystallography is a vast, highly specialized field, thus the following treatment is limited to an attempt to lay down sound but necessarily general guidelines and further detailed references should be consulted for each stage. A glossary of many of the technical terms we use is included at the end of the book. An assortment of addresses for programs, organizations, and data-bases is given in Section 9. The addresses of specialist suppliers are given in the appendix at the end of the book.

3. Crystallization

General reviews of protein crystallization techniques are provided by refs 5–7. Milligram quantities of pure (usually at least 95% pure), homogeneous protein are required. Preliminary trials may usefully commence given 1–2 mg, but a full survey of crystallization parameters and, hopefully, the eventual structure solution could require anything from a few milligrams to gram quantities of the protein. General points are:

- take note of information from the molecular biologists/protein chemists on factors influencing the stability and conformation of the protein in solution
- avoid repeated thawing and freezing of the stock of protein; lyophilization has proved deleterious to the subsequent crystallization of several proteins and is best avoided
- use protein solutions in the range 5–50 mg/ml or higher, generally the more concentrated the better, at least to start with
- limit the number of nucleation sites by filtering the protein solution, buffer solutions, etc., and also by cleaning any glass or plastic surfaces (air dusters as supplied in photographic shops for lens-cleaning work well)

The standard method for initial surveys is hanging-drop vapour diffusion (*Figure 1a*) since it is sparing of material and efficient in use of storage space. In this method any volatile components will equilibrate between a small volume drop and a larger volume reservoir. If water is the only component with a substantial vapour pressure, then the usual procedure is to have a precipitant present in the protein drop and also (at higher concentration) in the reservoir. Equilibration dehydrates and supersaturates the protein drop, increasing both the precipitant and protein concentrations. The basic requirements are a Linbro tissue culture multi-well plate (Flow Laboratories Inc.)

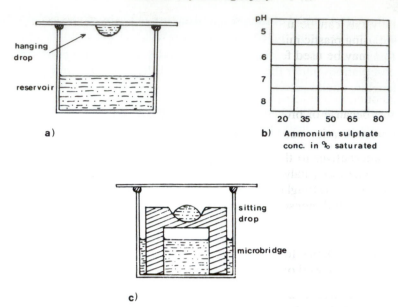

Figure 1. Crystallization methods: (a) hanging drop; (b) example of initial coarse survey of conditions; (c) sitting drop.

and glass (siliconized by treatment with dimethyldichlorosilane solution) or plastic cover slips.

Protocol 1. Hanging-drop vapour diffusion

1. Centrifuge protein to remove nucleation sites, precipitated protein, etc.
2. Make up 1 ml reservoirs containing buffer, precipitant, and additives in wells of tissue culture plate; grease rims of wells with vacuum grease.
3. Pipette 1 or 2 μl of reservoir on to cover slip and add 1 or 2 μl of protein to this drop (the ratio of these volumes will determine the protein concentration after equilibration).
4. Invert cover slip over well and check that seal is airtight.

Steps **2** and **3** in the above protocol are now commonly automated (8). The initial coarse survey (e.g. *Figure 1b*) is usually repeated over a finer grid in regions of interest (normally concentrations just below those which precipitate the protein). Generally, for protein crystal growth, one needs to be within 1 or 2% of the ideal concentration for a precipitant such as ammonium sulphate, but the requirements for polyethylene glycol (PEG) can be more flexible. Crystals appear over time-scales ranging from hours to months.

Microcrystals can be used for seeding fresh drops (9). A variation of this method using plastic microbridges (*Figure 1c*) (available from Crystal Micro-systems) may be used for sitting drops (3 μl plus 3 μl or greater) where:

- non-ionic detergents are used as additives (low surface tension)
- PEG or ethanol are the precipitating agents (hanging drops tend to increase in size because of problems with condensation)
- the crystallization conditions are established and need to be scaled up

Useful alternatives to these vapour diffusion techniques to bear in mind are the microdialysis (dialysis buttons from Cambridge Repetition Engineers) and microbatch (Douglas Instruments Ltd) methods.

With its multidimensional variables, crystallization remains an art, *Table 1* lists some of the currently fashionable mix-and-match alternatives. Ammonium sulphate, 2-methyl-2,4-pentanediol (MPD), and PEG 4000 are usually the precipitants of choice for initial attempts at crystallization; some examples of search schemes based on these precipitants are given in ref. 7. Many options,

---

Table 1. Examples of things to try in crystallization trials

Precipitants
Salts: ammonium sulphate, ammonium formate, sodium citrate
PEG: 400, 3000, 8000, 20000
Organic solvents: MPD, ethanol (at 4°C; difficult to avoid evaporation)
Mixtures: PEG + 0.5–1 M LiCl or NaCl, salts + 2–4% organic solvent

Additives
0.25–1% non-ionic detergent, e.g. β-octyl glucoside
Dioxane
Metal ions, e.g. Ca^{2+}, Zn^{2+}
Reducing agents: DTT
Glycerol

Variables
pH
Buffer type
Temperature: commonly try 5°C and 20°C
Gravity ('crystallization in space', a rather expensive parameter)

Co-crystallization partners
Inhibitors
Co-factors
Substrates
Fabs

Protein variants
Protease digestion
N- or C-terminal truncation
Different species/muteins
Different expression system (glycosylation, etc.)

---

including most of those in *Table 1*, are covered in refs 5–7. The application of an incomplete factorial approach to select a subset of trial conditions has been suggested (10) and is now used routinely in several laboratories (Hampton Research). Light scattering procedures are currently being developed as a possible monitor of potentially favourable crystallization conditions and may eventually aid in rationalizing the search (11). For several problem cases co-crystallization with antibodies (Fab fragments) has been successfully employed when the protein alone failed to crystallize (12). Non-ionic detergents have considerably broadened the scope of the field by facilitating the crystallization of several intact membrane proteins (13, 14). Heavily glycosylated proteins may also require special consideration, one option is to use neuraminadase treatment to reduce the level of glycosylation. Very recently the inclusion of glycerol as a co-solvent has been suggested as an aid in the crystallization of flexible proteins (15); however, high intramolecular flexibility often necessitates the use of truncated forms of the protein for crystallization.

Having obtained crystals two key questions must be faced:

(a) Are they salt crystals? These are usually distinguished by very high birefringence, high density, physical resilience, and a 'hard glassy appearance' arising from the large difference in refractive index between salt crystal and solvent; a few very intense diffraction spots at high resolution (often nothing below about 6 Å) provide the final proof.

(b) Do they give usable diffraction? Sadly, bona fide protein crystals which appear macroscopically well ordered (even birefringent) may not be sufficiently well ordered microscopically to diffract X-rays, or may diffract to only low resolution (less than 4 Å) or be unusable (multiple, twinned, etc.).

In our experience, additives such as β-octyl glucoside may be of help in eliminating problems involving disorder (16).

By measurement of the density of a protein crystal, values may be calculated for the fraction of the crystal volume occupied by solvent and the number of protein molecules in the asymmetric unit (17). This measurement may be carried out using a Ficoll density gradient (18). For crystals with high solvent content, seepage of the Ficoll molecules into the crystal necessitates a series of measurements taken over a time course from which the correct density at time zero may be extrapolated (19).

Protocol 2. Determination of crystal density over a time course

1. Make up solutions covering a range of densities in 5% steps from 60 to 10% (w/w) of Ficoll 400 (supplied by Pharmacia).

2. Layer in an ultracentrifuge tube; heating the solutions increases their fluidity and thus may be of help at this stage.

3. Calibrate using xylene/bromobenzene mixtures of known densities in the range 1.05–1.25 g/cm^3. Add to the gradient as small droplets and spin in a swinging bucket rotor. Plot positions along the gradient.

4. Add a crystal to the gradient and spin for 30 sec, measure the position (and hence the density) and repeat over a time-course of several minutes.

5. Fit measurements to a simple exponential function and extrapolate back to crystal density at time zero.

Assuming a typical partial specific volume for proteins of 0.74 cm^3/g, the following formula may be applied

$$M_p[\text{Da}] = 2.324(V/n)[\text{Å}^3](\rho_c[\text{g/cm}^3] - 1)$$

where V is the unit cell volume, n the number of asymmetric units per unit cell, ρ_c the crystal density, and M_p the mass of protein per asymmetric unit.

4. Data collection

Well-ordered crystals with dimensions of $0.1 \times 0.1 \times 0.1$ mm^3 (or less) may well be sufficient for data collection at a synchrotron. Considerably larger crystals are normally required for data collection on in-house rotating anodes. The protein crystal is usually mounted in a glass (relatively fragile) or quartz capillary tube of appropriate diameter (Astrophysics Research Ltd) as illustrated in *Figure 2*; detailed notes on the manipulation of crystals are given in ref. 20. Fragile plate-like crystals may be mounted in specially flattened capillary tubes (20). Where, for example, infusion of substrate is planned during the course of the data collection the crystal may be mounted in a flow cell (21). Very recently a method has been suggested whereby crystals are mounted suspended in a thin film confined within a thin-wire loop by surface tension (22); this may have advantages, particularly for the case of flash freezing. In some cases it is vital that the crystal is strictly maintained at the temperature at which it was grown. However, for crystals grown at room temperature, cooling by a few degrees during data collection in the intense X-ray beam of a synchrotron may improve crystal lifetime (crystals grown from high salt can often be cooled to −10°C with no ill effects). Flash freezing of crystals to liquid nitrogen temperatures is currently advertised as a means of greatly prolonging crystal lifetime (23). Potential users must weigh up this benefit against the frequently encountered difficulties of implementing this procedure for their particular crystals. At least one laboratory has found the advantages sufficient to merit establishing suitable conditions (choice of cryoprotectant, etc.) for every new project.

Time-resolved structural studies aimed at probing enzymic mechanisms require the data collection to be fast in comparison to the speed of the reaction. For some systems, fairly standard modes of data collection may be

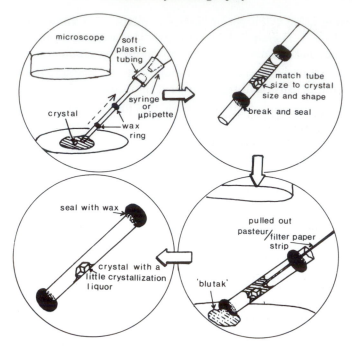

Figure 2. One method of mounting crystals.

sufficiently rapid (24), possibly facilitated by working at low temperature. Alternatively, very fast data collection using the Laue technique may be suitable (25). In general this is a particularly specialized area of data collection, thus no further discussion will be given here.

X-ray sources, detectors, and data collection methods are detailed in *Table 2*. Sealed tubes are the weakest X-ray sources and thus most current projects make use of a rotating anode or a synchrotron. In choosing the mode of data collection, the convenience of in-house data collection on a rotating anode must be balanced against the particular advantages of synchrotrons, namely:

(a) intense, highly collimated beam; essential for weakly diffracting crystals, very large unit cells, and weak high-resolution data

(b) short wavelength (less than 1 Å for some synchrotron X-ray beams compared to 1.54 Å from the traditional copper target on a rotating anode); gives reduced absorption effects and can be vital in enhancing the life times of certain crystals

(c) wavelength range and tunability; vital for Laue method of data collection and multiple wavelength anomalous dispersion (MAD) method of phasing

Alternatively, compared to copper, a gold rotating anode offers the advantage of X-rays with a shorter wavelength and, since characteristic gold emission

Table 2. Data collection

X-ray sources
Sealed tube
Rotating anode
Synchrotron

X-ray optics (in-house)
Nickel filter
Graphite monochromator
Franks mirrors

X-ray detectors
Linear proportional counters
Photographic film
Television area detectors with external phosphor: FAST
Multiwire proportional counters: Hamlin/Xuong, Xentronics
Imaging plates

Data collection methods
Diffractometry
Oscillation method
Laue (white radiation)

lines bracket the PtL_{III} absorption edge, allows some application of MAD methods in-house, as demonstrated by ref. 26. With the advent of detectors that work well at wavelengths below 1 Å it is time to reassess the relative merits of different anode materials. In-house data collection from crystals with very large unit cells (effective primitive cell greater than 200 Å) usually requires the focusing capabilities of Franks-type optics to give the required spot resolution (27). For the less demanding cases in-house we currently prefer the easier alignment and greater flux of an optical system based on a single, unbent, graphite monochromator and collimator. Again, with the new types of detectors becoming available, X-ray optics should be critically reassessed and the developments made at synchrotrons downloaded to the in-house systems.

At present the main choice of detector lies between photographic film, electronic area detectors, and imaging plates. Film remains a common choice at synchrotrons since it can match the data collection rate and provides sufficient spatial resolution to record the densely populated diffraction patterns from crystals with large unit cells at a moderate crystal to detector distance. Imaging plates offer high sensitivity and low background over a wide range of X-ray wavelengths, ideal for weakly diffracting crystals, high-resolution data, and the precise measurements at several wavelengths required for the MAD method. Given the potentially large active area of imaging plates and their respectable spatial resolution, these detectors can be expected to replace photographic film for most applications rapidly. The gas-filled multiwire and FAST electronic area detectors provide convenient and rapid data acquisition in-house. As a general rule, the bigger the useful area of the detector the better. This allows the crystal to detector distance to be increased, which is desirable both to improve the signal to noise ratio of the data and to ameliorate the limited spatial resolution of, in particular, many multiwire detectors. Note, however, that to get the most out of a large, high-resolution detector requires better X-ray optics than are currently used in-house.

Most common forms of data collection are now based on variations of the oscillation method (28). General points to consider are:

(a) Collimation should match beam size to the dimensions of the crystal or be smaller (note this is not the case for diffractometry).

(b) Crystal to detector distance must be sufficient to allow separation of diffraction spots and, having satisfied this criterion, is usually set to give the required resolution limit at the edge of the detector.

(c) Oscillation range:
   i. for electronic area detectors should be narrow (0.25° or less; as matches the typical rocking width of a reflection) to increase the signal to background ratio
   ii. for film is generally 1° or 2° (often as large as possible within the limits of increasing spot overlaps); reasons being, the labour of developing and scanning the films, and the loss of accuracy on addition of partials due to limitations in the data-processing algorithms and the inherent chemical fog
   iii. for imaging plates is currently similar to that used for film (to maintain a reasonable duty cycle). Indeed one, very successful, imaging plate installation employs the Weissenberg geometry to allow even greater angular ranges per image

(d) Length of exposure is governed by the balance between good counting statistics and limited crystal lifetime in the X-ray beam. Rapid data collection leading to a greater redundancy of data, although resulting in less impressive internal agreement between data, may provide a more complete data-set and reduce some systematic errors.

(e) Crystal orientation with respect to the oscillation axis is chosen classically to give most efficient data collection (i.e. minimum overall oscillation range for a full data-set); dependent on space group, unit cell dimensions, crystal morphology. However, as a routine data collection strategy on an in-house electronic area detector, it is often convenient to deliberately mis-set the crystal and collect over a total rotation range of 180°; this procedure yields complete data-sets but is not suitable when measurements of anomalous scattering are required (Bijvoet pairs should be collected on the same image).

12

When using the normal beam oscillation method, a cusp of data lying close to the rotation axis is, in principle, unmeasurable (for any one crystal orientation with respect to the oscillation axis). Note that:

(a) A short wavelength ameliorates this by 'thinning' the cusp.

(b) Symmetry in the reciprocal lattice often allows data related by symmetry to those in the cusp to be recorded if care is taken to align an evenfold rotation axis in the crystal to lie a few degrees away from the rotation axis (i.e. slightly mis-set the crystal).

The possible space groups for protein crystals are listed in *Table 3*. Crystal morphology very often gives some clue to the underlying symmetry and orientation of the crystallographic axes. Initial identification of space groups classically involves the use of a precession camera, but in our experience oscillation photographs at 0°, 90°, and, where appropriate, 30° or 45° about a crystallographic axis are usually sufficient to determine the unit cell dimensions and crystallographic point group (*Table 3*). Systematic absences indicating screw axes may be assessed from the final native data-set. We consider that the priority is to collect the data, not to waste potentially very valuable crystals characterizing the unit cell; the data may be processed in a lower symmetry space group and then checked for possible symmetry equivalents. Enantiomorphs (e.g. P3$_1$21 or P3$_2$21) cannot be distinguished until phase information is available, but this does not affect data collection strategies.

Data processing generally encompasses the following stages:

(a) indexing (assigning *hkl* to each diffraction spot); this usually now involves the use of autoindexing algorithms (29), crystal lifetime is not wasted in precise alignment to a known orientation while in the X-ray beam

(b) prediction of all recorded spot positions

(c) integration (now usually by profile fitting (30)); this involves measurement of the intensities at the predicted spot positions and estimation of appropriate background values for them

(d) scaling of the various observations (symmetry equivalents, etc.) to give a unique set, this allows some correction for crystal decay with time and variation in absorption effects for asymmetrically shaped crystals

Detector and source-dependent corrections will also be required at appropriate points in the above scheme. A specific correction for absorption effects may be appropriate in some cases (31). Post-refinement to give improved estimates of partiality for film (and imaging plate) data is useful, particularly in virus crystallography (32). Considerable care should be taken over the final scaling; typically, merging *R* factors on intensity defined as

$$R_{\text{merge}} = \frac{\Sigma_h \Sigma_j |I_{h,j} - <I_h>|}{\Sigma_h \Sigma_j <I_h>}$$

Table 3. Protein crystal space groups

System	Unit cell requirements	Point group	Space groups
Triclinic	none	1	P1
Monoclinic	$\alpha = \gamma = 90°$	2	P2, P2$_1$, C2
Orthorhombic	$\alpha = \beta = \gamma = 90°$	222	C222, P222, P2$_1$2$_1$2$_1$, P2$_1$2$_1$2, P222$_1$, C222$_1$, F222, I222[b], I2$_1$2$_1$2$_1$[b]
Tetragonal	$a = b$, $\alpha = \beta = \gamma = 90°$	4	P4 [P4$_1$, P4$_3$], P4$_2$, I4, I4$_1$
		422	P422, P42$_1$2, P4$_2$22, P4$_2$2$_1$2, [P4$_1$22, P4$_3$22], [P4$_1$2$_1$2, P4$_3$2$_1$2], I422, I4$_1$22
Trigonal	$a = b$, $\alpha = \beta = 90°$, $\gamma = 120°$	3[a]	P3, [P3$_1$, P3$_2$], R3
		32[a]	P321, [P3$_1$21, P3$_2$21], P312, [P3$_1$12, P3$_2$12], R32
Hexagonal	$a = b$, $\alpha = \beta = 90°$, $\gamma = 120°$	6	P6, [P6$_1$, P6$_5$], P6$_3$, [P6$_2$, P6$_4$]
		622	P622, [P6$_1$22, P6$_5$22], P6$_3$22, [P6$_2$22, P6$_4$22]
Cubic	$a = b = c$, $\alpha = \beta = \gamma = 90°$	23[a]	P23, P2$_1$3, F23, I23[b], I2$_1$3[b]
		432	P432, [P4$_1$32, P4$_3$32], P4$_2$32, F432, F4$_1$32, I432, I4$_1$32

[a] Space groups where symmetry of geometry of space group is greater than the underlying symmetry of the intensities (ambiguity in choice of index).
[] Enantiomorphs—not distinguishable on intensities alone.
[b] I222 versus I2$_1$2$_1$2$_1$ and I23 versus I2$_1$3 are indistinguishable in terms of symmetry and absences in reciprocal space. Beware, they will have different Harker sections in Patterson space!

Reflections observed in centred lattices

Centring	Conditions defining observed reflections
A centred	$k + l = 2n$
B centred	$h + l = 2n$
C centred	$h + k = 2n$
F (face centred)	$k + l = 2n$ and $h + l = 2n$ and $h + k = 2n$
I (body centred)	$h + k + l = 2n$
P	all reflections present
R (rhombohedral axes)	all reflections present
R (hexagonal axes)	$-h + k + l = 3n$

Reciprocal space conditions indicating screw axes

2$_1$ (for example along k), e.g. P2$_1$	$0k0$ only present for $k = 2n$
3$_1$, 3$_2$, e.g. P3$_1$21	$00l$ only present for $l = 3n$
6$_1$, 6$_5$, e.g. P6$_1$22	$00l$ only present for $l = 6n$
6$_2$, 6$_4$, e.g. P6$_2$22	$00l$ only present for $l = 3n$
6$_3$, e.g. P6$_3$22	$00l$ only present for $l = 2n$
4$_1$, 4$_3$, e.g. P4$_1$2$_1$2	$00l$ only present for $l = 4n$
4$_2$, e.g. P4$_2$2$_1$2	$00l$ only present for $l = 2n$

Table 3. *continued*

Reciprocal space mirror planes

monoclinic space groups (*b* axis unique)	perpendicular to *k*
orthorhombic space groups	perpendicular to *h*, *k*, and *l*
tetragonal space groups	perpendicular to *h*, *k*, (and *l* for point group 422) also at 45° to *h* and *k*
trigonal point group 3	none
trigonal P321 P3₁21 P3₂21 R32	at 30° to *h* and *k*
trigonal P312 P3₁12 P3₂12	along *h* and *k*
hexagonal point group 6	perpendicular to *l*
hexagonal point group 622	perpendicular to *l* and every 30° around *l* (starting in the *hl* plane)
cubic point group 23	perpendicular to *h*, *k*, and *l*
cubic point group 432	perpendicular to *h*, *k*, and *l* also at 45° to *h* and *k* (and all permutations thereof)

Additional notes

Cubic space groups have a threefold axis along the unit cell body diagonal.
Volume 1 of the *International tables for crystallography* (i.e. the old series) is invaluable but unfortunately out of print, see Section 9.

---

vary between about 3% for a very good, strong data-set and 11% for very weak data. For certain space groups (see *Table 3*) the symmetry of the geometry of the reciprocal lattice is greater than the underlying symmetry of the intensities (i.e. there is an ambiguity in the choice of index), for these cases certain independently indexed batches of data may require index manipulation before they can be scaled together. This may also be the case if the lattice exhibits pseudo-symmetry (e.g. two nearly equal cell dimensions) (33). Slippage of the crystal during data collection may require separate processing of sections of the data. Where practical, aim for:

- a reasonable amount of redundancy in the data to aid in correct scaling
- a fairly complete native data-set (over 80%), in particular missing reflections should be randomly distributed in reciprocal space

The importance of good quality data to all subsequent stages in a structure determination cannot be overemphasized; statistical measures may be massaged but the truth will out in the final electron density. Don't worry if the R_{merge} for a data-set increases a bit given a greater degree of redundancy, the final accuracy will be improved. The behaviour of R_{merge} as a function of resolution should be monitored; many workers feel that the maximum resolution claimed for an analysis should not exceed that for which R_{merge} reaches 25%. The completeness of the data is vital for the initial phasing and structure determination but for later difference maps is less important. For cases of non-crystallographic symmetry the fraction of data required for a difference

15

map may be estimated roughly from the degree of redundancy, e.g. for fivefold symmetry a (randomly sampled) fifth of a complete data-set should be adequate.

5. Phasing

Given multiple copies of the protein in the asymmetric unit, some valuable information on the molecular packing may be immediately available from calculations using just the native amplitudes.

(a) Check the self-rotation function for the orientation of any non-crystallographic symmetry axes; note that the alignment of non-crystallographic with crystallographic rotation axes will mask any such peaks, high-symmetry space groups will tend to yield noisier functions.

(b) Check the low resolution (e.g. 15–7 Å) native Patterson for simple translations between molecules in the asymmetric unit (e.g. a combination of crystallographic and non-crystallographic symmetry operators that leave the molecular orientation unchanged and produce a translation aligned with a unit cell axis).

This sort of basic information on the arrangement of the molecules in the unit cell can greatly aid an otherwise problematic phase determination. The three major methods for obtaining phase information will be dealt with separately in the following sections.

5.1 Single/multiple isomorphous replacement (SIR/MIR)

This method is the most generally applicable, it depends on the measurement of differences in the diffracted intensities that arise when heavy (usually metal) atoms are bound at a limited number of specific sites on the protein. Commonly used reagents are listed in *Table 4* and details of their use are discussed in ref. 34.

(a) *Note*: virtually all these reagents are very poisonous and should be handled accordingly.

(b) You can try co-crystallization, but generally start off by soaking native crystals in the heavy-atom solution.

(c) Preliminary soak conditions may be tested using small crystals to check for obvious cracking or total annihilation of the crystal, but suitable large crystals must be used for data collection.

(d) Typical concentrations of heavy atom compounds lie in the range 1 mM to 10 mM.

(e) Typical soak times range from hours to days.

Table 4. A selection of heavy atom compounds

Platinum compounds: K_2PtCl_4, $K_2Pt(NO_2)_4$, $K_2Pt(CN)_4$
Gold compounds: $KAu(CN)_2$, $NaAuCl_4$
Uranium compounds: $UO_2(NO_3)_2$, $UO_2(CH_3.COO)_2$
Lead compounds: $Pb(NO_3)_2$, $Pb(CH_3.COO)_2$, trimethyl lead acetate
Rare earths: lanthanide and samarium salts
Mercurials: p-chloromercuri benzene sulphonate (PCMBS), ethylmercury phosphate
   (EMP), mercuric acetate, mersalyl, 1,4-diacetoxy mercuri-2,3-dimethoxybutane
   (Baker's dimercurial)
Complex ions and clusters: K_2HgI_4 in excess KI, tetrakis(acetoxymercuri)methane
   (TAMM), $[[W_3O_2(O_2CCH_3)_6](H_2O)_3](CF_3SO_3)_2$ (WAC)

Certain chemical groups are traditionally good candidates for derivatiza-
tion, e.g. free sulphydryls, these are discussed in ref. 34. In addition, any
unique targets for heavy-atom binding to the protein should be carefully
assessed (e.g. a single, exposed disulphide bridge which can be reduced and
reformed incorporating a mercury atom (35)). Efforts have been made to
match the mass and symmetry of large, heavy-atom clusters to the special
nature of particular projects (36, 37). Proteins have been mutated to provide
theoretically favourable sites for heavy-atom binding (e.g. the creation of free
cysteines by site-directed mutagenesis, particularly of glutamine residues
(38)). An element of serendipity does, however, remain in obtaining a useful
heavy-atom derivative. Changes in colour and density of the crystal may
provide some guide as to whether heavy-atom binding has occurred but the
ultimate test remains X-ray data collection. On comparison of native and
putative derivative data:

- cell dimensions should show little change compared to those of the native
  (otherwise loss of isomorphism)

- mean differences should exceed those expected on the basis of random
  noise; one fully occupied mercury site per 17 kDa protein may be expected
  to yield mean differences in the structure factor amplitudes between native
  and derivative data-sets of about 25% (see *Figure 3*)

- the magnitude of the mean differences should not increase with resolution
  (indicative of non-isomorphism)

A wide range of putative derivatives (often between 10 and 50) may be
rapidly screened by collecting a small amount of diffraction data (at relatively
low resolution) for each, followed by full-scale data collection (to higher
resolution) for those that show promise. Particular caution should be exercised
in scaling together derivative data from different crystals, since heavy-atom
occupancies may differ, necessitating separate heavy-atom refinement for
each data set (39); ideally an entire heavy-atom data-set should be collected

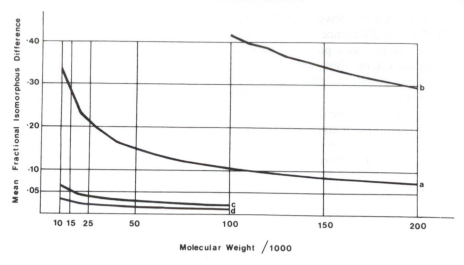

Figure 3. Mean fractional isomorphous difference (for structure factor amplitudes) verses molecular weight. Values are for wavelength 1.54 Å (CuKα) unless otherwise stated, and assume zero scattering angle and full occupancy for the heavy-atom site(s). (a) Single mercury (Hg) site; (b) $(Hg)_4$ cluster at very low resolution (the heavy-atom cluster diffracts coherently, i.e. as a single unit); (c) anomalous signal from single Hg site (Bijvoet difference); (d) anomalous signal from single Se site (Bijvoet difference) for wavelength 0.97 Å. Points to note: additional sites diffracting independently scale the above curves as (number of sites)$^{1/2}$; different heavy atoms (HA) scale the above curves as the ratio f_{HA} to f_{Hg}; the Bijvoet differences given for curves (c) and (d) are double the actual anomalous signal for a single reflection, since the quantity of interest is the difference between the amplitudes of $F(hkl)$ and $F(-h-k-l)$.

from just one crystal. Care should also be taken to obtain the best scaling of derivative to native data, an anisotropic scale factor is often particularly effective (39). Anomalous scattering signals, even from heavy atoms, are small (see Section 5.3 and *Figure 3*) but can be crucial in resolving phase ambiguities. Success will be critically dependent on the attainment of a high degree of accuracy in the data collection and scaling. Correlation coefficients provide a simple way of assessing the reliability of the measured anomalous differences (39).

For one or two sites in a low symmetry space group a few strong peaks on the Harker sections of the difference Patterson map may immediately be interpreted in terms of the heavy-atom positions (the interpretation of Patterson functions is described in the standard texts, e.g. ref. 4). In less obvious cases the comparison of difference Patterson maps calculated for independent resolution ranges or data-sets may aid in identification of significant peaks. In our experience the calculation of a simple correlation function on the peaks above 2σ between two such maps serves as a good measure of their useful information content (40). For particularly complex cases automatic Patterson

18

search procedures have been developed and can be very effective (e.g. ref. 40). Since a difference Patterson map is calculated using the square of the differences between pairs of numbers, a few rogue differences can have a disastrous effect, thus the rejection of outliers in the data (i.e. derivative reflections giving differences greater than $3 \times$ mean difference) is important. We also recommend origin removal as a standard procedure; large ripples generated around the origin peak by series termination errors may otherwise obscure genuine features in the map. Sharpening the observed structure factor amplitudes may be of help in difficult cases. However, it is important to bear in mind the difference between the complexity of a difference Patterson map when it arises from high symmetry and/or high non-crystallographic symmetry and straightforward noisiness indicative of a poor quality derivative which is unlikely to provide adequate phasing power. Always plot the Patterson peaks predicted by a given solution on the actual Patterson synthesis.

The coordinates of the heavy-atom sites should be improved, and values obtained for their occupancies and temperature factors, using heavy-atom refinement procedures. Heavy-atom refinement may be performed on phased structure factors or on amplitudes. Ultimately, if the problem is treated correctly, phased refinement should be the method of choice, since it allows all the available information to be used. However, for the case of a single derivative, amplitude refinement has much to recommend it and, in our experience, simple refinement against the Patterson map works well. If more than one derivative or anomalous scattering data are available, phased refinement should be superior. Centric reflections, because of the restrictions on their allowed phase angles, are particularly valuable at this stage. Unfortunately, heavy-atom refinement is inherently prone to severe problems arising from bias, especially for the acentric reflections. It is therefore important to check 'by eye' that the refinement actually appears to be improving the agreement with the difference Patterson maps in a sensible manner. After the major sites for one derivative have been established, the resultant phases may be used in difference Fourier syntheses to find additional minor sites in the same derivative or to solve other putative derivatives (but, again, check that the results seem sensible in the context of the original difference Patterson maps). Success in this is a very encouraging confirmation of the validity of the original solution. Further checks are provided by:

- good agreement with all major peaks on the original difference Patterson map
- sites deliberately omitted from the phasing showing up in Fourier difference maps
- interpretability of the electron density map for the protein (the ultimate test!)

The final mean figure of merit for the protein phases based on a good derivative should be 0.6 or better, but mean figures of merit of down to 0.3

indicate some phase information which, properly weighted, could be combined with that from other derivatives or may be improved by the methods of Section 6. The phasing power (mean heavy-atom structure factor amplitude/mean error of closure) should be monitored with respect to resolution; once this value falls below 1 it is generally assumed that no further phasing power is available from this derivative. An overall phasing power of greater than 1.5 is again indicative of a potentially useful derivative. A single isomorphous derivative has an inherent ambiguity in phase but, for favourable cases, can be sufficient for a structure determination, since this ambiguity may be resolved by the use of the anomalous scattering component (Section 5.3) and/or solvent flattening (Section 6). In the absence of anomalous scattering, Friedel pairs of reflections (hkl and $-h-k-l$) have equal intensities; therefore, even given multiple isomorphous derivatives, there will always be a choice of two, apparently equally valid, self-consistent sets ('hands') of heavy-atom positions (x, y, z or $-x, -y, -z$). Anomalous scattering information allows the correct choice of 'hand' to be made, failing this your electron density map, as like as not, will appear to possess left-handed helices; go back and swop the 'hand' (note: for enantiomorphic space groups this will also involve swopping to the 'mirror image' space group).

5.2 Molecular replacement (MR)

MR makes use of a known three-dimensional structure as an approximate model to provide starting phase angles for the observed structure factor amplitudes from the unknown protein structure (41, 42). This, of course, assumes that there is a close structural similarity. Thus the method is normally only applicable if the protein shows a high level of sequence identity (usually a minimum of 30% but preferably 50% or higher) to a protein of known three-dimensional structure. An exception is provided by virus families such as the Picornaviridae in which the capsid structures are similar despite very low sequence identity. The known structure of one of the components of a macromolecular complex may suffice if the known component comprises a significant percentage of the complex (over 33%). Where a family of homologous structures is available, it may be advantageous to construct an 'average' structure for use as the initial model (43).

The known (model) structure must be correctly placed in the unit cell of the unknown structure. If two structures are similar, then their Patterson functions will also be similar. For a crystalline protein, the Patterson function will be dominated, near the origin, by intramolecular vectors which are dependent on the orientation but not the position of the molecule in the crystallographic cell. Thus, analysis of Patterson functions near the origin allows the model structure to be orientated (by use of the rotation function). By consideration of intermolecular vectors it may then be positioned in the cell (by a translation search). In practice:

Protocol 3. Positioning the model structure in the cell

1. Calculate the cross-rotation function to indicate the correct orientation. This should be repeated for several resolution ranges to check the significance of the peaks; the integration radius should be varied over several trial values but should not exceed about half the expected dimensions of the unknown molecule. In particular, it must be smaller than the shortest cell dimension to avoid inclusion of the neighbouring origin peak. The model structure factors should be calculated for a P1 cell with cell dimensions much greater than the model (to avoid pollution of the model Patterson by intermolecular vectors). Sharpening the observed structure factors may help.

2. Place the model in the 'correct' orientation in the real cell; this requires particular care over orthogonalization changes. Check with rerun of cross-rotation function (relative rotations should now be zero). Since step **3** is very sensitive to the accuracy of the orientation, care should be taken to obtain the best possible value and, if several seem equally valid, each should be tried in **3**. Recently a simple strategy has been devised in which correlation coefficient refinement (based on maximizing the agreement between the observed and calculated Patterson coefficients) is applied to a large number of the best cross-rotation function solutions. This provides an automatic and reliable way of selecting and improving the most likely solutions (44).

3. Do a translation search to find the correct position in the cell, this may be judged by criteria based on either R factors or correlation coefficients. Correlation coefficients are generally more sensitive and reliable than R factors.

The presence of more than one copy of the unknown structure in the asymmetric unit obviously complicates the search unless the relative positions of the copies are known; including this information in the search model may be helpful (e.g. for dimers, trimers, or tetramers try an oligomeric model). The success of the search procedure may be crucially dependent on the exact model used (generally, where the method has been successful the r.m.s. deviations between the search model coordinates and the eventual refined crystal structure coordinates have been in the range 0.5–1 Å):

(a) for hinged structures, e.g. Fabs, break at hinge and treat two components separately (ref. 45 describes one possible strategy)

(b) remove additional domains, insertions, etc. known to be very different

(c) just Cα coordinates may occasionally (for cases of close similarity and good data) be sufficient but the result will be appreciably noisier; main-chain and even side-chain atoms can be built automatically on to the C^α coordinates by reference to existing data-bases

Criteria by which to judge the truth of the solution are:

(a) packs OK in unit cell (check on graphics system and/or with packing function)

(b) holds up over various resolution ranges

(c) given multiple subunits, is consistent with peaks in the self-rotation function

(d) if available, check with heavy-atom data, model phases used in a difference Fourier synthesis should reveal heavy-atom sites

The model is then subjected to rigid body refinement (45, 46), at first at low resolution; the initial R factor should be below 59% (random), often it is 50%. This value should then fall significantly during the refinement. Refinement may be extended to the secondary structure units, and ultimately full refinement carried out as in Section 8. *Warning*: because of the bias inherent in a method based on initial phases taken from a model, great care must be exercised in the refinement and inspection of the resultant maps (see Sections 7 and 8).

5.3 Anomalous scattering and mutiple wavelength anomalous dispersion (MAD)

Traditionally, anomalous scattering from a heavy-atom derivative has been used as supplementary information to resolve the phase ambiguity of a single isomorphous derivative. It must be stressed that as the effects are very small, accurate measurements are required even for this classic use of anomalous data. The statistics for the anomalous data in a heavy-atom refinement should discriminate between the two possible 'hands' of the heavy-atom sites, failure to achieve this is a clear indication that the anomalous data is of insufficient accuracy to contribute useful information to the phasing.

The use of anomalous scattering for the direct determination of the phase information is still non-standard and requires careful experimental planning, thus this section contains only a very superficial outline and interested readers must refer to the key references. For small proteins in which metal atom(s) provide measurable anomalous scattering and represent a substantial percentage of the normal scattering (at least 30%) the method of phase determination from partial structure resolved anomalous scattering may be attempted (47, 48). For larger proteins the MAD method requires the presence of a metal atom in the protein and, usually, the tunable wavelength capability of a synchrotron X-ray source (49). The result is essentially equivalent to perfect isomorphism. Very accurate measurements are required since the method depends on the values of very small differences in the diffracted amplitudes (*Figure 3*). This calls, for particular care in the data collection strategy:

(a) Measure Bijvoet pairs simultaneously or close in time (avoid errors from crystal decay).

(b) Measure Bijvoet pairs with equivalent beam path in crystal (avoid errors from differing absorption effects).
(c) Aim for very good counting statistics (image plates are the detector of choice at present because of fog on film and instabilities in electronic detectors).
(d) Measure blocks of data at several wavelengths about the appropriate absorption edge.

Particular care must be taken when scaling the data; local scaling is vital, outliers and weak data should be rejected from Patterson functions based on anomalous differences (49).

This method is of obvious use for metalloproteins. One route by which otherwise metal-free proteins may be labelled is by growing up bacteria to produce proteins incorporating selenomethionine (50); however, this does require your protein to be expressed in bacteria. Alternatively, a heavy-atom derivative may be solved as the native protein.

6. Phase improvement and extension

This stage is applicable when there is a redundancy of information in the asymmetric unit, either arising as a result of several copies of the protein (non-crystallographic symmetry), or more generally because of the crystal solvent content (which can range from 30% to 90%).

'Wang' solvent flattening is now a standard technique for phase improvement (51). Even for crystals with an average solvent content (50%) the constraint of level density in the solvent region is a powerful source of extra information, which can be sufficient to break the phase ambiguity for a single isomorphous derivative or critically improve the MIR phases to yield an interpretable map. Note that, in principle, any phases can be improved by this method; thus in a recent structure determination (52) the method was used to provide a dramatic improvement to MR phases (the initial electron density map was calculated using MR phasing from one component of a protein–protein complex; solvent flattening was applied to the corresponding difference map; this allowed 90% of the unit cell to be flattened, greatly enhancing the electron density for the unknown protein component). Points to watch are:

(a) Use a big enough convolution radius to 'blur' the map sufficiently when calculating the envelope.
(b) Do not overestimate solvent content as this will truncate the protein region, but do try to use all the available solvent.
(c) Limit truncation of negative regions within the protein to no more than about 3% of the total protein area.
(d) Iterate over several cycles of solvent flattening after each envelope calculation until there is no further phase change.

(e) Iterate envelope calculations until a smooth, stable boundary is established.

(f) Check that the envelope makes sense; heavy atoms at protein–solvent boundaries, continuity of protein regions, secondary structure components visible in electron density, any non-crystallographic symmetry visible in electron density.

(g) If the initial phases are particularly bad, it may help to start at low resolution (e.g. 6 Å) and gradually extend out to higher resolution.

(h) Beware, the figure of merit quoted by solvent flattening programs is artificially high (values of over 0.7 are usual) but hopefully there is some real improvement, judge only on the quality of the protein electron density (don't necessarily believe the convincing solvent density).

Where there are several copies of the molecule in the asymmetric unit (i.e. non-crystallographic symmetry), or more than one type of crystal, molecular averaging may also be used to improve the electron density map. In particular, when there is a high degree of redundancy in the asymmetric unit a set of low-resolution phases may be used as a start point from which *ab initio* phasing may be gradually extended to the limit of the native data (53). Be warned that this is a somewhat complex and (up to now) specialized procedure which places fairly heavy demands on computer time. Optimum strategies for real-space symmetry averaging are investigated in ref. 54. The interested reader should glean detailed advice from the successful (mainly virus) structure determinations alluded to in ref. 42. Our own experience, for an oligomeric protein, is documented in ref. 40.

In order to proceed with real-space averaging, one must establish:

• the orientation and position of the non-crystallographic symmetry axis/axes

• the boundary of the region over which non-crystallographic symmetry applies

Initial estimates for the orientation of a non-crystallographic symmetry axis may be obtained from self-rotation functions, heavy-atom sites, and inspection of existing electron density maps. This orientation will usually need to be refined by maximizing the correlation coefficient between the electron density at non-crystallographically related points in the initial map (if necessary, starting at extremely low resolution to improve the radius of convergence).

Attempts at phase extension must be carried out in very gradual steps in reciprocal space (in our experience a continuous series of steps of about a third of a reciprocal-lattice unit, suitable values may be calculated from the G function (42)). The resolution shell from 8 to 4 Å, which contains the information governing the general positioning of the secondary structural units, is often a particularly difficult region of reciprocal space to phase correctly.

Progress must be monitored carefully in terms of R factor (typical averaging R factors after convergence are 12–20%), correlation cofficients, and, ultimately, by inspection of electron density maps.

7. Map interpretation

For an electron density map calculated on the basis of good phases (mean figure of merit greater than or equal to 0.6):

(a) At 6 Å resolution the outline of the molecule is visible, α-helices appear as rods, β-sheets as slabs

(b) At 3 Å resolution, with luck, the polypeptide chain can be traced completely; amino acid side-chains are clearly visible and carbonyl bulges are present (allowing the planes of the peptide links to be determined).

(c) At 2.5 Å resolution usually about 50% of the side-chains can be correctly identified without knowledge of the amino acid sequence; carbonyl bulges on the polypeptide main-chain are clearly visible, as are tightly bound waters.

(d) At 1.5 Å resolution the individual atoms are almost resolved and the water structure is clear.

(e) At 1.2 Å resolution many hydrogen atoms are visible.

   With the aid of a known amino acid sequence an attempt to build an atomic model of the protein structure into the electron density may be made at about 3 Å resolution (for exceptionally good maps, 3.5 Å). Some people find a mini-map stack of contours on thin plastic sheets of help in establishing the initial main-chain trace, but the key tool for map interpretation is the computer graphics workstation, and new methods should soon completely eliminate the use of mini-maps (55). A skeletonized version of the electron density may be obtained automatically and then, in conjunction with the original density, reconnected/modified by hand at the graphics workstation to produce a rough 'sketch' of the structure on which to base detailed map interpretation/model building. Information input from data-bases of known protein structures provides a further, powerful aid for the rapid production of a stereochemically sound model (56). In the initial stages of a map interpretation the correspondence between distinctive stretches of the amino acid sequence and regions of electron density must be established. For this task lengths of α-helix or β-strand from the data-base, superimposed on the portions of secondary structure visible in the map, provide an invaluable scaffold and sense of scale. These regions then provide anchor points which, for an electron density map of sufficient quality, may be linked together to give an unambiguous interpretation of the polypeptide fold. Again 'ready-made' templates for β-turns, etc., drawn from a data-base, are useful at this stage to provide sound stereochemical guidelines, particularly for those less experienced in model building. Unfortunately, poor connectivity for the electron density in the (often relatively flexible) loops linking regions of secondary structure is a particularly common source of potential ambiguity in otherwise reasonable quality maps. For any proposed structure, check:

- chemical sense of heavy-atom positions
- conservation of core residues/secondary structure elements based on homologous structures and/or sequences
- polypeptide fold does not form a knot
- electron density for side-chains is compatable with sequence
- sensible environments for amino acid types, e.g. most hydrophobics buried in core of protein

It is crucial to bear in mind that the electron density map prior to its interpretation in terms of an atomic model provides the only truly unbiased data; everything after this stage will carry in it, to some degree, the assumptions of the model. Thus caution and keen critical awareness may be vital at this stage if very major errors are not to be perpetuated (57). The construction of a partial structure (for instance a protein domain), confined to a clearly interpretable region of the map, provides additional phase information. Phase combination (see Section 8) should thus yield a more fully interpretable map. This procedure, often used iteratively in combination with refinement, has proved to be a valuable strategy (for example, ref. 58). Particular caution must be exercised for structures solved by molecular replacement, since in this case the maps are never bias free. The omission of part of the model (e.g. a co-factor) throughout the molecular replacement procedure provides, by the subsequent quality of its electron density, an objective measure of the true information content of a map based on MR phases.

8. Refinement

The accuracy of the final protein structure is greatly improved by refinement (59), thus this procedure is now, rightly, the norm in virtually all structure determinations. However, beware, even with the use of tight stereochemical restraints, a full refinement of individual *xyz* coordinates and isotropic B factors for each of the (non-hydrogen) atoms of a protein is usually not meaningful at less than about 3 Å resolution because of the insufficient observation to parameter ratio (*Table 5*).

The radius of convergence of the least squares minimization on which refinement is based is rather small, hence the tendency for structures to stick in false minima and the requirement to return to the graphics for frequent

Table 5. Refinement—observation to parameter ratios

Resolution Å	FMDV	FMDV with NCS	TNF	PPb	β lac.
4.0	2.2	11.0	1.8	1.3	1.2
3.5	3.3	16.5	2.6	1.9	1.7
3.2	4.3	21.5	3.4	2.5	2.3
3.0	5.2	26.0	4.1	3.0	2.7
2.8	6.4	32.0	5.1	3.6	3.3
2.5	9.0	45.0	7.1	5.1	4.7
2.3	11.5	57.5	9.1	6.5	6.0
2.0	17.5	87.5	13.8	9.8	9.0
1.7	28.5	142.5	22.4	15.8	14.6
1.5	41.4	207.0	32.5	22.9	21.2
1.2	80.7	403.5	63.2	44.3	41.1
1.0	139.2	696.0	109.0	76.2	70.8

Values are given for: number of observable unique reflections (assumes 100% complete)/number of non-hydrogen atoms per asymmetric unit. Thus a value of 4 in the table would correspond to one observation per parameter in the case of free refinement of x, y, z, and B for each non-hydrogen atom.

The examples are drawn from our experience: FMDV, foot-and-mouth disease virus in space group I23 with over 80% solvent and RNA; FMDV with NCS, as above but with strict non-crystallographic symmetry constraints corresponding to an additional five-fold redundancy in the asymmetric unit; TNF, tumour necrosis factor in space group P3₁21 with 65% solvent; PPb, glycogen phosphorylase b in space group P4₃2₁2 with 50% solvent; β lac., β lactamase I in space group C2 with somewhat less than 50% solvent.

Points to note: the observation to parameter ratios are essentially independent of both the symmetry of the unit cell and the size of the protein; the ratios are strongly influenced by the crystal solvent volume; the ratios are strongly influenced by non-crystallographic symmetry, thus for FMDV with NCS the ratios at 3.0 Å are more favourable than they would be for PPb at 1.5 Å; in the case of stereochemically restrained refinement (e.g. X-PLOR, PROLSQ) the situation is improved somewhat since extra (chemical) observations are incorporated.

manual rebuilds. This problem has now been ameliorated by the use of restrained molecular dynamics to increase the radius of convergence (60); typically allowing a drop in the R factor from 48% to 25% to occur without any manual intervention. The general protocol is:

Protocol 4. Restrained molecular dynamics

1. Energy minimize freshly built structure to get rid of gross steric clashes, very bad stereochemistry, etc.

2. Estimate an overall B factor.

3. Heat model to 2000 or 3000 K for a restrained molecular dynamics run.

Protocol 4. *Continued*

4. Cool the model, usually 'slowly' over a period, of say, 1 picosecond.

5. Energy minimize final structure within new local (hopefully global) minimum.

6. Refine restrained isotropic B factors.

For a model based on molecular replacement, regions of high sequence homology (e.g. the core secondary structure) may be expected to be essentially correct (for the main-chain atoms to within 0.5 Å), thus strong restraints may be usefully applied to these regions in variations of the above protocol. A key criterion in refinement is the balance of weights between stereochemical and observational restraints. Wherever possible, non-crystallographic symmetry should be incorporated either as constraints (if exact, e.g. virus structures) or restraints. Scattering from the bulk solvent in the crystal produces a dramatic attenuation of the low resolution data (below 8 Å), thus a solvent correction term should be applied before incorporating these observations into the refinement. Multiple conformations for limited regions of the model (e.g. two possible orientations for a small inhibitor) may be modelled in some versions of the current refinement programs. This is likely to be important in providing an adequate description of proteins at high resolution.

The power of refinement methods (in particular molecular dynamics) to reduce the crystallographic R factor is open to abuse. Much of the burden of rebuilding faulty regions in the structure may have been shouldered by the computer but the crystallographer is still the one who has to decide if it is true. Recently, great concern has been expressed at the possibility of erroneous structures being refined by molecular dynamics to give plausible R factors. Once a model has been built, bias will always be present, hence it is important to build to the clearest possible MIR or MAD map to get things right at the beginning; unfortunately, for MR the model is there from the start and we know that it is not the right one. False MR solutions have been refined down to 30 or 27% R factors; certainly, substantial parts of a model are likely to require some revision at 25% R factor. Extension of the resolution (for example from 2.5 Å to 1.9 Å) can produce dramatic improvements in the quality of maps after refinement for a MR-based model, presumably mainly due to the much greater quantity of amplitude data pulling the model, and hence phases, into shape. For problems of bias arising in a structure determination based on MIR or MAD phases, combining the current refined phases with the original phases has proved to be worthwhile (61, 62). Omit maps often are of genuine assistance in indicating errors in the structure, but they rely for their power on a substantial portion of the remaining structure being correct and thus the mere absence of any contrary indications in an omit map should not be taken as absolute proof of a correct structure. The best sort of omit map is one for which the omitted portion has never been included in the model (e.g. a co-factor or inhibitor), clearly density for such a region is particularly reassuring, especially for MR-based structure determinations.

Points to watch to avoid disaster are:

(a) Refine and calculate the R factor on all the data (no sigma cut).

(b) Tightly restrain the stereochemistry (less than 0.02 Å r.m.s. deviation from ideal bond lengths).

(c) Do not include more parameters in the refinement than are strictly merited (*Table 5*).

(d) Do not put in lots of water molecules until the R factor is low and the refinement is at reasonably high resolution (but ultimately this depends on the quality of the phases and thus the clarity of the map).

(e) Virtually all main-chain phi psi angles should conform to the expected Ramachandran values.

(f) The distribution of B factors should make sense.

If all still looks good, well done, get writing! Please make the effort to deposit your coordinates with the Protein Data Base (see Section 9), this is the one piece of advice we'll try to stick to ourselves.

9. General information

9.1 Publications

International tables for crystallography compiled by the International Union of Crystallography. Sold and distributed by Kluwer Academic Publishers Group, PO Box 322, 3300 AA Dordrecht, The Netherlands.

Proceedings of the CCP4 Study Weekends enquiries on available titles to The Librarian, Daresbury Laboratory, Daresbury, Warrington WA4 4AD, UK.

World directory of crystallographers compiled by the International Union of Crystallography. Sold and distributed by Kluwer Academic Publishers Group, PO Box 322, 3300 AA Dordrecht, The Netherlands.

9.2 Crystallographic programs

CCP4 (general collection of crystallographic programs including data processing, molecular replacement and averaging programs), CCP4, Daresbury, Laboratory, Daresbury, Warrington WA4 4AD, UK.

GROPAT (Patterson solution), Drs Yvonne Jones or Dave Stuart, Laboratory of Molecular Biophysics, University of Oxford, South Parks Road, Oxford OX1 3QU, UK.

MERLOT (molecular replacement), Dr Paula Fitzgerald, Merck Sharp and Dohme Research Laboratories, R80M203, PO Box 2000, Rahway, NJ 07065, USA.

PROLSQ (refinement), Professor Wayne Hendrickson, Biochemistry and Molecular Biophysics Department, Columbia University, New York, NY 10032, USA.

PROTEIN (data processing and molecular replacement), Dr W. Steigemann, Strukturforschung II, MPI f. Biochemie, Am Klopferspitz, D-8033 Martin-sried, Germany.

SHELX (direct methods and Patterson solution), Professor George Sheldrick, Inst. f. Anorg. Chemie, Universität Tammannstr. 4, D-3400 Gottingen, Germany.

Wang (solvent flattening), Professor B.-C. Wang, Crystallography Department, University of Pittsburgh, Pittsburgh, PA 15260, USA.

X-PLOR (molecular dynamics refinement also molecular replacement), Professor Axel T. Brunger, The Howard Hughes Medical Institute and Department of Molecular Biophysics and Biochemistry, Yale University, 260 Whitney Avenue, PO Box 6666, New Haven, CT 06511, USA.

9.3 Contact addresses

Email network of crystallographers, Dr Martha M. Teeter, Department of Chemistry, Boston College, Chestnut Hill, MA 02167, USA, Bitnet: TEETER@BCCHEM Internet: TEETER@BCCHEM.BC.EDU.

International Union of Crystallography, 5 Abbey Square, Chester CH1 2HU, UK.

Protein Data Base, Chemistry Department, Brookhaven National Laboratory, Upton, New York 11973, USA.

Acknowledgements

We thank Anne Roper and Stephen Lee for assistance with the figures.

References

1. Wyckoff, H. W., Hirs. C. H. W., and Timasheff, S. N. (ed.) (1985). *Methods in enzymology,* Vols 114 and 115. Academic Press, San Diego.
2. Lipson, H. and Cochran, W. (1966). *The determination of crystal structures*. Bell and Sons, London.
3. Blundell, T. L. and Johnson, L. N. (1976). *Protein crystallography*. Academic Press, London.
4. Holmes, K. C. and Blow, D. M. (1966). *The use of X-ray diffraction in the study of protein and nucleic acid structure*. Wiley, New York.
5. MacPherson, A. (1982). *Preparation and analysis of protein crystals*. Wiley, New York.
6. Gilliland, G. L. and Davies, D. R. (1984). In *Methods in enzymology* (ed. W. B. Jacoby), Vol. 104, pp. 370–81. Academic Press, San Diego.
7. Ducruis, A. and Giegé, R. (ed.) (1992). *Crystallization of nucleic acids and proteins: a practical approach*. IRL Press, Oxford.

8. Cox, M. J. and Weber, P. C. (1987). *J. Appl. Cryst.*, **20**, 366–73.
9. Stura, E. A. and Wilson, I. A. (1990). *Methods: A companion to Methods in Enzymology*, **1**, 38–49.
10. Carter, C. W., Jr and Carter, C. W. (1979). *J. Biol. Chem.*, **254**, 12219–23.
11. Mikol, V., Hirsch, E., and Giege, R. (1990). *J. Mol. Biol.*, **213**, 187–95.
12. Laver, W. G. (1990). *Methods: A companion to Methods in Enzymology*, **1**, 70–4.
13. Michel, H. (ed.) (1991). *Crystallization of membrane proteins*. CRC Press, Cleveland, Ohio.
14. Garavito, R. M. and Picot, D. (1990). *Methods: A companion to Methods in Enzymology*, **1**, 57–69.
15. Sousa, R. and Lafer, E. M. (1990). *Methods: A companion to Methods in Enzymology*, **1**, 50–6.
16. Walker, N., Marcinowski, S., Hillen, H., Machtle, W., Jones, Y., and Stuart, D. (1990). *J. Crystal. Growth*, **100**, 168–70.
17. Matthews, B. W. (1968). *J. Mol. Biol.*, **33**, 491–7.
18. Westbrook, E. M. (1985). *Methods in enzymology* (ed. H. W. Wyckoff, C. H. W. Hirs, and S. N. Timasheff), Vol. 114, pp. 187–96. Academic Press, San Diego.
19. Bode, W. and Schirmer, T. (1985). *Biol. Chem. Hoppe-Seyler*, **366**, 287–95.
20. Rayment, I. (1985). In *Methods in enzymology* (ed. H. W. Wyckoff, C. H. W. Hirs, and S. N. Timasheff), Vol. 114, pp. 136–40. Academic Press, San Diego.
21. Petsko, G. A. (1985). In *Methods in enzymology* (ed. H. W. Wyckoff, C. H. W. Hirs, and S. N. Timasheff), Vol. 114, pp. 141–6. Academic Press, San Diego.
22. Teng, T. Y. (1990). *J. Appl. Cryst.*, **23**, 387–91.
23. Hope, H., *et al.* (1989). *Acta Cryst.*, **B45**, 190–9.
24. Hajdu, J., *et al.* (1987). *EMBO J.*, **6**, 539–46.
25. Schlichting, I., *et al.* (1990). *Nature*, **345**, 309–15.
26. Ryu, S. E., *et al.* (1990). *Nature*, **348**, 419–26.
27. Harrison, S. C. (1968). *J. Appl. Cryst.*, **1**, 84–90.
28. Arndt, U. W. and Wonacott, A. J. (ed.) (1977). *The rotation method in crystallography*. North Holland, Amsterdam.
29. Kabsch, W. (1988). *J. Appl. Cryst.*, **21**, 67–71.
30. Rossmann, M. G. (1979). *J. Appl. Cryst.*, **12**, 225–38.
31. Stuart, D. and Walker, N. (1979). *Acta Cryst.*, **A35**, 925–33.
32. Winkler, F. K., Schutt, C. W., and Harrison, S. C. (1979). *Acta Cryst.*, **A35**, 901–11.
33. Cavarelli, J., *et al.* (1991). *Acta. Cryst.*, **B47**, 23–9.
34. Petsko, G. A. (1985). In *Methods in enzymology* (ed. H. W. Wyckoff, C. H. W. Hirs, and S. N. Timasheff), Vol. 114, pp. 147–56. Academic Press, San Diego.
35. Ely, K. R., Girling, R. L., Schiffer, M., Cunningham, D. E., and Edmundson, A. B. (1973). *Biochemistry*, **12**, 4233–7.
36. O'Halleran, T., Lippard, S. J., Richmond, T. J., and Klug, A. (1987). *J. Mol. Biol.*, **194**, 705–12.
37. Ladenstein, R., Bacher, A., and Huber, R. J. (1987). *J. Mol. Biol.*, **195**, 751–3.
38. Nagai, K., Oubridge, C., Jessen, T. H., Li, J., and Evans, P. R. (1990). *Nature*, **348**, 515–20.
39. Stuart, D. I., Levine, M., Muirhead, H., and Stammers, D. K. (1979). *J. Mol. Biol.*, **134**, 109–42.
40. Jones, E. Y., Walker, N. P. C., and Stuart, D. I. (1991). *Acta Cryst.*, **A47**, 753–70.

31

41. Rossmann, M. G. (ed.) (1972). *The molecular replacement method,* International Science Review, No. 13. Gordon and Breach, New York.
42. Rossmann, M. G. (1990). *Acta Cryst.,* **A46,** 73–82.
43. Acharya, R., Fry, E., Stuart, D., Fox, G., Rowlands, D., and Brown, F. (1989). **337,** 709–16.
44. Brunger, A. T. (1990). *Acta Cryst.,* **A46,** 46–57.
45. Brunger, A. T. (1990). In X-PLOR (version 2.1) Manual, pp. 222–44. (Available from the author, at Yale University.)
46. Sussman, J. L. (1985). In *Methods in enzymology* (ed. H. W. Wyckoff, C. H. W. Hirs., and S. N. Timasheff), Vol. 115, pp. 271–303. Academic Press, San Diego.
47. Hendrickson, W. A. and Teeter, M. A. (1981). *Nature,* **290,** 107–13.
48. Stuart, D., *et al.* (1986). In *Structural biological applications of X-ray absorption, scattering, and diffraction,* pp. 315–28. Academic Press, San Diego.
49. Hendrickson, W. A., Smith, J. L., and Sheriff, S. (1985). In *Methods in enzymology* (ed. H. W. Wyckoff, C. H. W. Hirs, and S. N. Timasheff), Vol. 115, pp. 41–55. Academic Press, San Diego.
50. Hendrickson, W. A., Pahler, A., Smith, J. L., Satow, Y., Merritt, E. A., and Phizackerlay, R. P. (1989). *Proc. Natl Acad. Sci. USA,* **86,** 2190–4.
51. Wang, B. C. (1985). In *Methods in enzymology* (ed. H. W. Wyckoff, C. H. W. Hirs, and S. N. Timasheff), Vol. 115, pp. 90–111. Academic Press, San Diego.
52. Grutter, M. G., *et al.* (1990). *EMBO J.,* **9,** 2361–5.
53. Bricogne, G. (1976). *Acta Cryst.,* **A32,** 832–47.
54. Rayment, I. (1983). *Acta Cryst.,* **A39,** 102–16.
55. Jones, T. A., Zou, J.-Y., Cowan, S. W., and Kjeldgaard, M. (1991). *Acta Cryst.,* **A47,** 110–19.
56. Jones, T. A. and Thirup, S. (1986). *EMBO J.,* **5,** 819–22.
57. Branden, C.-I. and Jones, T. A. (1990). *Nature,* **343,** 687–9.
58. Kabsch, W., Mannherz, H.-G., Suck, D., Pai, E., and Holmes, K. C. (1990). *Nature,* **347,** 37–44.
59. Hendrickson, W. A. (1985). In *Methods in enzymology* (ed. H. W. Wyckoff, C. H. W. Hirs, and S. N. Timasheff), Vol. 115, pp. 252–70. Academic Press, San Diego.
60. Brunger, A. T., Kuriyan, J., and Karplus, M. (1987). *Science,* **235,** 458–60.
61. Rice, D. W. (1981). *Acta Cryst.,* **A37,** 491–500.
62. Stuart, D. and Artymiuk, P. (1985). *Acta Cryst.,* **A40,** 713–16.

2

Methods of structural analysis of proteins. Part 2—nuclear magnetic resonance

G. MARIUS CLORE and ANGELA M. GRONENBORN

1. Introduction

Over the past few years solution nuclear magnetic resonance (NMR) spectroscopy has been established as the only viable alternative to X-ray crystallography (see Chapter 1) for determining three-dimensional structures of proteins up to 10 kDa (1, 2). The main source of geometric information used to determine structures by NMR lies in short (<5 Å) approximate interproton distance restraints obtained from nuclear Overhauser enhancement (NOE) measurements, supplemented by torsion angle restraints derived from three-bond coupling constants. Methodological advances in the past year, particularly with respect to obtaining many more interproton distance restraints from the NOE data, coupled with stereospecific assignments and torsion angle restraints derived from systematic conformational search procedures, has led to a considerable increase in the precision and accuracy of the structures that can be obtained by NMR (3, 4). Recent results on five proteins, the C-terminal domain of cellobiohydrolase I (5), the homeodomain of the Antennapedia protein (6), interleukin-8 (7), the zinc finger domain of a human enhancer protein (8), and reduced human thioredoxin (9), indicates that atomic r.m.s. distributions about the mean coordinate positions of the order of 0.3–0.4 Å for the backbone atoms and 0.5–0.6 Å for the internal side-chains are readily attainable by NMR, and that the overall quality of these 'high-resolution' solution structures is comparable to that of ~2 Å resolution crystal structure (4, 10). In addition, recent developments in three- and four-dimensional NMR have expanded the horizons of three-dimensional (3D) protein structure determination by NMR to proteins up to 25 kDa, and potentially up to 40 kDa (4).

A number of detailed reviews have appeared on the theoretical basis of one- and two-dimensional NMR spectroscopy (11), on three- (2, 3, 4, 12) and four-dimensional (4) NMR spectroscopy, and on the application of NMR for determining 3D protein structures (1–4). Therefore, the present chapter will focus on the practical aspects of the NMR experiments employed in protein structure determination, as practised in the authors' laboratory.

2. General overview of protein structure determination by NMR

The initial goal in any protein structure determination by NMR involves the complete (or as complete as possible) assignment of all ^{1}H-resonances in the spectrum. This is achieved using two classes of experiment. The first involves demonstrating through-bond connectivities via scalar couplings. In general these are restricted to intra-residue connectivities, and are used mainly to delineate amino acid spin systems (cf. *Figure 1*). The second involves the application of the NOE experiment to demonstrate the proximity of protons close in space (<5 Å). Sequential resonance assignment proceeds by identifying sequential NOEs of the type NH(*i*)—NH(*i*+1), C$^{\alpha}$H(*i*)—NH(*i*+1) and C$^{\beta}$H(*i*)—NH(*i*+1), as well as some medium-range NOEs of the type NH(*i*)—NH(*i*+2,3), C$^{\alpha}$H(*i*)—NH(*i*+2,3,4), and C$^{\alpha}$H(*i*)—C$^{\beta}$H(*i*+3). Given a knowledge of the amino acid sequence, it is a conceptually simple task to make use of both these experiments in combination to assign the resonances sequentially, proceeding from one amino acid to the next along the polypeptide chain. The second step involves the delineation of elements of regular secondary structure on the basis of the pattern of NOEs involving the NH, C$^{\alpha}$H, and C$^{\beta}$H protons. In the case of β-sheets, very strong C$^{\alpha}$H(*i*)—NH(*i*+1) and weaker C$^{\beta}$H(*i*)—NH(*i*+1) sequential NOEs are observed, together with interstrand NOEs involving the NH and C$^{\alpha}$H protons. In the case of α-helices, consecutive stretches of medium to strong NH(*i*)—NH(*i*+1), weak to medium C$^{\beta}$H(*i*)—NH(*i*+1), and weak C$^{\alpha}$H(*i*)—NH(*i*+1) sequential NOEs are found, together with weak to medium C$^{\alpha}$H(*i*)—NH(*i*+2,3,4), C$^{\alpha}$H(*i*)—C$^{\beta}$H(*i*+3), and NH(*i*)—NH(*i*+2) NOEs. The third step comprises the assignment of as many NOE interactions as possible, particularly those involving residues that are far apart in the sequence but close together in space, which are crucial for determining the polypeptide fold. In the fourth step, stereospecific assignments of β-methylene protons and φ, ψ, and χ$_1$ torsion angle restraints are obtained on the basis of $^3J_{HN\alpha}$ and $^3J_{\alpha\beta}$ coupling constants and the intra-residue and sequential NOE data are obtained using conformational search procedures by matching up the experimental observables with the calculated ones present in the data-base, which can comprise either a systematic library of all φ, ψ, χ$_1$ conformational space in a 10° grid or a library of tripeptide fragments from high-resolution X-ray structures. Finally, the 3D structure of the protein can be calculated on the basis of the distance and torsion angle restraints by a variety of methods, such as metric matrix distance geometry (13), minimization in torsion angle space (14), restrained molecular dynamics (15), and simulated annealing (16, 17).

G. Marius Clore and Angela M. Gronenborn

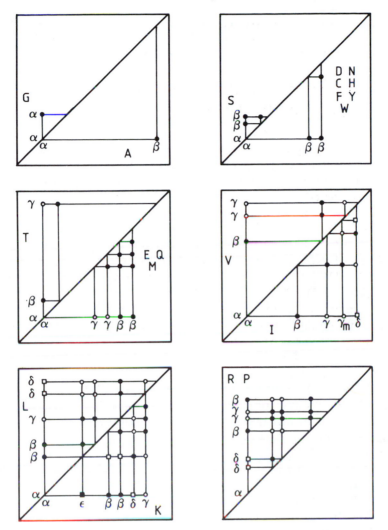

Figure 1. Patterns of through-bond scalar connectivities observed for different amino acids. In COSY spectra only direct connectivities (●) are observed, whereas in HOHAHA spectra direct (●) as well as single (○), double (□), and triple (■) relayed connectivities can be detected progressively as the mixing time in the pulse sequence is increased.

3. The NMR experiments

In this section we will outline the practical aspects of 2D, 3D, and 4D NMR experiments. The general relationship between these experiments is summarized in *Figure 2*. All 2D experiments comprise a preparation pulse; an evolution period, t_1, during which the spins are labelled according to their chemical shifts; a mixing period during which the spins are correlated with

35

2D NMR $P_a - E_a(t_1) - M_a - D_a(t_2)$

3D NMR $P_a - E_a(t_1) - M_a - E_b(t_2) - M_b - D_b(t_3)$

4D NMR $P_a - E_a(t_1) - M_a - E_b(t_2) - M_b - E_c(t_3) - M_c - D_c(t_4)$

Figure 2. Relationship between the pulse sequences for recording 2D, 3D, and 4D NMR spectra. Abbreviations: P, preparation; E, evolution; M, mixing; and D, detection.

one another; and, finally, a detection period, t_2. A series of experiments is recorded with different values of t_1, and 2D Fourier transformation with respect to both t_1 and t_2 results in the 2D spectrum. Higher-dimensional experiments can be devised by simply combining various 2D experiments. For example, a 3D experiment is obtained by combining two 2D experiments leaving out the detection period of the first experiment and the preparation pulse of the second. In this case, there are two evolution periods, t_1 and t_2, which are incremented independently, and an acquisition period, t_3. Three-dimensional Fourier transformation with respect to t_1, t_2, and t_3 yields the 3D spectrum. The relationship between 2D, 3D, and 4D NMR is illustrated schematically in *Figure 3*. References to the original articles describing the various pulse schemes are provided in the review articles cited in Section 1 (1–4, 11, 12).

3.1 Some basic experimental considerations

In general it is advisable to record spectra in the pure phase absorption mode in order to obtain as high a sensitivity and resolution as possible. To achieve this it is necessary to obtain quadrature detection in the indirectly detected dimension, which in the case of 2D NMR is the F_1 dimension. Four different methods may be used, as summarized in *Table 1*. For conventional 2D NMR, the time proportional phase incrementation (TPPI) method (18) is probably the most commonly used, and has the advantage that axial peaks are only observed at the edge of the spectrum. The States method (19), however, has

Figure 3. Schematic illustration of the progression and relationship between hetero-nuclear $^{13}C/^{15}N$-edited 2D, 3D, and 4D NOESY spectra. The closed circles represent NOE cross-peaks between NH and aliphatic protons. In the 2D spectrum, cross-peaks from 11 aliphatic protons to three NH protons at a single NH chemical shift are indicated. In the 3D spectrum, these peaks are spread into a third dimension according to the chemical shift of the directly bonded ^{15}N atoms, and in the figure the peaks are now distributed in three distinct ^{15}N planes, thereby removing the overlap associated with NH chemical shift degeneracy. The identity of the aliphatic protons, however, can still only be established on the basis of their ^1H chemical shifts. In the 4D spectrum, each ^{15}N plane of the 3D spectrum constitutes a cube composed of a series of slices at different ^{13}C chemical shifts. The identity of the originating aliphatic protons can now be established unambiguously on the basis of their ^1H and ^{13}C chemical shifts. (From ref. 32.)

G. Marius Clore and Angela M. Gronenborn

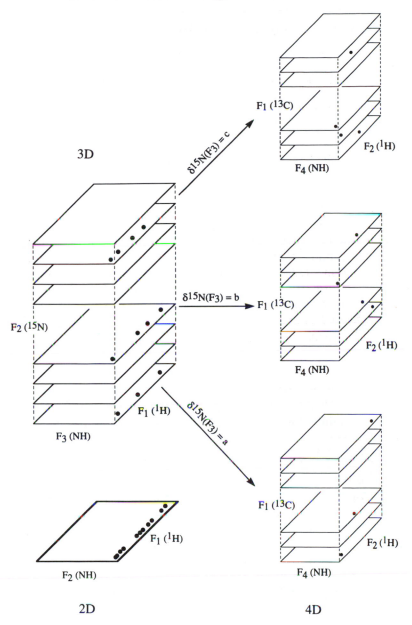

2D

4D

the advantage that the carrier position in the F_1 dimension can be shifted after data collection by applying a linear phase correction to the FIDs. Although axial peaks in the States method are observed in the centre of the spectrum, this does not present a problem in conventional 2D NMR as usually sufficient scans per increment are recorded to permit appropriate phase cycling for the

37

Table 1. Comparison of methods used to obtain quadrature detection in the indirectly detected dimension(s) in multi-dimensional NMR experiments (20)[a]

Method	Phase of preparation pulse Ψ	Acquisition receiver phase	Fourier transform	Axial peak
Redfield	$x(t_1)$	x	Real	Centre
	$y(t_1 + \Delta)$	x		
	$x(t_1 + 2\Delta)$	$-x$		
	$y(t_1 + 3\Delta)$	$-x$		
TPPI	$x(t_1)$	x	Real	Edge
	$y(t_1 + \Delta)$	x		
	$-x(t_1 + 2\Delta)$	x		
	$-y(t_1 + 3\Delta)$	x		
States	$x(t_1)$	x	Complex	Centre
	$y(t_1)$	x		
	$x(t_1 + 2\Delta)$	x		
	$y(t_1 + 2\Delta)$	x		
TPPI–States	$x(t_1)$	x	Complex	Edge
	$y(t_1)$	x		
	$-x(t_1 + 2\Delta)$	$-x$		
	$-y(t_1 + 2\Delta)$	$-x$		

[a] The time Δ equals $1/(2SW_1)$, where SW_1 is the spectral width in F_1. For the States and TPPI–States method, the x and y experiments are stored in separate locations to be processed as complex data. The TPPI–States method combines the different advantages of both the States and TPPI methods. Thus, shifting of the carrier position in the indirectly detected dimension(s) after data acquisition can be accomplished by simply applying a linear phase correction to the FIDs as in the States method. In addition, because the receiver phase is shifted by 180° for every t_1 increment, the axial signals invert their sign every t_1 increment as in the TPPI method, while the other signals remain unchanged relative to the regular States method. As a result, axial peaks in the TPPI–States method, like those in the TPPI method, are moved to the edge of the spectrum.

suppression of axial peaks. In 3D NMR, on the other hand, it is essential to use as few scans per increment as possible in order to ensure that the measuring time is confined to reasonable limits. Thus, in this case, a combination of the TPPI and States methods is the most desirable, offering the best of both methods (20).

The second consideration is the phase of the detected signal. In order to ensure flat baselines, the experimental conditions should be adjusted to ensure that the zero- and first-order phase corrections are either 0° and 0°, respectively, or 90° and 180°, respectively. The zero-order phase can be adjusted by varying the relative receiver phase, while the first-order phase can be adjusted by varying the delay between the last pulse of the experiment and the start of acquisition.

3.2 Two-dimensional homonuclear NMR

Although there are many experiments that can potentially be carried out, it is generally the case for proteins of up to 80–100 residues that the NMR spectra can be limited to four basic 2D homonuclear experiments: NOESY (21), P.COSY (22), PE.COSY (22, 23), and HOHAHA (24).

3.2.1 Observation of NOEs: NOESY

The pulse sequence of the NOESY experiment is given by

$$90_\psi - t_1 - 90_x - \tau_m - 90_\phi - Acq(t_2)$$

with phases $\psi = 4(x)$, $4(-x)$; $\phi = x$, y, $-x$, y; and Acq $= x$, y, $-x$, $-y$, $-x$, $-y$, x, y. The phase ϕ suppresses double and triple quantum coherence, while the phase ψ suppresses axial peaks.

For spectra in D_2O it is sometimes helpful to use very weak coherent pre-saturation to eliminate the residual HOD resonance. Spectra recorded in D_2O are used to observe NOEs between non-exchangeable protons.

To observe NOEs involving exchangeable protons (e.g. the backbone amides), spectra in H_2O must be recorded. Owing to limited dynamic range, it is therefore essential to suppress the massive water signal ($\sim$100 M). This can be achieved either by coherent pre-saturation or by the use of semi-selective excitation. Pre-saturation is easiest to implement but has the disadvantage that cross-peaks close to or degenerate with the water resonance are bleached out. Semi-selective excitation is accomplished by replacing the last 90_ϕ pulse in the NOESY experiment by a jump-return sequence $90_\phi - \tau - 90_{-\phi}$ where the delay, τ, is usually set to 80–90 μsec, and the carrier is placed at the frequency of the water resonance. In addition, it is useful to shift the phase of the preparation pulse by $45°$ to reduce the effects of radiation damping which may create an intense H_2O signal during data acquisition, particularly at shorter mixing times. In both cases the region of the spectrum that is used for analysis comprises the F_2(NH/aromatic)-F_1(NH/aromatic/aliphatic) region of the spectrum. To improve the baseline in the spectrum recorded with a semi-selective excitation sequence, it is advisable to zero the first point of the FIDs and apply a linear baseline correction to the FIDs prior to Fourier transformation in F_2 in order to reduce the wings of the water peak. This should then be followed by linear baseline correction of the NH region of the spectrum prior to Fourier transformation in F_1.

3.2.2 Observation of direct through-bond connectivities: P.COSY and PE.COSY

The P.COSY and PE.COSY spectra are used to detect direct through-bond connectivities via three-bond homonuclear $^1H-^1H$ J couplings. They differ from each other in so far that the P.COSY experiment yields cross-peaks with complete multiplet patterns, while the PE.COSY experiment yields cross-peaks with reduced multiplets. The latter is therefore employed for the accurate measurement of $^3J_{\alpha\beta}$ couplings. The pulse sequence for the 2D part of the experiment is given by

$$90_\psi - t_1 - \beta_x - Acq(t_2)$$

where β is $90°$ and $\sim35°$ in the P.COSY and PE.COSY experiments, respectively, and $\psi = x, -x$ and $Acq = x, -x$. In addition, a 1D FID with twice the number of data points and at least eight times the number of scans is recorded with $\beta = 0$. This 1D FID is then progressively left shifted to generate a $0°$ mixing pulse COSY which is then subtracted from the 2D time domain data set to yield the final 2D time domain data. The result of this procedure is that the dispersive components of the diagonal peaks are removed without affecting the cross-peaks. Hence, peaks close to the diagonal are readily observed. More traditional alternatives to the P.COSY and PE.COSY experiments are the DQF-COSY and E.COSY experiments, but their sensitivity is considerably lower by factors of two and at least four, respectively.

In the case of P.COSY and PE.COSY spectra recorded in water, the H_2O resonance is best suppressed by coherent pre-saturation.

3.2.3 Detection of multiple relayed through-bond connectivities: HOHAHA

The HOHAHA (and related TOCSY) experiment is used to demonstrate both direct and relayed through-bond connectivities, and constitutes an essential part in the delineation of amino acid spin systems. The basic pulse sequence for the experiment recorded in D_2O or in H_2O with pre-saturation is

$$90_\psi - t_1 - 90_\phi - SL_x - WALTZ17_y - SL_x - Acq(t_2)$$

with $\phi = 2(y), 2(-y)$; $\phi = x, -x$; and $Acq = 2(y), 2(-y)$. The $WALTZ17_y$ pulse train is given by the sequence $(270_y 360_{-y} 180_y 270_{-y} 90_y 180_{-y} 360_y$ $180_{-y} 270_y 270_{-y} 360_y 180_{-y} 270_y 90_{-y} 180_y 360_{-y} 180_y 270_y)$ $(270_{-y} 360_y 180_{-y} 270_y$ $90_{-y} 180_y 360_{-y} 180_y 270_{-y} 270_y 360_{-y} 180_y 270_{-y} 90_y 180_{-y} 360_y 180_{-y} 360_y 180_{-y}$ $270_y)(60_x)$, repeated n times. This pulse train transfers magnetization along the side-chain protons by anisotropic mixing of the ^1H spins. The duration of the mixing time is given by $n[(96 \times \tau_{90}) + \tau_{60}]$, and typical mixing times used range from 20 msec to 60 msec, depending on the extent of relayed magnetization transfer that one wishes to observe. At short mixing times, mainly direct connectivities are observed, and as the mixing time is increased so progressively single, double, and triple relayed connectivities are observed. The spin lock pulse (SL) is usually set to 1.5 msec. Note that relatively low radio-frequency (rf) power must be used for these experiments to prevent overheating the probe. Typically, the duration of the $90°$ pulse is $25–30$ μsec.

The pulse sequence for the experiment recorded in H_2O with a semi-selective excitation is given by (25)

$$90_\psi - t_1 - 90_{\phi 1} - SL_x - WALTZ17_y - SL_x - 90_{\phi 2} - \Delta - 90_{\phi 3} - \tau - 90_{-\phi 3} - Acq(t_2)$$

with $\psi = 4(-y), 4(-y)$; $\phi 1 = 8(x), 8(-x)$; $\phi 2 = 4(y), 4(-y)$; $\phi 3 = x, y, -x, -y$; and $Acq = x, y, -x, -y$. The delay Δ is set to half the HOHAHA mixing time to compensate for rotating frame Overhauser effects, and the delay τ is

40

usually set to 80–90 μsec. As in the case with the NOESY sequence with semi-selective excitation, the phase of the preparation pulse should be shifted by 45° to reduce radiation damping, and linear baseline correction should be applied to the FIDs prior to Fourier transformation in F_2.

3.3 Some useful two-dimensional heteronuclear experiments

As proteins get larger it becomes increasingly helpful to make use of uniformly labelled [^{15}N] and/or [^{13}C] protein, in order to effectively exploit hetero-nuclear experiments which allow the resolution of ambiguities arising from ^{1}H chemical shift degeneracy in the homonuclear 2D experiments. Such labelled proteins are readily obtained in microbial expression systems utilizing minimal medium with ^{15}NH$_4$Cl and [^{13}C$_6$] glucose as the sole nitrogen and carbon sources, respectively.

3.3.1 One-bond heteronuclear correlation

Two different experiments may be employed to measure ^{1}H-detected X-^{1}H correlations. The HMQC experiment (26) involves transfer of magnetization between ^{1}H and the directly bonded heteronucleus (^{15}N or ^{13}C) via multiple quantum coherence, while the 'Overbodenhausen' experiment (27, 28) employs single quantum coherence.

The sequence for the HMQC experiment is

$$
\begin{array}{l}
^1\text{H} \qquad 90_x - \Delta - \qquad\quad -t_1/2 - 180_\phi - t_1/2 - \qquad\quad - \Delta - \text{Acq}\,(t_2)\\[4pt]
^{15}\text{N} \qquad\qquad\qquad 90_\psi \qquad\qquad\qquad\qquad\qquad\qquad\qquad\quad 90_x
\end{array}
$$

with $\psi = x, -x$; $\phi = 2(x), 2(y), 2(-x), 2(-y)$; Acq $= 2(x, -x, -x,x)$; and the delay $\Delta \sim 1/(2J_{\text{NH}})$. In practice a slightly shorter value of Δ (4.5 msec) is employed to reduce net loss of magnetization via ^{15}N relaxation. The sequence for the 'Overbodenhausen' experiment is

$$
\begin{array}{l}
^1\text{H} \quad 90_x - \tau - 180_x - \tau - 90_y - t_1/2 - 180_x - t_1/2 - 90_x - \tau - 180_x - \tau - \text{Acq}(t_2)\\[4pt]
^{15}\text{N} \qquad\qquad 180_x \qquad 90_\psi \qquad\qquad\qquad\qquad\qquad\qquad 90_\phi \qquad 180_x
\end{array}
$$

with $\psi = x, -x, -x, x$; $\phi = x, x, -x, -x$; Acq $= x, -x, x, -x$; and τ set to slightly less (2.25 msec) than $1/(4J_{\text{NH}})$. The 180_x pulse in the centre of the evolution period should be a composite pulse (e.g. $90_x180_y90_x$).

The ^{1}H–^{15}N correlation experiment must be recorded in H$_2$O. The H$_2$O resonance can be suppressed using coherent pre-saturation. A particularly useful approach involves the application of a DANTE sequence for off-resonance irradiation, thereby permitting one to restrict the spectral width in the F_2 dimension to the NH region of the spectrum (29). The sequence typically used is $(P_x, P_y, P_{-x}, P_{-y})n$, where the length of the pulse P is given by $1/(4\Delta\nu)$ where $\Delta\nu$ is the frequency difference between the carrier and the water resonance.

41

In the case of ^{15}N, it is advisable to use the 'Overbodenhausen' experiment as the ^{15}N single quantum linewidths are significantly narrower than the multiple quantum ones, resulting in a large increase in resolution in the ^{15}N dimension (28). In the case of ^{13}C, however, there is little to choose between the two experiments. As the one-bond heteronuclear couplings are large (~95 Hz for ^{15}N, ~135 Hz for ^{13}C) both experiments are extremely sensitive, and can be recorded very rapidly. Indeed, good quality spectra can be obtained with only one scan per increment, making the ^{1}H–^{15}N correlation experiments an ideal technique for the measurement of NH exchange rates in the seconds to minutes range. Additionally, these spectra represent a finger-print of the protein conformation and can be used to assess quickly the structural integrity of mutant proteins.

3.3.2 Measurement of $^3J_{HN\alpha}$ couplings using ^{1}H–^{15}N correlation spectroscopy

$^3J_{HN\alpha}$ couplings are used to obtain ϕ torsion angle restraints by means of an empirical Karplus equation given by

$$^3J_{HN\alpha} = 6.4\cos^2(\phi - 60°) - 1.4\cos(\phi - 60°) + 1.9.$$

Although apparent values of $^3J_{HN\alpha}$ can be obtained from ^{1}H–^{1}H COSY spectra, a fundamental limitation of this experiment is that the minimum separation of the anti-phase components of the NH-C$^\alpha$H COSY cross-peaks is 0.58 times the NH linewidth. As a result, it is difficult to use COSY-type spectra to measure $^3J_{HN\alpha}$ coupling constants for proteins larger than about 10 kDa. This problem can be circumvented by making use of significantly narrower multiple quantum linewidths in the F_1 dimension coupled with procedures to correct for multiple quantum linewidths and dispersive con-tributions in order to measure these coupling constants from the F_1 splitting of the ^{15}N–^{1}H cross-peaks in an HMQC-J spectrum (30, 31).

The sequence of the HMQC-J spectrum is simply obtained by adding a $90°_y$ pulse after the last delay Δ in the HMQC sequence given above. The experi-ment should be recorded up to a total t_1 acquisition time of ~250 msec in order to obtain good digital resolution in F_1. The measured splittings taken directly from the F_1 cross-sections do not represent the actual couplings since the line shape of the multiplets is not purely absorptive and because the two peaks of the multiplet overlap. The dispersive component of the line shape leads to overestimation of $^3J_{HN\alpha}$ for large couplings, and the multiplet overlap causes underestimation and even disappearance of the splittings for small couplings. Consequently, an appropriate correction strategy must be em-ployed. This involves processing the spectrum with a Gaussian apodization function in F_1 using several different negative line-broadening parameters (e.g. −4, −6, −8, and −10 Hz) and a curve with a maximum at the end of the t_1 time domain data to optimize resolution. Splittings measured from the experimental spectra processed with the different parameters, along with initial guesses of the true $^3J_{HN\alpha}$ values and multiple quantum linewidths are then entered into a least-squares minimization program. Corrected values $^3J_{HN\alpha}$ are obtained by varying the input values of $^3J_{HN\alpha}$ and the multiple quantum linewidths (using, for example, Powell's non-linear optimization routine) to minimize the differences between the measured values of $^3J_{HN\alpha}$ for the different values of the line-broadening parameter and the corresponding calculated values of the apparent $^3J_{HN\alpha}$ splittings. The latter are obtained from the line shape simulated by Fourier transformation of a function describing the t_1 evolution of magnetization for the HMQC-*J* sequence, modified by application of a Gaussian apodization function. The time domain response for a multiplet centred at zero frequency is given by the expression: $\cos(2\pi.^3J_{HN\alpha}\Delta + t_1)e^{-t_1/T_{2,MQ}}$, where $T_{2,MQ}$ refers to the multiple quantum relaxation time and Δ is the delay time that allows for the efficient creation of multiple quantum coherence.

Experience with several larger proteins in a molecular weight range 15–20 kDa, in particular interleukin-1β, staphylococcal nuclease, and calmodulin, indicates that the HMQC-*J* experiment yields $^3J_{HN\alpha}$ coupling constants with an accuracy of ±0.5 Hz.

3.4 Three- and four-dimensional heteronuclear NMR spectroscopy

As the molecular weight increases above 10 kDa, two problems arise. The first is that the sensitivity of homonuclear *J* correlation experiments is reduced in larger proteins, owing to their slow tumbling which results in ^{1}H linewidths that are often significantly larger than many of the ^{1}H–^{1}H couplings. Secondly, and most importantly, extensive chemical shift overlap precludes interpretation of the 2D spectra. Both these difficulties can be overcome by the application of 3D and 4D heteronuclear NMR experiments. A summary of the various 3D and 4D experiments used in our laboratory is given in *Figure 4*, together with the phase cycling employed.

There are two fundamentally different approaches for increasing the resolution in protein NMR spectra. The conventional approach improves the resolution in 2D NMR spectra by increasing the digital resolution and by using strong resolution enhancement digital filtering functions at the expense of sensitivity. The alternative approach improves resolution by increasing the dimensionality of the spectrum and simultaneously yields important additional information about the system (i.e. ^{15}N and ^{13}C chemical shifts) (32). As this method makes use of large one-bond heteronuclear couplings, it is much less sensitive to wide linewidths associated with larger proteins. Because the resolution in 3D and 4D spectra is limited by digitization, spectra with equivalent resolution can be recorded at magnetic field strengths significantly lower than 600 MHz (e.g. the quality of 500 MHz multidimensional spectra is essentially the same as that of 600 MHz ones, although the signal-to-noise ratio is somewhat better at 600 MHz than at 500 MHz).

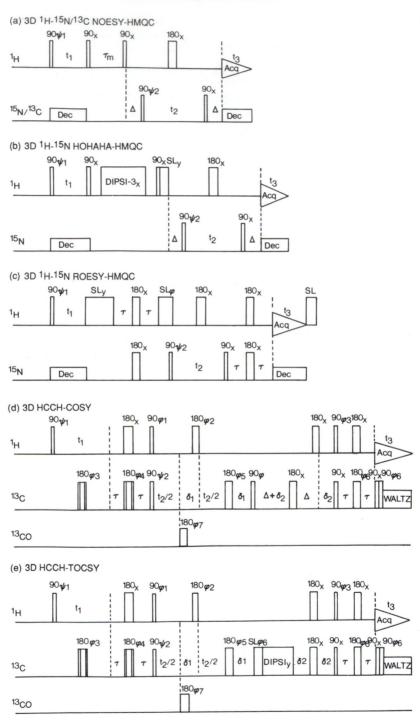

(a) 3D ^{1}H-^{15}N/^{13}C NOESY-HMQC

(b) 3D ^{1}H-^{15}N HOHAHA-HMQC

(c) 3D ^{1}H-^{15}N ROESY-HMQC

(d) 3D HCCH-COSY

(e) 3D HCCH-TOCSY

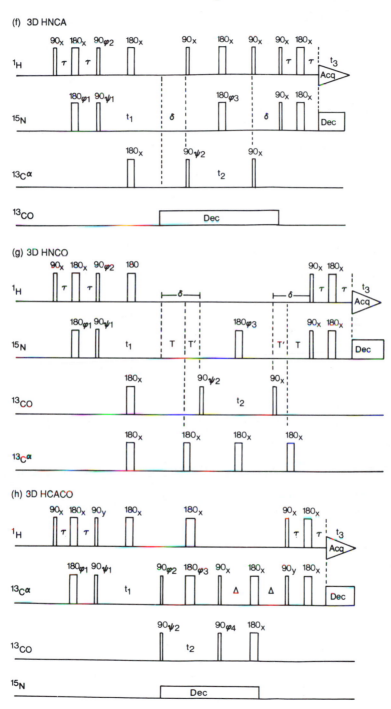

(f) 3D HNCA

(g) 3D HNCO

(h) 3D HCACO

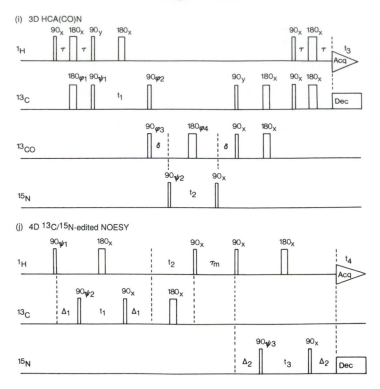

Figure 4. Pulse schemes for the most useful heteronuclear 3D and 4D NMR experiments. Phase cycling is as follows: 3D ^{1}H-^{15}N or ^{1}H-^{13}C NOESY HMQC (a) and 3D ^{1}H-^{15}N HOHAHA–HMQC (b): $\psi 1 = x, -x$; $\psi 2 = 2(x), 2(-x)$; Acq = $x, -x, -x, x$. 3D ^{1}H-^{15}N ROESY–HMQC (c): $\psi 1 = x, -x$; $\psi 2 = 2(x), 2(-x)$, $\phi = 4(x), 4(-x)$; Acq = $(x, -x, -x, x)$ $(-x, x, x, -x)$. 3D HCCH–COSY (d) and HCCH–TOCSY (e): $\phi 1 = y, -y$; $\phi 2 = 4(x), 4(y), 4(-x), 4(-y)$; $\phi 3 = 8(x)$, $8(-x)$; $\phi 4 = 2(x), 2(-x)$; $\phi 5 = 2(x), 2(y), 2(-x), 2(-y)$; $\phi 6 = 4(x), 4(-x)$; $\phi 7 = 8(x), 8(y)$; $\psi 1 = 16(x)$; $\psi 2 = 32(x)$; Acq = $2(x, -x, -x, x)$ $2(-x, x, x, -x)$. 3D HNCA (f) and HNCO (g): $\psi 1 = x$; $\psi 2 = 2(x), 2(-x)$; $\phi 1 = x, -x$; $\phi 2 = y, -y$; $\phi 3 = 4(x), 4(y), 4(-x), 4(-y)$; Acq = $x, 2(-x), x, -x, 2(x), -x$. 3D HCACO (h): $\psi 1 = x$; $\psi 2 = 2(x), 2(-x)$; $\phi 1 = x, -x$; $\phi 2 = y, -y$; $\phi 3 = 4(x), 4(-x)$; $\phi 4 = 8(x)$, $8(-x)$; Acq = $2(x, -x, -x, x), 2(-x, x, x, -x)$. 3D HCA(CO)N (i): $\psi 1 = x$; $\psi 2 = 4(x), 4(-x)$; $\phi 1 = x$, $-x$; $\phi 2 = y, -y$; $\phi 3 = 2(x), 2(-x)$; $\phi 4 = 8(x), 8(y), 8(-x), 8(-y)$; Acq = $x, 2(-x), x, -x, 2(x), 2(-x)$, $2(x), -x, x, 2(-x), x$. 4D ^{13}C/^{15}N-edited NOESY (j): $\psi 1 = 4(x)$, $\psi 2 = 2(x, -x)$, $\psi 3 = 2(x), 2(-x)$; Acq = $x, -x, -x, x$. Quadrature in the indirectly detected dimensions is achieved by changing the phases labelled ψ_i in an independent manner using the TPPI–States method (cf. *Table 1*). Typically, for 3D experiments we record a data set comprising 32 complex points in the heteronuclear dimension(s), 128 complex points in the indirectly detected ^{1}H dimension, and 512–1024 real points in the ^{1}H acquisition dimension. For 4D experiments, we typically record 8–16 complex points in the heteronuclear dimensions, 64 complex points in the indirectly detected ^{1}H dimension, and 256–512 real points in the ^{1}H acquisition dimension. The data are then processed with one zero-filling in all dimensions.

3.4.1 Assignment of spin systems by heteronuclear three-dimensional NMR

The strategy to identify spin systems using heteronuclear 3D NMR comprises two stages. In the first step the NH and ^{15}N chemical shifts of each residue are correlated with the corresponding ^{13}C$^\alpha$ and C$^\alpha$H proton chemical shifts using 3D ^{1}H–^{15}N HOHAHA–HMQC (33; *Figure 4b*) and triple resonance ^{1}H–^{15}N–^{13}C$^\alpha$ (HNCA) (34; *Figure 4f*) spectroscopy, respectively. The first experiment relies on transfer of magnetization between NH and C$^\alpha$H spins via the three-bond homonuclear intra-residue $^3J_{HN\alpha}$ coupling, while the second involves transfer of magnetization via the one-bond heteronuclear $^1J_{NC\alpha}$ intra-residue coupling. In the HOHAHA–HMQC experiment, aliphatic ^{1}H chemical shifts evolve during the period t_1. Transfer of magnetization originating on aliphatic protons to the corresponding intra-residue NH protons is achieved via anisotropic mixing of ^{1}H magnetization in the next step, using, for example, a WALTZ17 sequence (see Section 3.2.3). Magnetization, which now resides on the NH protons, is transferred to the directly bonded ^{15}N spins by means of multiple quantum coherence. ^{15}N chemical shifts evolve during the period t_2. Magnetization is then transferred back from the ^{15}N spins to the directly bonded NH protons via multiple quantum coherence and detected during the acquisition period t_3. The HNCA experiment, on the other hand, transfers magnetization originating on an NH proton to its directly bonded ^{15}N spin via an INEPT sequence, followed by the evolution of ^{15}N chemical shifts during the period t_1; subsequent application of 90° pulses to both ^{1}H and ^{13}C$^\alpha$ spins establishes three-spin NH–^{15}N–^{13}C$^\alpha$ coherence, and evolution of solely ^{13}C$^\alpha$ chemical shifts during the period t_2 is ensured by refocusing of ^{1}H and ^{15}N chemical shifts through the application of ^{1}H and ^{15}N 180° pulses at the midpoint of the t_2 period. Magnetization is then transferred back to the NH protons by simply reversing the above procedure. It is important to realize that both types of spectra must be recorded in order to correlate the C$^\alpha$H and ^{13}C$^\alpha$ chemical shifts of a given residue unambiguously, as for larger proteins with many degenerate C$^\alpha$H chemical shifts, the ^{13}C$^\alpha$ chemical shifts cannot simply be determined by recording a 2D ^{1}H–^{13}C shift correlation spectrum.

With the C$^\alpha$H and ^{13}C$^\alpha$ resonances of reach residue correlated, the amino acid spin systems can be delineated using the HCCH–COSY (*Figure 4d*) and HCCH–TOCSY (*Figure 4e*) experiments (35–37). Both experiments are based on analogous principles, making use of the well-resolved one-bond ^{1}H–^{13}C (~140 Hz) and ^{13}C–^{13}C (30–40 Hz) J couplings to transfer magnetization, thereby circumventing the problems associated with conventional methodologies (e.g. ^{1}H–^{1}H COSY, HOHAHA, etc.) which rely on poorly resolved ^{1}H–^{1}H couplings (3–12 Hz). The pulse scheme for the two experiments is naturally comparable. In the first step, ^{1}H magnetization from a proton is transferred to its directly bonded ^{13}C nucleus via the $^1J_{CH}$ coupling in an INEPT-type manner. In the second step, ^{13}C magnetization is transferred between ^{13}C neighbour(s) via the $^1J_{CC}$ coupling. In the case of the HCCH–COSY experiment, this is achieved by a 90° ^{13}C COSY mixing pulse so that magnetization exchange only occurs between two directly bonded ^{13}C atoms; in the HCCH–TOCSY experiment, on the other hand, isotropic mixing of ^{13}C spins is achieved using a DIPSI-3 pulse train such that both direct and multiple-relayed magnetization transfer occurs along the carbon chain. (The DIPSI-3 sequence is given by $(245_x395_y250_x275_y30_x230_y360_x 245_y370_x340_y350_x260_y270_x30_y225_x365_y255_x395_y)(245_y395_x250_y275_x30_y230_x 360_y245_x370_y340_x350_y260_x270_y30_x225_y365_x255_y395_x)$ repeated for a duration of 20–25 msec.) Ultimately, ^{13}C magnetization is transferred back to ^{1}H via the $^1J_{CH}$ coupling and detected during t_3. The final result is a 3D spectrum in which each $^1H(F_1)$–$^1H(F_3)$ plane has an appearance similar to that of a 2D ^{1}H–^{1}H COSY or HOHAHA/TOCSY experiment, but is edited by the ^{13}C chemical shift of the ^{13}C nucleus directly bonded to the ^{1}H at the diagonal position from which the magnetization originates. Further, in contrast to the 2D correlation experiments, the cross-peaks in each plane do not occur symmetrically on either side of the diagonal. Consider the case where magnetization is transferred from proton A to proton B. In the plane corresponding to the ^{13}C chemical shift of the ^{13}C nucleus directly bonded to proton A where magnetization originates, a correlation is observed between the diagonal peak at $(F_1,F_3) = (\delta_A,\delta_B)$ and a cross-peak $(F_1,F_3) = (\delta_A,\delta_B)$ in one half of the spectrum. The symmetrical correlation between the diagonal peak at $(F_1,F_3) = (\delta_B,\delta_B)$ and the cross-peak at $(F_1,F_3) = (\delta_B,\delta_A)$ appears in the plane corresponding to the ^{13}C chemical shift of the ^{13}C nucleus directly bonded to proton B (*Figure 5*). By this means unambiguous checks on the assignments are afforded at each step of the process, which are made all the easier as the ^{13}C chemical shifts for different carbon types are located in

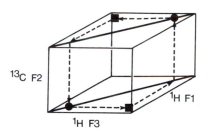

Figure 5. Schematic diagram of two slices of an HCCH spectrum at different $^{13}C(F_2)$ chemical shifts, showing the diagonal and cross-peaks expected for a simple two-spin system. The diagonal peaks are represented by circles and the cross-peaks by squares. Note that each slice is asymmetric about the diagonal, and cross-peaks only appear in the F_3 dimension. This is due to the fact that magnetization originating from a proton attached to a given ^{13}C nucleus is only visible in the F_2 slice corresponding to the ^{13}C chemical shift of this particular ^{13}C nucleus. The corresponding cross-peak at the other side of the $F_1 = F_3$ diagonal is found in the slice taken at the ^{13}C frequency of the destination carbon. (From ref. 37.)

characteristic regions of the ^{13}C spectrum, with little overlap between them. This characteristic of ^{13}C chemical shifts also enables one to make use of extensive folding to reduce the $^{13}C(F_2)$ spectral width to about 20 p.p.m., thereby increasing the available resolution in the $^{13}C(F_2)$ dimension. The HCCH–COSY experiment is particularly useful for identifying Gly, Ala, Thr, and Val spin systems, as well as for amino acids of the AMX type (e.g. Ser, Cys, Asn, Asp, His, Tyr, Phe, and Trp). For longer side-chains, the HCCH–TOCSY experiment has to be used, as it is often difficult to locate connectivities between $C^\beta H$ and $C^\gamma H$ protons, $C^\gamma H$ and $C^\delta H$ protons, and $C^\delta H$ and $C^\varepsilon H$ protons, as their shifts are frequently similar and sometimes degenerate. Complete Leu, Ile, Lys, Arg, and Pro spin systems are readily identified using this technique, in contrast to the situation in conventional ^1H–^1H spectroscopy, as even for smaller proteins of 50–80 residues the ratios of linewidths to ^1H–^1H three-bond couplings are unfavourable for efficient magnetization transfer.

The success and practical applicability of these 3D techniques has been demonstrated by the results on interleukin-1β, a protein of 153 residues (17.4 kDa), where complete ^1H, ^{15}N, and ^{13}C resonance assignments were obtained using the 3D HNCA, HCCH–COSY, and HCCH–TOCSY experiments described above (37).

3.4.2 Conventional sequential assignment using three-dimensional heteronuclear NMR

As pointed out in Section 2, conventional sequential assignment relies on identifying NOE connectivities between adjacent residues involving the NH, $C^\alpha H$, and $C^\beta H$ protons. The same approach can be used in heteronuclear 3D NMR, making use of the 3D ^1H–^{15}N HOHAHA–HMQC experiment described above (Section 3.4.1) to identify intra-residue connectivities between NH and $C^\alpha H$ protons, and the 3D ^1H–^{15}N NOESY–HMQC experiment (*Figure 4a*) to identify through-space connectivities (33). This procedure has been used to obtain complete sequential assignments of the polypeptide backbone of interleukin-1β (153 residues, 17.4 kDa) (38–39). The two experiments are very similar and differ in the mixing sequence (HOHAHA versus NOE mixing). Thus, the 3D ^1H–^{15}N NOESY–HMQC spectrum has the same appearance as the $NH(F_2)$–$^1H(F_1)$ region of a conventional ^1H–^1H NOESY spectrum, spread out in a series of slices according to the ^{15}N chemical shifts. Sequential assignment can thus proceed in a relatively straightforward manner by hopping from one pair of HOHAHA/NOESY planes to another pair, connecting them via either $C^\alpha H(i)$–$NH(i+1)$ or $NH(i)$–$NH(i+1)$ NOEs in a manner analogous to that employed in the analysis of 2D spectra. An alternative and simple method of analysis is readily conceived by selecting strips of data from each slice containing cross-peaks arising from individual amide resonances, thereby eliminating the empty space that is present in the 3D spectrum and the redundancy that is caused by the fact that a series of cross-peaks from a single amide NH may appear in more than one slice of the 3D spectrum (38). In this manner, all the information present in the 3D ^{1}H–^{15}N NOESY–HMQC spectrum can be reduced to a relatively small number of 2D plots.

In addition, the ^{1}H–^{15}N ROESY–HMQC experiment (*Figure 4c*; 40) is useful for distinguishing chemical exchange effects from NOEs (as these two processes result in cross-peaks of opposite sign) and for circumventing problems associated with spin-diffusion. This experiment is particularly valuable for identifying bound water molecules (40).

3.4.3 Sequential assignment via well-resolved one-bond J couplings using triple resonance heteronuclear three-dimensional NMR spectroscopy

In addition to the conventional sequential assignment approach which relies on the use of the NOE, heteronuclear 3D NMR can be employed to assign sequentially the backbone NH, ^{15}N, ^{13}C$^\alpha$, ^{13}CO, and C$^\alpha$H resonances by means of heteronuclear one-bond couplings using a number of triple resonance experiments without any need for a knowledge of spin sytems (34). This approach relies on five 3D experiments, two of which, the ^{1}H–^{15}N HOHAHA–HMQC and the HNCA experiment, have already been described above with respect to spin-system identification (Section 3.4.1). The other three experiments are the 3D HNCO (*Figure 4g*), HCACO (*Figure 4h*) and HCA(CO)N (*Figure 4i*) experiments (34). The connectivities observed in these five experiments are illustrated in *Figure 6*, from which it is readily apparent that there will be at least two, and often three, independent pathways for determining any given sequential connectivity.

The HNCO experiment (*Figure 4g*) correlates the NH and ^{15}N chemical shifts of residue i with the ^{13}CO shift of the preceding residue via the one-bond $^1J_{NCO}$ coupling (~15 Hz), thereby providing sequential connectivity information. In this experiment, magnetization originating from NH protons is transferred to the directly bonded ^{15}N spins using an INEPT sequence, following which ^{15}N magnetization evolves exclusively under the influence of the ^{15}N chemical shift as a result of ^{1}H, ^{13}CO, and ^{13}C$^\alpha$ decoupling through the application of 180° pulses at the midpoint of the t_1 interval. During the delay, δ, ^{15}N magnetization becomes anti-phase with respect to the polarization of the carbonyl spin of the preceding residue via the $^1J_{NCO}$ coupling, and the subsequent 90° ^{13}CO pulse converts this magnetization into ^{15}N–^{13}CO two-spin coherence. Evolution of ^{13}CO chemical shifts then occurs during the period t_2, and the contributions of the ^{15}N chemical shift and ^{1}H–^{15}N J coupling on the one hand, and of the ^{13}C$^\alpha$–^{13}CO J coupling on the other, are removed by the application of 180° ^{15}N and ^{13}C$^\alpha$ pulses, respectively, at the t_2 midpoint. Magnetization is finally transferred back to the NH protons by reversing the transfer steps and is detected during t_3.

The HCACO experiment (*Figure 4h*) correlates the intra-residue C$^\alpha$H,

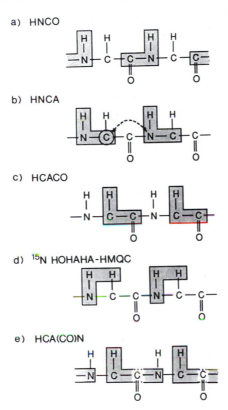

a) HNCO

b) HNCA

c) HCACO

d) ^{15}N HOHAHA-HMQC

e) HCA(CO)N

Figure 6. Connectivities observed in 3D triple resonance correlation experiments that can be used for sequential backbone assignments via well-resolved one-bond heteronuclear J couplings (from ref. 34).

$^{13}C^{\alpha}$, and ^{13}CO shifts. Magnetization is transferred from $C^{\alpha}H$ protons to the directly bonded $^{13}C^{\alpha}$ spins via an INEPT sequence, and $^{13}C^{\alpha}$ magnetization evolves during the period t_1 under the influence of $^{13}C^{\alpha}$ chemical shift as well as the $^{13}C^{\alpha}$–^{13}CO and $^{13}C^{\alpha}$–$^{13}C^{\beta}$ J couplings, while the $C^{\alpha}H$–$^{13}C^{\alpha}$ J coupling is removed by the 180° ^{1}H pulse applied at the midpoint of t_1. Transfer of magnetization occurs next in a COSY-like manner from $^{13}C^{\alpha}$ spins to ^{13}CO spins via the $^{1}J_{C_{\alpha}CO}$ coupling by the application of simultaneous 90° $^{13}C^{\alpha}$ and ^{13}CO pulses. Evolution of ^{13}CO chemical shifts occurs during t_2, and the effects of $^{13}C^{\alpha}H$–^{13}CO and $^{13}C^{\alpha}$–^{13}CO couplings are removed by the application of 180° ^{1}H and $^{13}C^{\alpha}$ pulses at the midpoint of t_2. ^{13}CO magnetization is then transferred back to $^{13}C^{\alpha}$ by the application of simultaneous $^{13}C^{\alpha}$ and ^{13}CO pulses, at which point the $^{13}C^{\alpha}$ magnetization is anti-phase with respect to the ^{13}CO spin state. This anti-phase magnetization is removed during the subsequent interval 2Δ, and finally magnetization is transferred all the way back to the $C^{\alpha}H$ spins by a reverse-INEPT sequence and detected during t_3.

The HCA(CO)N experiment (*Figure 4i*) provides a second source of sequential information by correlating the $C^\alpha H$ and $^{13}C^\alpha$ shifts of residue i with the ^{15}N shift of residue $i+1$. The experiment is very similar to the HCACO one, except that magnetization transferred to the ^{13}CO spin in the HCACO experiment is subsequently transferred to the ^{15}N spin of residue $i+1$. This is achieved by simply including an interval $\delta \sim 0.3/J_{NCO}$ (18 msec) after the end of the t_1 period so that the ^{13}CO magnetization becomes anti-phase with respect to that of the directly bonded ^{15}N spin. The subsequent ^{15}N 90° pulse generates two-spin $^{15}N-^{13}CO$ coherence which evolves during t_2 under the influence of the ^{15}N chemical shift only, as a 180° ^{13}CO pulse refocuses the effects of both ^{13}CO chemical shifts and $^{13}CO-^{13}C^\alpha J$ coupling. At the end of the t_2 period, magnetization is transferred back to the $C^\alpha H$ protons by reversing the transfer steps and is detected during t_3.

Finally, yet a third sequential connectivity pathway is afforded by the HNCA experiment (34), which, in addition to the intra-residue NH(i)–^{15}N(i)–$^{13}C^\alpha$(i) correlations, may reveal a number of weaker inter-residue NH(i)–^{15}N(i)–$^{13}C^\alpha$($i-1$) correlations which arise via the small (≤ 7 Hz) two-bond $^2J_{NC\alpha}$ inter-residue coupling.

These triple resonance experiments have been used to obtain complete backbone ^1H, ^{13}C, and ^{15}N assignments for calmodulin, a protein of 16.7 kDa (34), and should prove very powerful for proteins up to ~ 25 kDa.

3.4.4 Identification of long-range NOE connectivities by heteronuclear three- and four-dimensional NMR

The key to determining the protein fold lies in the identification of NOEs between residues far apart in the sequence but close together in space. This step can only be accomplished once complete (or almost complete) resonance assignments are available. It is often the case, particularly for larger proteins, that the assignments of a large number of such tertiary NOE cross-peaks in conventional 2D $^1H-^1H$ NOESY spectra is ambiguous due to severe resonance overlap. This can be resolved by the application of 3D and 4D heteronuclear edited NOESY spectroscopy.

i. NOEs between non-exchangeable protons

In the case of non-exchangeable protons (which are invariably attached to carbon atoms), ambiguities arising from ^1H chemical shift overlap can be resolved by recording a 3D $^{13}C-^1H$ NOESY–HMQC spectrum in which NOEs between the protons are spread out according to the chemical shift of the ^{13}C spin directly bonded to the originating proton (41, 42). The pulse scheme is identical to that of the 3D $^1H-^{15}N$ NOESY–HMQC experiment except, of course, that ^{13}C pulses replace the ^{15}N pulses. Just as in the case of the HCCH–COSY and HCCH–TOCSY 3D spectra described above (Section 3.4.1), the spectrum at each ^{13}C slice of the 3D $^1H-^{13}C$ NOESY–HMQC is asymmetric about the diagonal. Thus, an NOE originating on proton A and ending on proton B will be seen in the ^{13}C slice corresponding to the ^{13}C spin attached to proton A, while the corresponding NOE from proton B to proton A is seen in the ^{13}C slice corresponding to the ^{13}C spin attached to proton B. As a result, numerous cross-checks are available to resolve ambiguities and render cross-peak assignments essentially foolproof.

ii. NOEs involving protons attached to nitrogen

In the 3D ^{1}H–^{15}N NOESY–HMQC experiment, NOEs between NH protons and aliphatic protons are spread into a third dimension by the chemical shift of the directly bonded ^{15}N atoms. While, as discussed in Section 3.4.2, this 3D experiment effectively removes, in all but a very few cases, chemical shift degeneracy associated with the NH protons, it leaves ambiguities associated with severe overlap of the aliphatic protons unaffected. Thus, even if a cross-peak connecting an aliphatic and amide proton is well resolved in the 3D spectrum, it is frequently not possible, with the exception of cases involving the C$^\alpha$H resonances, to identify conclusively the aliphatic proton involved on the basis of its ^{1}H chemical shift. This problem can be overcome by intro-ducing a fourth dimension comprising the ^{13}C chemical shift of ^{13}C spins directly bonded to the aliphatic protons (32). Such a ^{15}N/^{13}C-edited experi-ment is designed by combining three separate 2D experiments, namely ^{1}H–^{13}C HMQC, ^{1}H–^{1}H NOESY, and ^{1}H–^{15}N HMQC sequences. Transfer of magnetization between protons and the directly bonded ^{15}N or ^{13}C hetero-nucleus is achieved by means of multiple quantum coherence, while transfer between protons occurs via through-space NOE effects. The transfer of magnetization from aliphatic protons thus follows the pathway:

$$^1H \xrightarrow{^1J_{CH}} {}^{13}C \xrightarrow{^1J_{CH}} {}^1H \xrightarrow{NOE} NH \xrightarrow{^1J_{NH}} {}^{15}N \xrightarrow{^1J_{NH}} NH.$$

The pulse scheme for the 4D ^{13}C/^{15}N-edited NOESY experiment is shown in *Figure 4j*. The chemical shifts of ^{13}C, ^{1}H, and ^{15}N evolve during the variable time periods t_1, t_2, and t_3 which are incremented independently, and the NH signal is acquired during the acquisition period t_4. There are three key aspects of practical importance concerning this 4D experiment. First, the number of peaks in this 4D spectrum is the same as that present in the corresponding ^{15}N/^{13}C edited 3D and 2D spectra, so that the extension to a fourth dimension increases the resolution without a concomitant increase in complexity. Secondly, the through-bond transfer steps are highly efficient as they involve one-bond heteronuclear couplings (90–130 Hz) which are much larger than the linewidths. As a result, the sensitivity of the experiment is high and can easily be performed on a 1–2 mM sample of uniformly labelled ^{15}N/^{13}C protein. Thirdly, extensive folding can be employed to maximize resolution in the ^{13}C(F_1) dimension so that each ^{13}C coordinate corresponds to a series of ^{13}C chemical shifts separated by intervals of approximately 20 p.p.m. This does not complicate the interpretation of the 4D spectrum in any way since all

^{13}C resonances would have been previously assigned using the 3D HNCA, HCCH–COSY, and HCCH–TOCSY experiments described in Section 3.4.1, and the appropriate ^{13}C chemical shift is easily ascertained on the basis of the ^{1}H chemical shift of the aliphatic proton from which the magnetization originates.

iii. Other heteronuclear four-dimensional NMR experiments for assignment of long range NOEs

It is clear that a number of other 4D experiments can be performed that are useful for the computer automated assignment of long-range NOEs. For example, the 4D ^{1}H–^{13}C–HMQC–^{1}H–^{1}H–NOESY–^{1}H–^{13}C–HMQC experiment permits straightforward assignment of NOEs in a completely automated manner as each cross-peak is labelled by the ^{1}H and ^{13}C chemical shifts of both the originating and destination CH groups (43). Further, these experiments permit one to observe NOEs between protons with completely degenerate chemical shifts, providing the corresponding shifts of the directly bonded heteronucleus are different, as the diagonal peaks only occur in very limited regions of the spectrum where $\delta(F_1) = \delta(F_3)$.

Acknowledgements

We thank first and foremost Dr Ad Bax, as well as Drs Paul Driscoll, Mitsuhiko Ikura, and Lewis Kay for many valuable discussions and fruitful collaborations in the development of the heteronuclear 3D and 4D NMR experiments described here. The work in the authors' laboratory was supported in part by the Intramural AIDS Directed Anti-Viral Program of the Office of the Director of the National Institutes of Health.

References

1. Wüthrich, K. (1986). *NMR of proteins and nucleic acids.* John Wiley, New York.
2. Clore, G. M. and Gronenborn, A. M. (1989). *CRC Crit. Rev. Biochem. Mol. Biol.,* **24,** 479.
3. Gronenborn, A. M. and Clore, G. M. (1990). *Anal. Chem.,* **62,** 2.
4. Clore, G. M. and Gronenborn, A. M. (1992). *Ann. Rev. Biophys. Biophys. Chem.,* **20,** in press.
5. Kraulis, P. J., *et al.* (1989). *Biochemistry,* **28,** 7241.
6. Qian, Y. Q., Billeter, M., Otting, G., Müller, M., Gehring, W. J., and Wüthrich, K. (1989). *Cell,* **59,** 573.
7. Clore, G. M., Appella, E., Yamada, E., Matsushima, K., and Gronenborn, A. M. (1990). *Biochemistry,* **29,** 1689.
8. Omichinski, J., Clore, G. M., Appella, E., Sakaguchi, K., and Gronenborn, A. M. (1990). *Biochemistry,* **29,** 9324.
9. Forman-Kay, J. D., Clore, G. M., Wingfield, P. T., and Gronenborn, A. M. (1990). *Biochemistry,* **30,** 2685.

10. Wüthrich, K. (1989). *Acc. Chem. Res.,* **22,** 36.
11. Ernst, R. R., Bodenhausen, G., and Wokaun, G. (1987). *Principles of nuclear magnetic resonance in one and two dimensions.* Clarendon Press, Oxford.
12. Fesik, S. W. and Zuiderweg, E. R. P. (1990). *Quart. Rev. Biophys.,* **23,** 97.
13. Havel, T. F., Kuntz, I. D., and Crippen, G. M. (1983). *Bull. Math. Biol.,* **45,** 665.
14. Braun, W. (1987). *Quart. Rev. Biophys.,* **19,** 115.
15. Clore, G. M., Brünger, A. T., Karplus, M., and Gronenborn, A. M. (1986). *J. Mol. Biol.,* **191,** 523.
16. Nilges, M., Clore, G. M., and Gronenborn, A. M. (1988). *FEBS Lett.,* **229,** 317.
17. Nilges, M., Clore, G. M., and Gronenborn, A. M. (1988). *FEBS Lett.,* **239,** 129.
18. Drobny, G., Pines, A., Sinton, S., Weitekamp, D., and Wemmer, D. (1979). *Faraday Div. Chem. Soc. Symp.,* **13,** 49.
19. States, D. J., Haberkorn, R. A., and Ruben, D. J. (1982). *J. Magn. Reson.,* **48,** 286.
20. Marion, D., Ikura, M., Tschudin, R., and Bax, A. (1989). *J. Magn. Reson.,* **85,** 393–9.
21. Jeener, J., Meier, B. H., Bachman, P., and Ernst, R. R. (1979). *J. Chem. Phys.,* **71,** 4546.
22. Marion, D. and Bax, A. (1988). *J. Magn. Reson.,* **80,** 528.
23. Mueller, L. (1987). *J. Magn. Reson.,* **72,** 191.
24. Bax, A. (1989). In *Methods in enzymology* (ed. N. J. Oppenheimer and T. L. James), Vol. 176, p. 151. Academic Press, San Diego.
25. Bax, A., Sklenar, V., Clore, G. M., and Gronenborn, A. M. (1987). *J. Am. Chem. Soc.,* **109,** 7188.
26. Bax, A., Griffey, R. G., and Hawkins, B. L. (1983). *J. Magn. Reson.,* **55,** 301.
27. Bodenhausen, G. and Ruben, D. J. (1980). *Chem. Phys. Lett.,* **69,** 185.
28. Bax, A., Ikura, M., Kay, L. E., Torchia, D. A., and Tschudin, R. (1990). *J. Magn. Reson.,* **86,** 304.
29. Kay, L. E., Marion, D., and Bax, A. (1989). *J. Magn. Reson.,* **84,** 72.
30. Kay, L. E. and Bax, A. (1990). *J. Magn. Reson.,* **86,** 110.
31. Forman-Kay, J. D., Gronenborn, A. M., Kay, L. E., Wingfield, P. T., and Clore, G. M. (1990). *Biochemistry,* **29,** 1566.
32. Kay, L. E., Clore, G. M., Bax, A., and Gronenborn, A. M. (1990). *Science,* **249,** 411.
33. Marion, D., *et al.* (1989). *Biochemistry,* **29,** 6150.
34. Ikura, M., Kay, L. E., and Bax, A. (1990). *Biochemistry,* **29,** 4659.
35. Bax, A., Clore, G. M., Driscoll, P. C., Gronenborn, A. M., Ikura, M., and Kay, L. E. (1990). *J. Magn. Reson.,* **87,** 620.
36. Bax, A., Clore, G. M., and Gronenborn, A. M. (1992). *J. Magn. Reson.,* in press.
37. Clore, G. M., Bax, A., Driscoll, P. C., Wingfield, P. T., and Gronenborn, A. M. (1990). *Biochemistry,* **29,** 8172.
38. Driscoll, P. C., Clore, G. M., Marion, D., Wingfield, P. T., and Gronenborn, A. M. (1990). *Biochemistry,* **29,** 3542.
39. Driscoll, P. C., Gronenborn, A. M., Wingfield, P. T., and Clore, G. M. (1990). *Biochemistry,* **29,** 4668.

40. Clore, G. M., Bax, A., Wingfield, P. T., and Gronenborn, A. M. (1990). *Biochemistry*, **29,** 5671.
41. Ikura, M., Kay, L. E., Tschudin, R., and Bax, A. (1990). *J. Magn. Reson.*, **86,** 204.
42. Zuiderweg, E. R. P., McIntosh, L. P., Dahlquist, F. W., and Fesik, S. W. (1990). *J. Magn. Reson.*, **86,** 210.
43. Clore, G. M., Kay, L. E., Bax, A., and Gronenborn, A. M. (1991). *Biochemistry*, **30,** 12.

3

Molecular sequence data-bases and their uses

MARK S. BOGUSKI, JAMES OSTELL, and DAVID J. STATES

1. Introduction

There are many aspects of nucleic acid and amino acid sequences pertinent to protein engineering. For example, promoters and other types of non-coding regulatory DNA and RNA can be studied and manipulated with the purpose of designing transcription and translation units for maximal expression of natural or mutated proteins in heterologous systems. However, this topic is covered elsewhere in this volume (Chapters 13–15) and for present purposes we consider nucleic acid sequences only as additional sources of conceptually translated protein sequences. The study of natural sequence variation in protein families also has obvious implications for protein design and engineering but will not be dealt with here (see Chapter 14).

This chapter describes

- the design and implementation of the GenInfo Backbone (GIBB) sequence data-base by the National Center for Biotechnology Information at the US National Library of Medicine (1)
- new, network-based sequence similarity search tools and new methods for assessing the statistical significance of sequence similarities
- various methods of multiple sequence alignment for the identification and analysis of protein sequence motifs as well as for protein modelling

The practical use of data-base searching and sequence alignment tools will be demonstrated in a detailed example.

2. Data-bases in molecular biology

2.1 Overview

Some major sequence data-bases, particularly GenBank, EMBL, and NBRF/PIR, have been discussed extensively elsewhere, including a previous volume in this series (2, 3). Important characteristics of these data-bases are included in *Table 1*. For a brief historical account of the development of genetic sequence data-bases, the reader should consult ref. 4. A more recent effort, that deserves special mention, is the OWL composite protein sequence data-base that was designed and constructed as part of the ISIS (Integrated Sequences/Integrated Structures) project of the British Protein Engineering Club (5). We also include a discussion of SWISS-PROT because of its unique features, high quality of annotation, and its explicit cross-references to the Brookhaven data-base of protein crystallographic structures.

2.2 OWL

OWL (not an acronym but rather a symbol of the collaboration between Birkbeck College and Leeds, both of which contain an owl in their heraldic crests) is a meta-database of protein sequences whose entries are consolidated from eight primary source data-bases of protein and (translated) nucleic acid sequences (6). These include NBRF/PIR, SWISS-PROT, GenBank, NBRF/PIR NEW, NEWAT86, PSD-Kyoto, NRL–3D, and Brookhaven PDB. OWL was designed primarily to address a number of practical problems associated with having to search multiple-source data-bases that have different file formats and considerable degrees of redundancy among their sequence entries. The design of OWL was a compromise: sequence redundancy is removed but supplementary annotation from the 'lower priority' data-bases (see below) is lost.

During the construction and updating processes of OWL:

(a) Source data-bases are first converted to a common format that is a software-compatible extension of the NBRF/PIR format and, in the case of GenBank, amino acid translations are performed using the programs of Fickett (7, 8) based upon information in the GenBank features tables.

Table 1. Some major data-bases for molecular/structural biology

Name	Contents	Availability
GenInfo	Nucleic acid and protein sequences with MEDLINE abstracts from published articles and by direct submission from authors; GenBank and NBRF/PIR data incorporated in ASN.1 Sponsor: National Center for Biotechnology Information, NIH	Network, CD-ROM geninfo@ncbi.nlm.nih.gov
GenBank[a]	Nucleic acid sequences from published articles and by direct submission from authors; translations of coding regions separately available (GenPept). Sponsors: National Institute of General Medical Sciences, NIH by contract to Intelligenetics, Inc. and Los Alamos National Laboratory	Network, CD-ROM, magnetic tape gos@genbank.bio.net gb-sub%life@lanl.gov (for direct submission)

Table 1. *continued*

Name	Contents	Availability
EMBL	Nucleic acid sequences from published articles and by direct submission from authors Sponsor: European Molecular Biology Organization	Network, CD-ROM, magnetic tape datalib@embl.bitnet datalib@embl-heidelberg.de
NBRF/PIR	Protein sequences from published articles Sponsors: National Library of Medicine, NIH by grant to the National Biomedical Research Foundation/Protein Identification Resource	Network, CD-ROM, magnetic tape pirmail@gunbrf.bitnet
SWISS-PROT	Protein sequences from published articles and translations of coding regions from EMBL Sponsors: Amos Bairoch, University of Geneva	Network, CD-ROM, magnetic tape bairoch@cmu.unige.ch datalib@embl-heidelberg.de
OWL	Protein sequences consolidated from NBRF/PIR, SWISS-PROT, GenBank (translation), NEWAT86, PSD-Kyoto, NRL_3D, and PDB Sponsor: University of Leeds; Science and Engineering Council, UK; Protein Engineering Initiative	Network, CD-ROM, magnetic tape uig@daresbury.ac.uk
PDB	Protein and nucleic acid three-dimensional structures by direct submission from authors of crystallographic and NMR data as well as models	Network, magnetic tape pdb@bnlchm.bitnet
NRL–3D	Protein sequence and structure data-base derived from PDB and searchable in the NBRF/PIR environment Sponsors: Office of Naval Research; US Army Medical Research and Development Command	Network, CD-ROM, magnetic tape pirmail@gunbrf.bitnet
CCSD	Complex Carbohydrate Structural Database containing complex carbohydrate, glycolipid, and glycopeptide sequences and annotations from published articles Sponsors: Complex Carbohydrate Research Center at the University of Georgia; US Department of Energy; NIH; European Economic Community	IBM PC floppy diskettes CarBank@uga.bitnet

Current releases at the time of this writing include: GenBank 67.0 (55 169 276 bases in 43903 loci), NBRF/PIR 27.0 (7 620 668 residues in 26798 sequences), SWISS-PROT 17.0 (6 524 504 residues in 20024 sequences), PDB July 1990 (583 coordinate sets representing about 115 megabytes of data), and CCSD (approx. 4000 entries). To provide some idea of the redundancy among PIR, SWISS-PROT, and translated GenBank (GenPept), a composite collection of all three with exact matches removed contains 12 363 598 residues in 43036 sequences (W. Gish, personal communication).
[a] As of October 1992 NCBI will assume responsibility for distribution of Genbank.

(b) This translation is followed by procedures to correct the translation of flagged anomalous codons and to produce separate protein sequence entries from GenBank entries that contain multiple open reading frames.

(c) The source data-bases have priorities defined in terms of the judged quality of sequence validation and annotation. The annotation in PIR and SWISS-PROT is equally comprehensive but PIR was arbitrarily chosen as the highest priority data-base and is used as the main data-base against which lower priority data-bases are sequentially compared. This comparison process uses a heuristic sequence alignment protocol to eliminate redundant or trivially different sequence entries.

(d) In the final steps, all non-redundant entries are merged into the composite data-base and a number of validation procedures are carried out.

The construction and updating process is semi-automated and modular, providing for scientific judgements by human experts to be applied at each stage. The NBRF/PIR file format permits a wide range of pre-existing sequence analysis software to be used with OWL. In addition, specially designed information retrieval software has been developed (9).

2.3 SWISS-PROT

SWISS-PROT (10) is another example of a composite database of protein sequences and is built by merging entries from PIR with translated entries from EMBL using a set of automated tools for maintenance and annotation. (Some sequence data comes directly from the published literature.) The goals of SWISS-PROT include:

- minimal redundancy
- maximal annotation
- integration with other data-bases

One unique feature of the SWISS-PROT effort is the compilation of a set of defined protein sequence motifs in a linked data-base called PROSITE (11). These PROSITE features are used as the basis for the automated annotation of SWISS-PROT. By using such an automated annotation procedure, global updates and consistency checks may be performed, and the data-base is easily maintained in a machine-readable form. Manual supervision of the automated documentation procedure is necessary, but the automated tools serve a critical role in facilitating data handling. The actual specification of the PROSITE entries is accomplished through an electronic collaboration between Amos Bairoch and scientists who are subspecialists in particular sequence families.

In addition to the PROSITE links, SWISS-PROT entries also maintain cross-references to the PIR, EMBL, and Brookhaven data-bases as well as a human retrovirus/AIDS data-base (12), the on-line version of McKusik's

Medelian inheritance in man (13), and the REBASE collection of type 2 restriction enzymes (14). Finally, an ENZYMES data-base provides links to the standard IUB Enzyme Commission numbers.

3. The GenInfo Backbone data-base project

3.1 Introduction

The National Center for Biotechnology Information (NCBI) was established in November 1988 by the US Congress (Public Law 100-607) in recognition of the essential and expanding role of bio-informatics in a wide range of research and development activities connected with human health and disease (1). The NCBI's legislative mandate includes the creation of automated systems for storing and analysing knowledge about molecular biology, biochemistry, and genetics; research into advanced methods for data retrieval, and analysis and facilitation of the use of data-bases and software by biomedical researchers; and co-ordination of efforts to gather biotechnolgy information world-wide. As part of a broad programme of activities to meet these goals, the NCBI is developing a new generation of nucleic acid and protein sequence data-bases.

3.2 Limitations of current data-bases

In spite of the best efforts at sequence annotation, entries in current data-bases remain essentially detached from their intellectual wellspring—the scientific literature. Furthermore, efforts to provide comprehensive, in-depth annotation, as well as some model of biological classification, hinder the timely entry of new data and its release to the scientific community in computer-readable form. To most end-users, the present sequence data-bases represent an electronic Tower of Babel, due largely to historical circumstances of their development such as different levels and sources of funding, differences in the perceived needs of various segments of the scientific community, inconsistent nomenclature, and conflicting standard terminologies. It is now clear that scientists need an up-to-date and integrated information resource.

The GenInfo Backbone data-base project was designed to address these issues. This new initiative takes advantage not only of experience gained from the previous generation of molecular sequence data-bases but also of its association with the US National Library of Medicine, which has a 25-year history of collecting, indexing, and distributing computerized biomedical information.

The US National Library of Medicine indexes approximately 325 000 articles from 3500 journals per year for inclusion in the MEDLINE® bibliographic data-base. Those articles that contain molecular sequence data are routed to sequence indexing specialists in NLM's Division of Library Operations who enter factual information and sequences into the GenInfo Backbone.

The Backbone is designed to perform two functions:

(a) The first is the inclusion of a broad range of sequence data, some of which is not currently represented in other sequence data-bases. This breadth of coverage is made possible by access to the full journal scanning stream currently in place at the NLM. In addition, to capture sequence data present in journals not traditionally indexed by NLM (such as those dealing primarily with plant science), a bibliographic file that will complement the MEDLINE data-base is under development (The US National Agricultural Library has assisted in the identification of candidate journal titles in its data-base, AGRICOLA). The US Patent Office will also be a source of GenInfo sequence data.

(b) The second major goal is to provide a means by which molecular sequences can be used as retrieval keys to the scientific literature. Molecular sequences are concise, standard, and readily accessible. They provide an ideal complement to free-text-based retrieval methods.

Despite the great promise of direct (electronic) submission of sequence data to public data-bases, the biological context of this data will remain firmly rooted in the scientific literature, and it is unreasonable to expect that direct submission of data will, in the foreseeable future, be comprehensive without a robust literature-based entry stream. The annotation available through published papers frequently exceeds that provided in direct submissions, and there is no substitute for the peer review process in maintaining the scientific quality of the data. Furthermore, the frequency of new experiments on, and interpretations of, previously published sequences is only increasing and literature-based sequence collections provide important pointers to this information. Author-direct submission and the scientific literature (electronic or otherwise) will both serve as important sources of sequence data for some time to come.

3.3 The GenInfo Backbone (GIBB) data-base

3.3.1 The Backbone concept

An ideal molecular sequence data-base should provide:

- an up-to-date and accurate archive of all publically available data
- a knowledge base that dynamically expands and adapts as new concepts and data become available
- a substrate upon which experts in a variety of fields may apply their specialized knowledge
- a stable and standardized resource for which software tools may be written with confidence

Achieving these multiple and conflicting goals would be impossible in a single data-base, but they can be met by a set of related, loosely coupled data-bases that share a very well-defined syntax for referencing overlapping sets of information, and a common stable set of reference data (*Figure 1*). The GenInfo Backbone is a simple data-base containing the MEDLINE abstract and literature citation and minimally annotated molecular sequence data. It is designed to function as a stable and standardized foundation upon which more detailed and specialized data-bases can be built and from which customized user views can be computed (see Section 3.3.3 below).

3.3.2 Major characteristics

The GenInfo Backbone is an integrated information resource containing not only nucleic acid and protein sequence data but also annotation from the articles in which the sequences were published, the MEDLINE abstract of the publication, and taxonomic data. The GIBB is meant to be an accurate transcription of sequence data as presented in scientific publications and does nto attempt to model biology or impose interpretations of the data apart from those of the author(s) of those articles. Once the sequences have been accurately transcribed, they remain as permanent data-base records. This ensures stability of the data-base contents. Each data-base entry also includes data elements that will serve as defined links to other data-bases, currently in existence and yet to be developed (these include: journal citation, MEDLINE unique identifier, McSH® (Medical Subject Heading) indexing terms, NBRF/PIR retrieval codes, GenBank name/accession number, Enzyme Commission number/name, gene symbol, gene product, organism name/taxonomic level, and the sequence itself). As such, the Backbone is meant to provide a minimal but well-defined substrate upon which to build richer data representations, and to act as a crossroads of information resources relevant to molecular biology and biotechnology.

i. Annotation of new sequence entries

All newly published nucleic acid and protein sequences are annotated with only those features that are widely accepted and understood, non-controversial, and unambiguous, such as organism, type of molecule, and coding region(s). Interpretative features such as promoters or enhancers are not included. (It is envisioned that specialized data-bases will be created that link their entries back to GenInfo. In fact, a data-base of transcription factors and the DNA sites to which they bind (15), can be mapped to Backbone entries via MEDLINE unique identifiers—numbers that are assigned to every article that NLM indexes.) No attempt is made to second-guess or re-evaluate an author's assertion about a sequence. When possible, an author's terms are mapped to standard vocabularies (such as officially sanctioned gene names). Also, translations of coding regions (using the appropriate genetic codes) are checked automatically, but this does not lead to replacement of the translation given by the author if there is a discrepancy; the discrepancy is noted and the published translation stands as it is. All sequence annotations in the

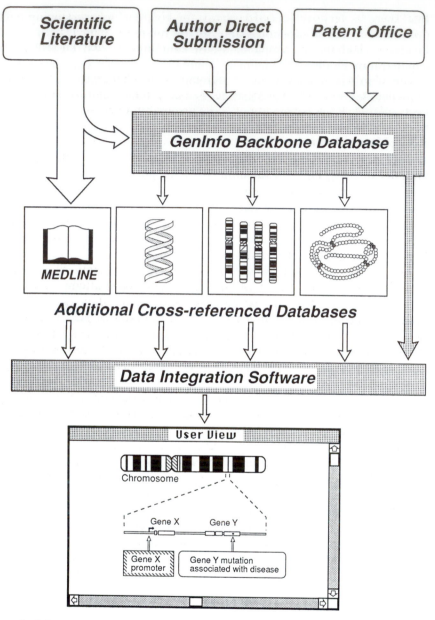

Figure 1. Schematic diagram showing the flow of information used to build the GenInfo Backbone (GIBB), to link GIBB to specialty data-bases, and to create a customized, end-user view.

Backbone are *features* defined as structured blocks of information, specific to each feature type and associated with one or more sequence intervals or locations. This design ensures uniform structure, explicit knowledge, and explicit scope.

The GenInfo Backbone has a number of mechanisms for representing sequence relationships. For example, a coding region feature explicitly connects a DNA/RNA coding region to its protein product. Another example is that in which the sequences of the exons of a gene are reported without the sequences of the introns. In this case the exon sequences are entered separately but then grouped into a segmented sequence to maintain the linear integrity of the genetic information. If the lengths but not the sequences of the introns are given, these introns become 'virtual' sequences that can be grouped and annotated like any other. Likewise, sequences that cannot be entered, due to the style or quality of their published presentation, are also represented and annotated as virtual sequences so that their parent publications are not lost from the data-base.

Additionally, GIBB does not merge sequence variants or overlapping sequences into a single entry as other data-bases have attempted to do. Of course, non-redundant, special-purpose views of the data can and will be built. This design philosophy leaves the full reservoir of published information available for construction of specialized data-bases and also minimizes the need for expert judgements at the data entry level. Higher-level views and interpretations of the data can subsequently be carried out by domain experts or sequence curators.

The literature-based model used in GenInfo facilitates the handling of sequence errors. Transcription and keyboard errors arising during data entry are minimized by the use of duplicate entry and verification with periodic internal data audits. When such errors are identified, the original entry is corrected to reflect the published data. In cases where subsequent work reveals the presence of actual sequencing errors, and these are communicated to GenInfo either in the form of published errata or directly submitted reports, a revised entry is added to the data-base. Through this mechanism, it is possible to support users who need to track errors and revision histories while providing an automated non-redundant view of the data for users interested only in the best available sequence. In either case, it is important to emphasize that molecular sequence data is experimental data and errors will be present in any data-set.

ii. Neighbouring based retrieval techniques

One of the primary GIBB CD-ROM distribution media, the Entrez CD-ROM (see Note added in proof), includes a unique set of statistical neighbouring indices. Separate tables were compiled for protein sequences, nucleic acid sequences, and text comparisons. The protein and nucleic acid sequence similarities were computed using sequence alignment. Sequences were preprocessed to remove low entropy and repetitive sequence elements (States and Claverie, in preparation). These filtered sequences were then searched against the full data-base using the BLAST alignment tool, a calculation requiring several days on a high-performance parallel compute server. For each sequence, a list of its best significant neighbours was compiled and stored in compressed form on the CD-ROM, allowing the user to rapidly move through sets of similar sequences.

The text neighbouring is based on the title and abstract information from MEDLINE entries associated with the sequence data. Term lists and relevance weighting for these terms were calculated using a relevance pair model (16). As with the sequence neighbour tables, lists of the closest neighbours for every citation were stored on the Entrez CD-ROM in compressed format. The user can readily shift between sequence and text neighbouring to explore the data-base and identify relevant information.

3.3.3 Building richness through linkage

The GIBB itself is a relatively simple data-base in terms of structure and content, for all of the reasons discussed above. Yet it is specifically designed to facilitate the automated construction of richer and more flexible world views (e.g. *Figure 2*). This is made possible by having a stable, well-specified structure and through the use of explicit links. For example, links to the published literature are present in the form of precisely defined bibliographic citations and MEDLINE unique identifiers that are codes used to specify individual articles within the MEDLINE system. There is an explicit link between nucleic acid coding regions and protein sequences. There are emergent links between sequences via 'homology' searches. There will be direct links to the full text of US Patent Office documents. NCBI is currently working, in conjunction with the US Naval Research Laboratory and the Brookhaven data-base so that GIBB sequence entries can be related directly to known crystal structures. If it is not so already, the importance of such a linkage for protein modelling and engineering will become clearer in the subsequent example section (Section 6).

3.3.4 Implementation-independent data exchange

In computational biology, as in many other fields, output from one computer program or data-base may become input for another. The lack of data interchange standards has hindered scientific productivity because it has forced computing biologists and software developers to expend a great deal of time and effort to adapt their programs continuously to cope with a large variety of *ad hoc* data formats. To address this problem, the NCBI has chosen to specify object models for programs and data-bases in Abstract Syntax Notation 1 or ASN.1 (17). ASN.1 is a flexible, data description language and is one of a series of international standards (known as the Open Systems Interconnection or OSI) for communication among computer applications

Mark S. Boguski, James Ostell, and David J. States

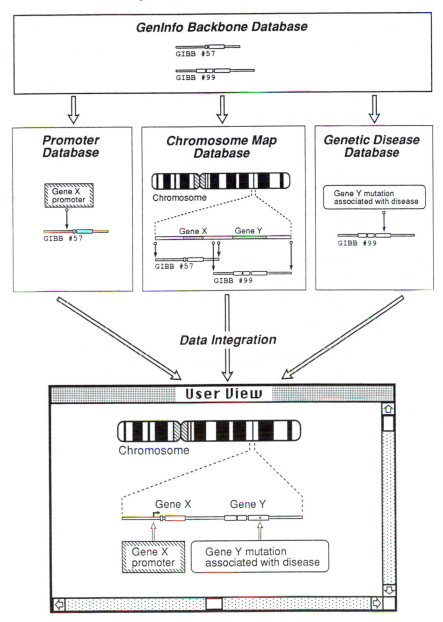

Figure 2. Schematic representation of mapping specialized data structures from independent data-bases (middle row of boxes) to GenInfo Backbone (GIBB) sequences (top box) using standard data structures for sequence location (arrows ended with circles). A possible user view created by integrating the specialized data through the common GIBB coordinates is presented at the bottom.

that has been adopted by the US and other governments and many major commercial vendors. ASN.1 has a rich syntax capable of representing complex biological objects. Furthermore, commercial software tools are available to analyse ASN.1 specifications for syntactic correctness and for encoding and decoding ASN.1 to C language structures.

Simply speaking, ASN.1 specifies *what* information is to be exchanged, not *how* it is to be exchanged. The implementation of a sequence data-base (whether it be relational, object-oriented, or flat file) is unimportant so long as one can retrieve defined elements from it. Furthermore, it is immaterial if the implementation changes because one can still retrieve the same data objects one could before the change. The use of ASN.1 encourages modularity and interoperability in the design of software tools. The GenInfo Backbone uses the ASN.1 standard, and other major data-base groups are considering providing their data in ASN.1 form (ASN.1 renditions of GenBank, PIR, and SWISS-PROT will be provided on the GenInfo CD-ROM; an ASN.1 version of the Complex Carbohydrate Structure Database (18) is being prepared).

The GenInfo Backbone (or just Backbone) was designed from the beginning to communicate explicitly biological sequence objects in this standard form. The great advantage of the explicit specification is that specification allows the sequence (and other) data to be mapped to or retrieved from any type of data-base model. GenInfo is currently implemented as a *relational data-base* which offers a number of distinct advantages over other data-base systems (19, 20) (all prior data/bases have been implemented as 'flat files' although recent efforts have been made to retrofit some of these with relational technology). Briefly, these advantages include simplicity, reflected both in a single data structure, TABLES (at the logical level), and a small number of non-procedural operators that operate on entire sets of values; Structured Query Language (SQL) support; the VIEW mechanism that allows users to tailor the data-base to their own requirements by defining virtual tables in terms of one or more base tables; and a sound theoretical base of data representation. Other key properties of relational data-bases are physical and logical data independent, for both users and user programs, and dynamic data definition, which means that new tables and columns can be added at any time without having any untoward effects on the logic of existing programs.

The well-defined and robust properties of the relational data-base model translate into simplified data-base design, ease of installation, operation and application development, distributed data-base support, and extendability. A number of relational data-base management systems are commercially available, and the relational model is overwhelmingly the dominant data-base implementation in the commercial applications. It is important to emphasize, however, that the internal format of the data-base is not of great concern to the user. For the user, it is the definition of the data-base which tells what is available and, ultimately, how data can be accessed.

68

Object-oriented data-bases (21) may represent the next step in data-base evolution and their new capabilities, such as rich data modelling constructs, direct support for inference, and novel data types (graphic images), would appear to make them ideal for representing biological entities and relationships. However, this technology is presently not mature enough for a production system. Nevertheless, because of its explicit specification of sequence objects, the GenInfo Backbone can take advantage of new software and hardware technologies as they become available.

3.3.5 Access and distribution

The GenInfo data-base is intended as a public, non-commercial, and international resource. Retrieval software, using message-passing technology, is being written to operate over the Internet and experimental implementations are already in use at a number of institutions. (Internet is a US computer network that connects most major colleges, universities, research institutions, and government laboratories (22). Network services are based on TCP/IP (Transmission Control Protocol/Internet Protocol) and include remote login (*telnet*) and file transfer (*ftp*). The need for, and new opportunities engendered by, high-speed electronic networks are discussed by Palca (23).) Messages, for example, will incorporate requests for particular records or for sequence similarity searches. Contents of the GenInfo Data Repository (information generated by NCBI as well as that contributed by outside groups) may be downloaded by anonymous *ftp*. In addition to this online service, the GenInfo Backbone and Data Repository files, along with information retrieval and sequence analysis tools, will be regularly distributed on CD-ROM for Apple Macintosh and IBM-compatible Microsoft Windows systems (see Note added in proof).

4. Sequence similarity searching

4.1 Introduction

The value of molecular sequence data-bases depends not only on their role as simple repositories of information but also on their properties as computable objects for information retrieval and analysis. The volume of sequence data has long surpassed human information processing capacity, even for simple tasks such as searching for related sequences. Automated tools can greatly extend human capabilities, but to utilize them, it is important to define goals and objectives.

In the case of similarity searching, several issues commonly arise:

- relevance (biological context)
- sensitivity (finding all relevant matches)

- specificity (avoiding irrelevant matches such as those that occur by random chance)
- efficiency

In an ideal world, it would not be necessary to choose between these criteria. Unfortunately, the models currently available to describe sequence similarity are only approximately correct. As a result, sequence string matches may or may not correspond to matches in biological function. The formalism of sequence level pattern matching permits the application of a variety of statistical approaches to data-base searching. Inherent in the use of these limited models is a statistical balance between the criteria of sensitivity and specificity. High thresholds for statistical significance will improve specificity at the expense of missing some relevant matches. While decreasing the threshold for significance will include such matches and increase the sensitivity, the number of false positives will increase. The larger and more diverse the data-base being searched, the more difficult it will be to achieve both specificity and sensitivity simultaneously using only symbolic pattern matching methods.

The relevance of finding a pattern match in sequence searching may depend on the context in which the search is being run and the source of the data. The identification of relevant matches can be optimized through the integration of sequence data with data such as bibliographic information and molecular structures, and by using complex query strategies to access multiple facets of these data. As described above, the annotated data-base is an essential tool in such integration and has been a major focus of work at the NCBI.

Finally, efficiency is an issue. A highly sensitive and specific algorithm would be irrelevant if it could not be implemented to run in a short time on available computing systems. Defined in this way, efficiency is a complex issue influenced by many aspects of the algorithm and computing system in question. High-speed microprocessors, parallel processing, and specialized co-processors may aid in some phases of a search, but to achieve high performance it is necessary to avoid bottlenecks in all aspects of the computation.

4.2 Efficient search algorithms

Identifying the best sequence similarity requires a search of a *query sequence* against all possible positions in the target sequence. Attempting similarity searches by direct enumeration of all possible alignments in the presence of possible insertions or deletions is a combinatorially complex problem. The application of dynamic programming techniques (24) to this problem limits the computation time to order N^2, greatly increasing the speed of the computation. However, without specialized computing hardware, even dynamic programming approaches are too slow for routine data-base searching. Hash table or neighbourhood searches provide an additional acceleration which does make routine searching feasible (25). These methods are based on the construction of a table showing where short sequence segments may appear in the key target sequence. If a match is found in the table, it can immediately be evaluated and extended by going to the sequence itself. There is an approximate nature to these methods in that they may miss alignments with extensive regions of low similarity that lack any localized regions of high similarity. Such alignments are, in practice, unusual enough that the approximate methods have proven to be very useful.

4.3 Evaluation of statistical significance

The algorithm employed in a sequence search may have a significant impact on its discriminatory ability. For example, searches based on translated protein sequence are frequently able to recognize significant matches that are not apparent by analysing the nucleic acid sequence directly. (This difference in sensitivity is due to codon degeneracy and to the fact that DNA has only a four-character alphabet. The protein alphabet, of course, has 20 characters and each of the amino acids can be assigned a number of values that represent different physical, chemical, or evolutionary properties. The most commonly used scoring system for amino acid similarities is the PAM250 or Mutation Data Matrix (26) although PAM120 and PAM200 matrices may be superior for particular applications (27).) Several search strategies based on complete or approximate alignment of sequences, including the presence of insertions and deletions, have been utilized widely (3). The statistical characteristics of these algorithms may be empirically characterized by Monte Carlo methods employing randomized sequences (it must be noted that for a wide range of similarity functions, the distribution of maximal scores is an extreme value distribution (28) and alignment scores are usually not significant until they are 5–10 standard deviations above the mean). However, a rigorous derivation of the significance of observed scores has not been worked out.

Tests for statistical significance assume that the unrelated sequences in a data-base are well represented by a random character stream. In data-bases of biologically derived sequence, there are many clear exceptions to this assumption. Gene duplication and rearrangement is an important biological process which may result in the appearance of highly related sequences in quite different biological contexts (29). Amino acid usage is not uniform and varies from source to source. Higher-level correlates are present between pairs of amino acids or nucleotides, and the role of peptide sequences as triggers for biological events, such as post-translational modification, may lead to anomalous selection pressures on particular sequences (30).

With the ever-expanding size of sequence data-bases, the balance between search sensitivity and specificity becomes increasingly critical, and the rigorous application of statistical formalism can help to optimize performance. Recently, Karlin and Altschul described rigorous tests for the significance of ungapped sequence alignments (31). This approach, when implemented in an efficient neighbourhood search algorithm (32), provides a basis for very rapid sequence similarity searching while retaining discriminatory behaviour.

Ultimately the purpose of sequence data-base searching is the identification of macromolecules with related structures and functions. Symbolic pattern matching is a limited model for the physical processes of protein folding and active site chemistry. The limitations of the symbol string model for protein structure necessarily lead to the presence of a 'grey area' where pattern matches may or may not correlate with structural and functional similarity.

Using any of the data-base search algorithms, a number of matches are likely to be reported. For analysis and modelling work to proceed, one or more of these must be above the threshold of randomly expected matches. In the case of BLAST (see below), a probability statistic is reported directly. For other methods, such as FASTA (33), the number of randomly observed hits with a given alignment score seems to follow an extreme value distribution, and the logarithm of the number expected will decrease linearly with the score. By plotting this relationship empirically, a cut-off score can be determined above which no random matches would be expected (34). This cut-off may be verified by randomly permuting the target sequence and searching against the data-base; no matches with scores above the cut-off should be observed (3).

The set of sequences chosen for more detailed analysis is obviously a critical choice that will affect subsequent model building. In particular, including a sequence which is not truly related into a data-set for analysis may lead to false conclusions. On the other hand, excluding truly related sequences will ignore relevant data. The use of a cut-off score criterion (as described above) in data-base searches is often found to be overly restrictive, and a number of related sequences will be missed. Rather than simply lower the score cut-off, it is often productive to search the data-base again using the sequences already identified as being related to the sequence of interest. This process can be repeated to both add and delete sequences to/from the working group of related sequences. When a match against the data-bases is found in several different searches, it is much more likely to represent a true positive that should be included in the analysis. Conversely, when a sequence matches only one of a set of related target sequences, it is more likely to be a false positive and can be excluded. A related strategy has been implemented in a program called BLAST3 (35).

Having obtained a set of sequences for analysis, it is next necessary to consider the extent to which they represent independent data. For example, the protein sequence data-bases contain several hundred copies of quite closely related immunoglobulins. If such a large number of closely related sequences are present, this is obviously a source of statistical bias. The problem has been treated in detail using a hierarchical tree clustering method to derive weights for each of the constituent sequences (36). An alternative, and simpler, approach is to exclude all but one member of a group of closely related proteins because the marginal contribution of each one of the additional closely related sequences will be small. As will be discussed below, alignment and analysis of families of multiple sequences may be computationally demanding. The fewer the sequences that need be considered, the more tractable calculations become.

4.4 Advantages of network-based services

Data-base searching is typically a 'burst' activity in which brief periods of computation are interspersed with long periods of idle time during which the search results are being read and analysed by the investigator. The use of centralized computing facilities may help to optimize search efficiency. In establishing centralized resources, provisions can be made for the availability of faster hardware than the individual investigator or laboratory could afford to purchase or maintain themselves. Each user then receives faster response to their queries, with little contention by all users for the same computational resource. With the high-speed local and wide-area networks now available throughout much of the US and Europe, a computation service such as this need not be located in the same building or even in the same city.

In addition to using faster processors, a centralized resource may be able to maintain entire data-bases resident in memory *between* searches, thus avoiding disk I/O delays entirely. Memory-resident data-bases also permit parallel processors to be used. Network-based search services, such as NCBI's implementation of the BLAST program (32), utilize these concepts.

Access to the most recent data is critical in many areas of biomedical research. Network-based search servers reduce the number of computation sites that need to be kept updated with the most current data (GenBank conducts an on-line service that provides Internet access to similarity searching with FASTA and text word searching with the IRX information retrieval system) (37, 38). The networks can also be used as distribution media, with master servers providing updates to remote slave servers automatically.

5. Multiple sequence alignment

It is now often the case that if one related sequence can be identified, then several other significant matches will also be found. The alignment of an entire family of proteins may provide more information than the alignment of any pair of those sequences. Relationships among multiple sequences may improve the accuracy of alignment, and the increased sample size improves the chance that a representative range of the permissible sequence variation will be observed.

Mathematically, multiple sequence alignment represents a generalization of the simple pairwise sequence alignment. Dynamic programming can be supplied in more than two dimensions, but the complexity of the computation grows exponentially with the number of sequences being considered. In this setting, bounding techniques become increasingly valuable. The Carillo–Lipman algorithm (39) uses pairwise alignments to limit the volume of the N-dimensional space that needs to be searched in the full alignment, greatly improving overall speed. This algorithm has been implemented in the program MSA (40) with appropriate weighting to circumvent the problem of having a small number of very closely related sequences dominate the alignment (36).

An alternative, approximate algorithm employs iterative improvement to self-consistency. In this approach, a trial multiple sequence alignment is used as a basis for realigning each of the component sequences until each is in an optimal alignment. Although the globally optimal alignment is guaranteed to be self-consistent in this sense, the fact that an alignment is self-consistent does not ensure that it represents the global optimum. In practice, local minima are frequently present. This method has been widely described and utilized (see references cited by Lipman *et al.* (40)) because it is straight-forward and allows large numbers of sequences (>100) to be aligned.

Incorporating chemical and biological knowledge into multiple sequence alignment can greatly improve the reliability of results, but this may require manual intervention. An important advance in this area is the development of a hybrid approach: the Multiple Alignment Construction and Analysis Work-bench (MACAW) provides statistically based alignments on segments of a multiple alignment while allowing the user to interactively define, lock, and shift different portions of the data (41). Prior knowledge not expressible in the linear sequence may be used to guide the alignment process and ensure that active sites and other features are preserved, but the basic alignment procedure remains automated with rigorous statistical criteria applied. Because the user may limit the automated alignment to small blocks of sequences, computational demands are manageable, and good response times have been achieved with personal computer-based implementations (MACAW runs under Microsoft® Windows on IBM-compatible personal computers).

6. Data-base searching and multiple sequence alignment for protein modelling: an example

6.1 Introduction

To illustrate and expand on the principles described in the foregoing sections, we work through many of the essential steps of sequence identification and alignment that are usually antecedent to the construction of a detailed molecular model. The example we have chosen is that of human apolipoprotein D (apo-D) which is a 169-residue serum glycoprotein that belongs to the α_2-microglobulin (or lipocalin) protein superfamily. (Protein superfamily names are assigned in the PIR data-base according to the member that was dis-covered first. In an effort to be more descriptive of function, Pervaiz and

Brew (42) coined the term 'lipocalin' to describe the fact that proteins in this family carry lipophilic ligands in a calyx-shaped binding pocket. However, according to Lindsay Sawyer (personal communication), the coinage 'lipo-calycin' is more etymologically correct.) Several short reviews (42, 43, 44) describing this protein family have previously appeared and a detailed molecular model has been developed (45).

For the purposes of molecular modelling, the ultimate goal of a *sequence data-base search* is to find a matching sequence that has two essential characteristics:

(a) that it is similar enough to the query sequence that one can assume both proteins have the same general structure or fold (of course there are cases of convergently evolved proteins whose structures are superimposable yet share no significant similarity in primary sequence; obviously, sequenced-based modelling is not relevant in this context); and

(b) that the matching sequence corresponds to a known 3D structure whose coordinates are obtainable.

In the latter case, it is helpful to use a sequence data-base, such as SWISS-PROT, that contains explicit cross-references to a protein structure data-base (Brookhaven PDB), or NRL–3D (46, 47) which contains only those primary sequences that correspond to Brookhaven entries. Of course, there are many protein crystal structures whose coordinates have not been deposited in a public data-base and one must rely on the scientific literature to find these examples (GIBB index sequences will facilitate this process).

Regarding the issue of sufficient similarity, Chothia and Lesk (48) have provided some helpful guidelines. Note, in the present context, that retinol-binding protein and bilin-binding protein are two members of the lipocalin superfamily that share only about 15% aligned sequence identity, yet Huber *et al.* (49) were able to build a preliminary model of bilin-binding protein based upon the retinol-binding protein coordinates. Indeed, so few 3D structures are actually available (relative to the number of available sequences) that modelling in the face of only distant similarity can be justified on these grounds.

6.2 FASTA

Partial results of a FASTA search of the NBRF/PIR data-base using apo-D as a query sequence are shown in *Figure 3*. Note the magnitude of the score (868) for the apo-D self-comparison and the fact that several other sequence matches appear to be clearly distinguishable from the background noise (*Figure 3*). In fact the next eight matches after apo-D itself are bona fide members of the lipocalin family; one then encounters the first false positive match (the rabbit ryanodine receptor). Two additional members of the lipo-calin family (insecticyanin and androgen-dependent epididymal protein) are

```
lphud.aa, 169 aa vs /usr/ncbi/db/fasta/pir library
using protein matrix

7620668 residues in 26798 sequences
 statistics exclude scores greater than 72
 mean initn score:  24.1 (8.07)
 mean init1 score:  23.4 (6.53)
 5262 scores better than 28 saved, ktup: 2, variable pamfact
 joining threshold: 27  scan time: 0:01:28

The best scores are:                                    initn init1   opt

LPHUD  Apolipoprotein D precursor - Human                 868   868   868
A26958 Apolipoprotein D precursor - Human                 868   868   868
CUWOI  Insecticyanin - Tobacco hornworm                   138    80   212
A27786 Plasma retinol-binding protein - Human            119    80   144
S06278 Plasma retinol-binding protein - Human            119    80   146
VAHU   Plasma retinol-binding protein precursor - Human  119    80   146
VART   Plasma retinol-binding protein (pRBP) precursor   118    81   139
VARB   Plasma retinol-binding protein (pRBP) - Rabbit    117    77   138
A26969 Retinal retinol-binding protein precursor - Chi   112    67   167
A32202 Prostaglandin-D synthase - Rat #EC-number 5.3.    105    76    85
B35041 Ryanodine receptor - Rabbit                        88    40    40
S00819 Insecticyanin - Cabbage butterfly                  84    84   208
S04200 NAD+ ADP-ribosyltransferase - Mouse #EC-number     80    47    49
S04654 Ryanodine receptor - Rabbit                        76    40    40
S01830 Hypothetical protein p69 - Mycoplasma hyorhinis     75    38    50
A22735 Cytochrome-c oxidase polypeptide I - Emericella     75  - 33    33
A35041 Ryanodine receptor - Human                          75    38    42
SQRTAD Androgen-dependent epididymal 18.5K protein pre     73    73    93
GNWVJE Genome polyprotein - Japanese encephalitis viru     73    43    46
JT0337 Penton-base protein - Mastadenovirus h5             72    38    42


A26958 Apolipoprotein D precursor - Human                 868   868   868
 100.0% identity in 169 aa overlap


          -----/self-alignment omitted/-----


CUWOI Insecticyanin - Tobacco hornworm                    138    80   212
 31.2% identity in 170 aa overlap


                10        20        30        40        50
LPHUD   QAFHLGKCPNPPVQENFDVNKYLGRWYEIEKIPTTFEN-GRCIQANYSLMENGKIKVLN
        . :. : X:.    ..::.. . : :.::.:.X . :: :.:. :.:.   .::   .
CUWOI   GDIFYPGYCPDVKPVNDFDLSAFAGAWHEIAKLPLENENQGKCTIAEYKY--DGKKASVY
                10        20        30        40        50

        60        70        80        90        100       110
LPHUD   QELRADGTVNQIEGE---ATPVNLTEPAK--LEVKFSWFMPSAPYWILATDYENYALVYS
        ... ..:. .::.  :. .. :...:  .. ::.  . .  :.::::::.:::. :.
CUWOI   NSFVSNGVKEYMEGDLEIAPDAKYTKQGKYVMTFKFGQRVVNLVPWVLATDYKNYAINYN
        60        70        80        90        100       110

           120       130       140       150       160
LPHUD   CTCIIQ-LFHVDFAWILARNPNLPPETVDSLKNILT--SNNIDVKKMTVTDQVNCPKLS
        :.. . : :::!... : .: . ..::. :. ::..:. .:
CUWOI   CDYHPDKKAHSIHAWILSKSKVLEGNTKEVVDNVLKTFSHLIDASKFISNDFSEAACQYS
        120       130       140       150       160       170

VAHU Plasma retinol-binding protein precursor - Human    119    80   146
 28.1% identity in 146 aa overlap
```

```
                    10        20        30        40
LPHUD         QAFHLGKCPNPPVQENFDVNKYLGRWYEI-EKIPTTFENGRCIQAN
              :.::::  ...  :  ::.  .:  :...     :  :.
VAHU      MKWVWALLLLAAWAAAERDCRVSSFRVKENFDKARFSGTWYAMAKKDPEGLFLQDNIVAE
            10        20        30        40        50        60

           50        60        70        80        90
LPHUD     YSLMENGKIKVLNQE-LRADGTVNQ-IEGEATPVNLTEPAKLEVKFSW----FMP--SAP
          .:.  :.:.....  ...  :.  .  .  .:  ..  ..:::...:.  :   :..  ..
VAHU      FSVDETGQMSATAKGRVRLLNNWDVCADMVGTFTDTEDPAKFKMKY-WGVASFLQKGNDD
            70        80        90       100       110

          100       110       120       130       140       150
LPHUD     YWILATDYENYALVYSCTCIIQL---FHVDFAWILARNPN-LPPETVDSLKNILTSNNID
          .X:..::..::. ::X   ...:       .........:.:: ::::. .  ...
VAHU      HWIVDTDYDTYAVQYSCR-LLNLDGTCADSYSFVFSRDPNGLPPEAQKIVRQRQEELCLA
          120       130       140       150       160       170

               160
LPHUD     VKKMTVTDQVNCPKLS

VAHU      RQYRLIVHNGYCDGRSERNLL
          180       190


B35041 *Ryanodine receptor - Rabbit                    88    40    40
   31.8% identity in 22 aa overlap

               30        40        50        60        70        80
LPHUD     YLGRWYEIEKIPTTFENGRCIQANYSLMENGKIKVLNQELRADGTVNQIEGEATPVNLTE
                            X:. .  ...::  ......  ::X
B35041    VFAQPIVSRARPELLHSHFIPTIGRLRKRAGKVVAEEEQLRLEAKAEAEEGELLVRDEFS
          3340      3350      3360      3370      3380      3390


S01830 Hypothetical protein p69 - Mycoplasma hyorhinis  75    38    50
   26.9% identity in 26 aa overlap


A22735 Cytochrome-c oxidase polypeptide I - Emericella  75    33    33
   100.0% identity in 3 aa overlap


A35041 *Ryanodine receptor - Human                      75    38    42
   71.4% identity in 7 aa overlap


SQRTAD Androgen-dependent epididymal 18.5K protein pre  73    73    93
   18.6% identity in 167 aa overlap


Library scan:  0:01:28  total CPU time:  0:01:30
```

Figure 3. Edited output of the FASTA program (25, 33). The sequence for human apolipoprotein D (signal peptide excluded) was extracted from the NBRF/PIR data-base (release 27.0) and then used as a query sequence to search this data-base. The top 20 matches are shown along with alignments for selected matched sequences. The complete execution (with the k-tup parameter set to 2) required 90 sec of CPU time on a Silicon Graphics 4D/280 computer using a single processor (this performance is slightly faster than on a Sun Microsystems Sparcstation 1+). A search comparable in sensitivity (k-tup = 1) to the BLASTP search of *Figure 4* required 266 CPU sec.

found among the next nine matches but their scores are not clearly distinguishable from background noise and would require additional statistical tests, such as Monte Carlo analysis, to support a conclusion of significant similarity. However there are two clues in the data-base search results that indicate the possible significance of these matches. These are:

(a) a large increase in the optimized score ('opt') over the initial score ('init1') that results when the program attempts to extend the local similarity for insecticyanin; and

(b) a moderate degree of sequence identity (18.6%) that extends over a large overlap (167 residues) for androgen-dependent protein—the same or even a greater degree of identity over a small region is not persuasive (e.g. ryanodine receptor) unless one is dealing with a very specific local sequence pattern, such as an active site.

6.3 Basic Local Alignment Search Tool (BLAST)

Partial results of a BLAST search, using the same data-base and query sequence, are shown in *Figure 4*. BLAST output includes the Karlin score ('Score'), the expected number ('Expect') of matches to the query in the data-base of this size, the data-base *P*-value ('P-value'), the pairwise *P*-value ('P'), and the significant subsequence alignment(s). The pairwise *P*-value indicates the likelihood that the query sequence and the matched sequence are related by chance based on *target frequencies* (31) of amino acids derived from representative sequences (one from each superfamily) in the PIR data-base. The data-base *P*-value equals the pairwise *P*-value times the total data-base size, divided by the length of the query sequence. The data-base *P*-value thus gives a better indication of how surprising it would be to find such a match in a data-base of the size under consideration. The significance of the matches displayed in *Figure 4* is unquestionable. Note that for most of the matched data-base sequences, two significant local similarities (motifs) were identified and these were in the same relative locations in all matched proteins.

Neither FASTA nor BLAST identifies, in a single search, every known member of the lipocalin family (in particular, β-lactoglobulin was absent). (To illustrate how useful sequence annotation and classification can be, simple text word searches of PIR and SWISSPROT with 'alpha-2u-globulin' and 'lipocalin', respectively, identified 20 and 36 entries, respectively, for proteins that belong to this superfamily. These results were considerably more comprehensive than simple sequence similarity searching alone.) A very useful strategy for exhaustive analysis (previously noted in Section 4.3) employs iterative searching using as sequential query sequences all significant matches obtained in the initial search. This process is repeated until all of the most distantly related sequences are found. Alternatively, one can search the data-base for *multiple alignments* as opposed to pairwise alignments (35). This technique often allows one to detect distant relationships that are more easily

Mark S. Boguski, James Ostell, and David J. States

National Center for Biotechnology Information

Experimental GENINFO(R) BLAST Network Service

Query: LPHUD Apolipoprotein D - Human | 950.0 8.0 1.0 1.0 1.0

Database: Protein Identification Resource 27.0 (complete), Dec. 31, 1990
           26,798 sequences; 7,620,668 total residues.

		Smallest Poisson	
Sequences producing High-scoring Segment Pairs:	High Score	Probability P(N)	N
LPHUD Apolipoprotein D precursor - Human	950.0 8.0 1.0 ...	921	8.4e-130	1
A26958 Apolipoprotein D precursor - Human	921	8.4e-130	1
A26969 Retinal retinol-binding protein precursor - Chicken	78	4.2e-08	2
CUWOI Insecticyanin - Tobacco hornworm	950.0 9.0 1.0 1....	70	8.0e-08	2
VART Plasma retinol-binding protein (pRBP) precursor - R...	68	1.0e-05	2
VARB Plasma retinol-binding protein (pRBP) - Rabbit	95...	64	3.5e-05	2
S00819 Insecticyanin - Cabbage butterfly	75	6.6e-05	2
A27786 Plasma retinol-binding protein - Human	68	7.0e-05	2
S06278 Plasma retinol-binding protein - Human	68	7.0e-05	2
VAHU Plasma retinol-binding protein precursor - Human	...	68	7.6e-05	2
A30013 Plasma retinol-binding protein precursor - African ...	61	0.0023	2
A32202 Prostaglandin-D synthase - Rat #EC-number 5.3.99.2	67	0.028	1
SQRTAD Androgen-dependent epididymal 18.5K protein precurs...	67	0.028	1
D26890 Alpha-2u-globulin V precursor - Mouse	50	0.12	2
HCHU Alpha-1-microglobulin/inter-alpha-trypsin inhibitor...	61	0.20	1
A25303 Alpha-1-microglobulin precursor - Human	61	0.20	1
JL0107 IgG Fc receptor precursor, type III-2 (NK cells) - ...	58	0.46	1
JU0284 IgG Fc receptor precursor, type III-1 (polymorphonu...	58	0.46	1
S00758 Surface glycoprotein CD16 precursor - Human	58	0.46	1
A32933 IgG receptor CD16 (clone HL12) - Human (fragment)	58	0.46	1

>LPHUD Apolipoprotein D precursor - Human | 950.0 8.0 1.0 1.0 1.0
       Length = 189

  Score = 921, Expect = 8.4e-130, P-value = 8.4e-130, length = 169

            -----/self-alignment omitted/-----

>A26969 Retinal retinol-binding protein precursor - Chicken
       Length = 196

  Score = 78, Expect = 0.00068, P-value = 0.00068, length = 17

Query: 98 YWILATDYENYALVYSC 114
             YW++ TDY+NYA+ Y+C
Sbjct: 126 YWVIDTDYDNYAITYAC 142

  Score = 65, Expect = 0.057, Poisson P-value = 4.2e-08, length = 20

Query: 12 PVQENFDVNKYLGRWYEIEK 31
             +V +NFD ++Y G+WY + K
Sbjct: 30 SVKDNFDPKRYAGKWYALAK 49

>CUWOI Insecticyanin - Tobacco hornworm | 950.0 9.0 1.0 1.0 1.0
       Length = 189

  Score = 70, Expect = 0.010, P-value = 0.010, length = 16

```
Query:    99 WILATDYENYALVYSC 114
             W+LATDY NYA+ Y+C
Sbjct:   104 WVLATDYKNYAINYNC 119
```

Score = 64, Expect = 0.080, Poisson P-value = 8.0e-08, length = 31

```
Query:     3 FHLGKCPNPPVQENFDVNKYLGRWYEIEKIP 33
             F  G CP+    ++FD++ + G W EI K+P
Sbjct:     4 FYPGYCPDVKPVNDFDLSAFAGAWHEIAKLP 34
```

Score = 42, Expect = 1.4e+02, Poisson P-value = 0.00030, length = 24

```
Query:   126 AWILARNPNLPPETVDSLKNILTS 149
             AWIL+++  L +T + + N+L +
Sbjct:   132 AWILSKSKVLEGNTKEVVDNVLKT 155
```

>A32202 *Prostaglandin-D synthase - Rat #EC-number 5.3.99.2
        Length = 188

Score = 67, Expect = 0.029, P-value = 0.028, length = 15

```
Query:    13 VQENFDVNKYLGRWY 27
             VQ NF+ +K+LGRWY
Sbjct:    30 VQPNFQQDKFLGRWY 44
```

Score = 41, Expect = 2.0e+02, Poisson P-value = 0.39, length = 13

```
Query:   101 LATDYENYALVYS 113
             + TDY++YA ++S
Sbjct:   120 VETDYDEYAFLFS 132
```

>SQRTAD Androgen-dependent epididymal 18.5K protein precursor - Rat | 950.0 3.0
        1.0 1.0 1.0
        Length = 184

Score = 67, Expect = 0.029, P-value = 0.028, length = 18

```
Query:    12 PVQENFDVNKYLGRWYEI 29
             +V  +FD++K+LG WYEI
Sbjct:    19 AVVKDFDISKFLGFWYEI 36
```

>D26890 Alpha-2u-globulin V precursor - Mouse
        Length = 180

Score = 50, Expect = 9.4, P-value = 1.0, length = 16

```
Query:    14 QENFDVNKYLGRWYEI 29
             ++NF+V+K  G+W+ I
Sbjct:    25 RQNFNVEKINGKWFSI 40
```

Score = 43, Expect = 1.0e+02, Poisson P-value = 0.12, length = 11

```
Query:   100 ILATDYENYAL 110
             IL TDY+NY +
Sbjct:   110 ILKTDYDNYIM 120
```

>HCHU Alpha-1-microglobulin/inter-alpha-trypsin inhibitor precursor - Human |
      950.0 5.0 1.0 1.0 1.0
      Length = 352

```
  Score = 61, Expect = 0.22, P-value = 0.20, length = 20

Query:      10 NPPVQENFDVNKYLGRWYEI 29
               N  VQENF++++  G+WY++
Sbjct:      28 NIQVQENFNISRIYGKWYNL 47


>JL0107 IgG Fc receptor precursor, type III-2 (NK cells) - Human
        Length = 254

  Score = 58, Expect = 0.62, P-value = 0.46, length = 22

Query:      79 LTEPAKLEVKFSWFMPSAPYWI 100
               L++P  LEV  +W +  AP W+
Sbjct:      96 LSDPVQLEVHIGWLLLQAPRWV 117




Parameters:
  E = 19., S = 48, T = 11, X = 20, W = 3, M = PAM120
  E2 = 0.080, S2 = 36

Statistics:
  Lambda = 0.338, Lambda/ln2 = 0.488, K = 0.179, H = 0.927
  Expected/Observed high score = 56/921
  # of residues in query:  169
  # of neighborhood words in query:  6933
  # of exact words scoring below T:  0
  Database:  Protein Identification Resource 27.0 (complete), Dec. 31, 1990
  # of residues in database:  7,620,668
  # of word hits against database:  3,251,514
  # of failed hit extensions:  2,604,529
  # of excluded hits:  646,894
  # of successful extensions:  91
  # of HSPs reported:  48
  # of overlapping HSPs discarded:  43
  # of sequences in database:  26,798
  # of database sequences with at least one HSP:  34
No. of states in DFA:  558 (55 KB)
Total size of DFA:  130 KB (150 KB)
Time to generate neighborhood:  0.11u 0.00s 0.11t
No. of processors used:  8
Time to search database:  6.67u 0.23s 6.90t
Total cpu time:  7.37u 2.17s 9.54t
```

Figure 4. Edited output of the BLASTP program (32). The query sequence and data-base searched were exactly the same as in *Figure 3*. The top 20 matches are shown along with significant local alignments for selected matched sequences. The BLAST family of programs have been parallelized and this computation required only 7.37 CPU sec on a Silicon Graphics 4D/280 computer when all eight processors were used (the CPU time increased to about 48 sec when the program was restricted to one processor).

distinguished from background noise and, indeed, the BLAST3 program (35) identifies β-lactoglobulin as being related to apo-D (not shown).

6.4 Considerations for further analyses

At this point, data-base searching has revealed several possible candidates upon which to base an apo-D model, including the insecticyanins and retinol-binding proteins. Inspection of the data-base annotation for each of these sequence entries indicates that none of these proteins are represented in the Brookhaven data-base. However, examination of the scientific literature shows that the two insecticyanins (49, 50, 51), human plasma retinol-binding protein (52), and bovine β-lactoglobulin (53) all have published crystal structures (some of these articles contain sequence data qualifying as GIBB index sequences and would be retrievable by sequence 'homology' searches). These proteins bind small hydrophobic ligands in a pocket, or calyx, formed by eight antiparallel β-strands arranged as two stacked orthogonal sheets. Despite this common architecture, however, three different intrachain disulphide bonding patterns are seen (54) and one needs to keep this in mind when interpreting sequence alignments.

Assuming all coordinate sets are available from the authors, which structure would be most appropriate to use for the apo-D model? Important issues to consider are:

(a) the degree of *global* sequence similarity to apo-D

(b) the presence of sequence motifs (local similarities) that are conserved among all family members and thus must be accommodated in any model

(c) important structural features, such as the locations of ½ cystine residues, salt bridges, and allowable surface loops and turns

FASTA search results and pairwise sequence alignments can address the first issue. BLAST searches and MACAW multiple alignments can be used to locate and quantify significant sequence motifs. Multiple alignment using dynamic programming with data weighting (MSA) can then be used on a refined subset of the sequences to obtain precise estimates for the locations of insertions, deletions, and other sequence features that must be accounted for in a model.

6.5 Sequence motifs and optimal alignments

MACAW analyses of 12 representatives of the lipocalin superfamily are shown in *Figure 5*. Two highly conserved regions of local similarity are seen

---

Figure 5. Screen dumps from MACAW program, version 1.0 (41) during the course of analysis of 12 lipocalin family members: APOD, human apo-D (SWISS-PROT code APD$HUMAN); ICYA, tobacco hornworm insecticyanin (ICYA$MANSE); RETB, human retinol-binding protein (RETB$HUMAN); LACB, bovine β-lactoglobulin (LACB$BOVIN); MUP, rat major urinary protein (a.k.a. α-2-microglobulin) (MUP$RAT); AN18, rat androgen-dependent epididymal protein (AN18$RAT); OLFA, frog olfactory protein (OLFA$RANPI); HC, human α-1-microglobulin/inter-α-trypsin inhibitor (HC$HUMAN); CO8G, human complement component 8 γ-chain (CO8G$HUMAN); A1AG, human α-1-acid glycoprotein (orosomucoid) (A1AG$HUMAN); PP14, human pregnancy-associated endometrial protein (PP14$HUMAN). (a) Analysis of motif I as described in the text; (b) analysis of motif II as described in the text.

(these conserved local similarities have previously been noted by other authors (42, 43, 44)):

(a) 20–24 residue span (motif I) containing an invariant Gly–X–Trp tripeptide (*Figure 5a*); and

(b) a 9–12 residue span (motif II) containing a nearly invariant Thr–Asp–Tyr tripeptide (*Figure 5b*).

(a)
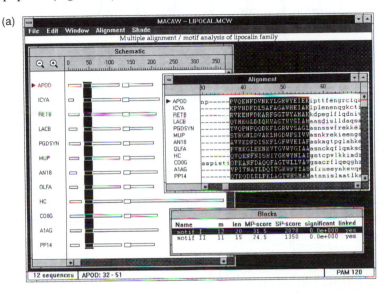

(b)
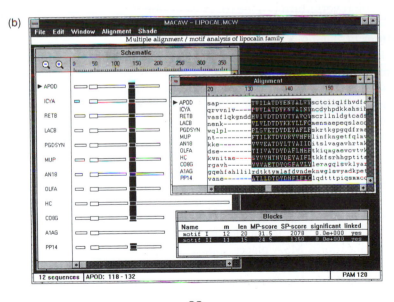

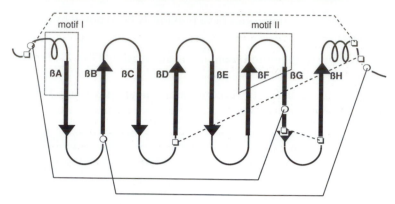

Figure 6. Schematic diagram of lipocalin family structure. β-strands A through D and E through H form two orthogonal β-sheets. The antiparallel β-strands are usually connected by reverse turns, although short α-helices served this function in certain family members (see *Figure 5*). NH$_2$- and COOH-terminal α-helices are indicated by spirals. Dotted lines connecting boxes indicate the *three* intrachain disulphide bonds present in human retinol-binding protein (Cys4/Cys160, Cys70/Cys174, Cys120/Cys129; numbering for mature protein, i.e. minus signal peptide). Solid lines connecting circles indicate the *two* intrachain disulphide bonds present in insecticyanin (Cys9/Cys119, Cys43/Cys175; mature protein). The boxes labelled motif I and motif II are as described in the text.

The statistical significance of these motifs is unquestionable. Note that these sequence motifs were explicitly identified by the BLAST search as consistently matched regions of local similarity (*Figure 4*). Motif I encompasses the NH$_2$-terminal α-helix and first β-strand; motif II corresponds to the distal portion of the sixth β-strand, the proximal portion of the seventh β-strand, and their connecting reverse turn (*Figure 6*). A clear implication of the conservation of motif II is that this turn is crucial and should not be the location of arbitrary insertions or deletions during modelling.

Preliminary analysis with MSA (not shown) indicated that a global multiple alignment, within reasonable bounds, does not exist among apo-D and the four other related sequences for which crystal structures are available. Thus the most distantly related protein, β-lactoglobulin, was dropped and the analysis repeated, results of which are shown in *Figure 7*. Note that the pattern of conserved cysteine residues is identical between apo-D and in-secticyanin but differs significantly from retinol-binding protein. In the two cases where ½ cystines are aligned among all three sequences, both have very different ½ cystine pairs in retinol-binding protein versus insecticyanin. This fact, in addition to the higher degree of sequence identity between apo-D and the insecticyanins (*Figure 3*) indicates that insecticyanin would be superior to retinol-binding protein for modelling an apo-D structure.

In two locations, MSA places apparently unfavourable gaps that interrupt β-strands F and G (*Figures 6* and *7*). The alignment model can be further refined by adding the second insecticyanin sequence (BBP$PIERB) and

(a)

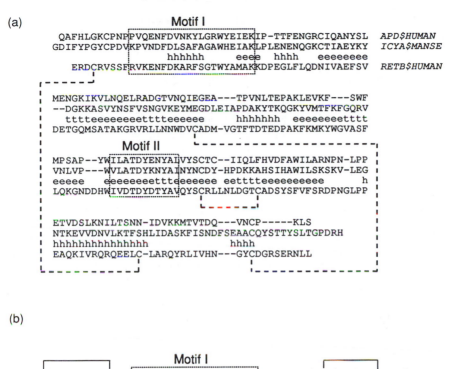

(b)

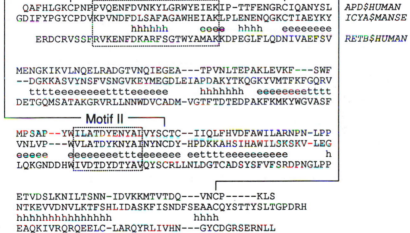

Figure 7. Annotated multiple alignments. Apo-D, insecticyanin, and retinol-binding protein were simultaneously aligned using the MSA program (40). The sequence alignments in panels (a) and (b) are exactly the same. The secondary structure annotations (h = α-helix, e = β-strand, t = reverse turn) are based upon the interpretation of Holden *et al.* of the crystal structure of insecticyanin (50). Motifs I and II are as described in the text. (a) Disulphide bonds in retinol-binding protein (see legend to *Figure 6*); (b) disulphide bonds in insecticyanin.

excluding retinol-binding protein. This results in a better multiple alignment that places gaps only at surface loops or turns (45).

7. Summary

We have described the GenInfo Backbone data-base and compared its design and characteristics with those of other sequence data-bases. We have also discussed general principles of homology-based sequence similarity searching and alignment as a prelude to molecular modelling. Finally, we have shown how to apply these principles in a detailed example.

Acknowledgements

We thank Stephen Altschul, Dennis Benson, Jean-Michel Claverie, Warren Gish, David Lipman, Greg Schuler, Fred Thompson, and John Wootton for valuable suggestions on and contributions to the manuscript.

Note added in proof

The *Entrez: Sequences* CD-ROM, briefly mentioned in the text, is now available for Apple Macintosh, Microsoft Windows and X-windows/Unix systems for an annual subscription fee of $57 US and includes six bi-monthly updates. For more information and a free subscription to the NCBI Newsletter, write to the National Center for Biotechnology Information, National Library of Medicine, National Institutes of Health, Bldg. 38A, Room 8N-803, 8600 Rockville Pike, Bethesda, MD. 20894 USA. Alternatively contact NCBI via electronic mail at info@ncbi.nlm.nih.gov. A detailed introduction to the Entrez system, as well as further examples of data-base searching and multiple alignment, may be found in ref. 55.

References

1. Benson, D., Boguski, M., Lipman, D. J., and Ostell, J. (1990). *Genomics*, **6**, 389–91.
2. Bishop, M. J., Ginsburg, M., Rawlings, C. J., and Wakeford, R. (1987). In *Nucleic acid and protein sequence analysis: a practical approach*, (ed. M. J. Bishop and C. J. Rawlings), pp. 83–146. IRL Press, Oxford.
3. Doolittle, R. F. (1990). In *Methods in enzymology* (ed. R. F. Doolittle), Vol. 183, pp. 3–49. Academic Press, San Diego.
4. Smith, T. F. (1990). *Genomics*, **6**, 702–7.
5. Akrigg, D., *et al.* (1988). *Nature*, **335**, 745–6.
6. Bleasby, A. J. and Wootton, J. C. (1990). *Prot. Engng*, **3**, 153–9.
7. Fickett, J. W. (1986). *TIBS*, **11**, 190.
8. Fickett, J. W. (1986). *TIBS*, **11**, 382.

9. Bleasby, A. J. and Wootton, J. C. (1990). *Prot. Engng,* **3,** 153–60.
10. Bairoch, A. and Boeckmann, B. (1992). *Nucl. Acids Res.,* in press.
11. Bairoch, A. (1992). *Nucl. Acids Res.,* in press.
12. Myers, G., Rabson, A. B., Josephs, S. F., Smith, T. F., Berzofsky, J. A., and Wont-Staal, F. (1990). Theoretical Biology and Biophysics Group T-10, Los Alamos National Laboratory.
13. McKusick, V. A. (1990). *Mendelian inheritance in man. Catalogs of autosomal dominant, autosomal recessive, and X-linked phenotypes.* Johns Hopkins University Press, Baltimore.
14. Roberts, R. (1990). *Nucl. Acids Res.,* **18,** 2331–66.
15. Ghosh, D. (1990). *Nucleic Acids Res.,* **18,** 1749–56.
16. Wilbur, W. John (1992). *Proc. 55th Annual Meeting of the American Soc. Information Science.* Pittsburgh, USA.
17. Rose, M. T. (1990). *The open book: a practical perspective on OSI.* Prentice Hall, Englewood Cliffs, NJ.
18. Doubet, S., Bock, K., Smith, D., Darvill, A., and Albersheim, P. (1989). *TIBS,* **14,** 475–7.
19. Date, C. J. (1986). *Relational database: selected writings.* Addison-Wesley, Reading, Massachusetts.
20. Date, C. J. and Warden, A. (1990). *Relational database writings 1985–1989.* Addison-Wesley, Reading, Massachusetts.
21. Kim, W. and Lochovsky, F. H. (1989). ACM Press, New York.
22. Comer, D. (1988). *Internetworking with TCP/IP: principles, protocols, and architecture.* Prentice Hall, Englewood Cliffs, NJ.
23. Palca, J. (1990). *Science,* **248,** 160–2.
24. Kruskal, J. B. (1983). In *Time warps, string edits, and macromolecules: the theory and practice of sequence comparison,* (ed. D. Sankoff and J. B. Kruskal), pp. 1–53. Addison-Wesley, Reading, MA.
25. Pearson, W. R. (1990). In *Molecular evolution: computer analysis of protein and nucleic acid sequences,* (ed. R. F. Doolittle), pp. 63–110. Academic Press, San Diego.
26. Dayhoff, M. O. (1978). *Atlas of protein sequence and structure.* National Biomedical Research Foundation, Washington, DC.
27. Altschul, S. F. (1992). *J. Mol. Biol.,* in press.
28. Altschul, S. F. and Erickson, B. W. (1988). *Bull. Math. Biol.,* **50,** 77–92.
29. Gilbert, W., Marchionni, M., and McKnight, G. (1986). *Cell,* **46,** 151–3.
30. Mott, R. F., Kirkwood, T. B., and Curnow, R. N. (1989). *CABIOS,* **5,** 123–31.
31. Karlin, S. and Altschul, S. F. (1990). *Proc. Natl Acad. Sci. USA,* **87,** 2264–8.
32. Altschul, S. F., Gish, W., Miller, W., Myers, E. W., and Lipman, D. J. (1990). *J. Mol. Biol.,* **215,** 403–10.
33. Pearson, W. R. and Lipman, D. J. (1987). *Proc. Natl Acad. Sci. USA,* **85,** 2444–8.
34. Collins, J. F. and Coulson, A. F. W. (1990). In *Molecular evolution: computer analysis of protein and nucleic acid sequences,* (ed. R. F. Doolittle), pp. 474–87. Academic Press, San Diego.
35. Altschul, S. F. and Lipman, D. J. (1990). *Proc. Natl Acad. Sci. USA,* **87,** 5509–13.
36. Altschul, S. F., Carroll, R. J., and Lipman, D. J. (1989). *J. Mol. Biol.,* **207,** 647–53.
37. Benson, D., Huntzinger, R., Johnson, D., and Kulkarni, R. (1988). Lister Hill Center, National Library of Medicine.

38. Harman, D., Benson, D., Fitzpatrick, L., Huntzinger, R., and Goldstein, C. (1988). In *RIAO 88 User-oriented content-based text and image handling*, pp. 840–8.
39. Carrillo, H. and Lipman, D. J. (1988). *SIAM J. Appl. Math.*, **48**, 1073–82.
40. Lipman, D. J., Altschul, S. F., and Kececioglu, J. D. (1989). *J. Mol. Biol.*, **86**, 4412–15.
41. Schuler, G. D., Altschul, S. F., and Lipman, D. J. (1990). *Proteins*, **9**, 180–90.
42. Pervaiz, S. and Brew, K. (1987). *FASEB J.*, **1**, 209–14.
43. Sawyer, L. (1987). *Nature*, **327**, 659.
44. Godovac-Zimmerman, J. (1988). *TIBS*, **13**, 64–6.
45. Peitsch, M. C. and Boguski, M. S. (1990). *The New Biologist*, **2**, 197–206.
46. Namboodiri, K., Pattabiraman, N., Lowrey, A., and Gaber, B. P. (1988). *J. Mol. Graphics*, **6**, 211–12.
47. Pattabiraman, N., Namboodiri, K., Lowrey, A., and Gaber, B. P. (1990). *Protein Seq. Data Anal.*, **3**, 387–405.
48. Chothia, C. and Lesk, A. M. (1986). *EMBO J.*, **5**, 823–6.
49. Huber, R., *et al.* (1987). *J. Mol. Biol.*, **198**, 423–34.
50. Holden, H. M., Rypniewski, W. R., Law, J. H., and Rayment, I. (1987). *EMBO J.*, **6**, 1565–70.
51. Huber, R., *et al.* (1987). *J. Mol. Biol.*, **198**, 499–513.
52. Newcomer, M. E., *et al.* (1984). *EMBO J.*, **3**, 1451–4.
53. Papiz, M. Z., *et al.* (1986). *Nature*, **324**, 383–5.
54. Cowan, S. W., Newcomer, M. E., and Jones, T. A. (1990). *Proteins Struct. Func. Genet.*, **8**, 44–61.
55. Boguski, M. S. (1992). *J. Lipid Res.*, **33**, 957–74.

4

Design of protein structures

CHRIS SANDER, MICHAEL SCHARF,
and REINHARD SCHNEIDER

1. Introduction

1.1 The protein design cycle

The design of protein structures, whether by variation of natural proteins or from scratch, is an essential first step in the development of newly engineered proteins. Designs based on theoretical and computer methods enter the protein design cycle of repeated steps of experimental testing and improvement, much as in other engineering disciplines. Success in protein design depends on how well we make use of this cycle and on how well we understand the principles of protein folding, both in terms of the underlying molecular physics and in terms of knowledge extracted from the data-bases of macromolecular sequences and structures.

1.2 Two examples: redesign, *de novo* design

Here we illustrate two basic approaches to protein design, one incremental, a redesign project, the other bolder, a *de novo* design. The first example is the re-engineering of the loop connections in a naturally occurring protein of simple architecture, a bundle of four α-helices, Rop protein. The second example is the design from scratch of a geometrically regular sandwich of two four-stranded β-sheets, called Shpilka. We describe some of the methods used in these designs, but the selection is far from exhaustive.

1.3 Two tools: secondary structure and interface preference parameters

Of the many theoretical, data-based, and statistical tools, we present two in detail, namely two types of statistical preference tables for amino acid residue types in specific structural positions in a protein. The first uses structural states in terms of secondary structure type, position in a secondary structure segment, and solvent accessibility. The second, in terms of contacts between secondary structure segments and with water. These parameters can be used to help answer questions such as 'which residue types are preferred in solvent-exposed positions at the C-terminal end of an α-helix?' or 'what is the typical residue composition of a helix–sheet interface?'

1.4 Aspects not covered here

This chapter focuses on purely structural aspects, leaving out the important questions of biological function or activity. Also not covered here are: protein design using non-natural scaffolds, design of metal binding sites, the design of enzymatically active sites, and the design of membrane channels. Readers interested in further details of the theoretical and computer methods used in protein design are referred to recent reviews (1, 2) and may request from the authors detailed written reports from protein design workshops held in 1986 and 1990, under the auspices of the European Molecular Biology Organization (3, 4, 5).

2. Redesigning natural proteins: example, an α-helix bundle

2.1 The effect of mutations on protein structure

As the problem of designing protein structures from first principles is a very difficult one, we are well advised to devote some effort to a simpler problem, that of understanding the effects of amino acid sequence changes on a known protein structure. In which way does replacement of certain amino acids affect the conformation of the native (time-averaged) structure? In which way do point mutations affect the stability of the native protein structure relative to an unfolded or denatured state?

2.2 Nature's experience

Nature has accumulated vast experience, apparently by trial and error, of testing point mutations. Just think of the many variations in sequence in immunoglobulin molecules tested daily. Natural selection operates by selecting for functional properties, but structural integrity is a prerequisite for the proper function of many proteins. So the thousands of natural sequences known to be homologous to proteins of known structure can be considered the result of a giant series of evolutionary experiments in testing the effect of sequence changes on protein structure. We merely have to draw on the data-bases of protein sequences and structures to learn from these experiments (6, 7, 8, 9).

2.3 The protein engineer's experience

In comparison, the number of deliberately engineered mutations tested in the laboratory is very small, and often very little direct information about the details of structural changes is available. However, the data-base of experi-

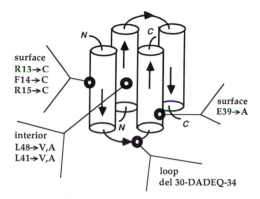

surface
R13→C
F14→C
R15→C

surface
E39→A

interior
L48→V,A
L41→V,A

loop
del 30-DADEQ-34

Figure 1. Cartoon of Rop protein based on the wild-type crystal and NMR structures (44, 45). Mutations on the protein *surface*, in the protein *interior*, and in *loop* regions have explored the stability of this fold with respect to sequence changes (12). Because of the dimeric symmetry of the helix bundle, each mutation appears twice in the three-dimensional structure, once in the front and once in the back subunit. In the figure, each mutation is only shown once, for the front subunit. The main result is that this four-helix bundle is very resistant to sequence changes on the solvent-exposed surface and in loop regions, and quite sensitive to mutations in the protein core (contact interface between the helices). Thus the bundle is an excellent scaffold for protein redesign experiments. The amino acid sequence of wild-type Rop is, in one-letter code, MTKQEKTALN MARFIRSQTL TLLEKLNELD ADEQADICES LHDHADELYR SCLARF.

mentally determined structures of mutant proteins is increasing steadily. One of the richest sets of mutant crystal structures, of phage T4 lysozyme, comes from the group of Brian Matthews (10, 11). Here, we illustrate some of the principles used in redesigning protein structure with a particularly simple example, that of the four-helix bundle, Rop (*Figure 1*).

2.4 Regions of protein structure

Just as a mechanical engineer would, protein engineers have developed certain concepts of the constructive elements of a protein, based on physical principles and on the analysis of natural proteins. A principal distinction is between the *framework*, which consists of hydrogen-bonded secondary structure segments, and *loops*, which connect these segments. In Rop protein, the framework consists of the four α-helices. Another important distinction is between the exposed surface, or protein *exterior*, defined as those parts of the protein accessible to direct contact by solvent molecules in the time-averaged native structure, and the protein *interior*, not in direct contact with solvent. In Rop protein, each helix has an interior and an exterior face, most of the backbone is in the interior and most of the loop residues are exposed to solvent. A third major distinction is made between the protein *core*, defined as those regions which are the most conserved in sequence and structure in the evolution of a protein family, and the *variable regions*, which are the most flexible in conformation and the most variable in sequence. In the Rop four-helix bundle, the mutually contacting interior faces of the helices constitute the core. These distinctions are useful not only in describing protein structure but also in formulating design rules. We now illustrate some of these rules by example.

2.5 Mutations in the protein core

In our example protein, Leu41 and Leu48 in the core of Rop protein were each changed to Val or Ala (*Figure 1*) (G. Cesareni and M. Kokkinidis, personal communication). The expectation was that replacing one non-polar side-chain by another would not lead to major structural changes but that the introduction of a hole in the hydrophobic protein core would reduce the stability of the native state. The crystal structure (M. Kokkinidis, personal communication) confirmed that the deletion of methyl groups in these interior side-chains simply leaves a hole in the protein interior without significant structural rearrangement. Calorimetric measurements (H. J. Hinz and P. Weber, personal communication) confirmed that the introduction of holes in the protein interior destabilizes the native structure relative to unfolded states.

2.6 Mutations on the exposed protein surface

As the primary interactions of exterior residues are with solvent molecules, their contribution to details of three-dimensional (3D) structure are expected to be minimal. On the other hand, solvation properties, such as a tendency for protein–protein aggregation, are expected to depend on surface residues, although perhaps not in a position-specific way. In the Rop example, replacement of many residues on the exposed surface of α-helices had no apparent effect of the native structure of the protein (detailed 3D structures were, however, not determined). Examples are the mutants D28A, D32F, R13C, R15C, F14C, E39A, D59H (*Figure 1*) (12) (although the function of some of these mutants was impaired).

2.7 Mutations in loops

Residues in loop regions also are primarily exposed to solvent and are observed to be variable in evolution, but they may also have a crucial role in determining the folding pathway. In one experiment, Cesareni and colleagues (12) deleted five amino acids (30-DADEQ-34; *Figure 1*) from the loop connecting the two helices of the Rop monomer in order to test the possibility of the formation of a single long helix. The deletion was chosen such that the long helix would have a continuous strip of hydrophobic residues on one helix face, as analysed in a helical wheel projection. The outcome of the experiment underscored the considerable conformational flexibility of loop regions and the persistence of the core structure formed by the interface between the four helices of the bundle: in the crystal structure, the helical bundle was well formed, except that a few residues at the ends of the two helices simply unravelled and formed a new type of loop in place of the deleted five residues (12).

2.8 Re-engineering protein topology

Mutation experiments on Rop and many other globular proteins have established the fact that the protein surface as well as loop regions are, in general, the most tolerant of mutations while the core is much more sensitive. Physically, the difference is primarily due to the fact that interactions involving side-chains in the protein core are highly specific and vary little in time, while interactions involving solvent are mostly non-specific and fluctuate strongly in time.

These results suggest the simple hypothesis that protein folding is dominated by interactions in the core of the protein. Let us call this the 'core hypothesis of protein folding'. According to this hypothesis, both the formation of the correct tertiary structure and the stability of folded proteins may be explained in terms of tight packing and specific interactions in the protein interior; the surface of the protein merely needs to have average solvation properties and loops merely need to provide connections of the appropriate length and flexibility.

To test one aspect of this hypothesis, the question asked was: to what extent could the loops connecting the four helices of the Rop bundle be rearranged without negatively affecting the correct fold? In a first experiment, the chain threading of this small protein was modified in a simple but radical fashion: deletion of one loop and addition of two new loops would turn the wild-type dimeric bundle into an artificial left-handed monomeric bundle, with an identically packed hydrophobic core but rather different topology of loop connections (*Figure 2*).

The synthetic gene for the correspondingly redesigned left-handed monomeric bundle (LmRop) was overexpressed and subsequent nuclear magnetic resonance spectroscopy (NMR) experiments yielded sufficient distance constraints (NOEs) to prove that helix–helix packing in the topologically rearranged bundle was preserved, in complete agreement with the core hypothesis of protein folding (ref. 13 and S. C. Emery *et al.*, submitted). The results of this type of experiment indicate that the protein engineer has considerable latitude in re-engineering loop regions in this type of protein, not only in length and amino acid sequence, but also in topology (14).

The successful modifications of Rop protein illustrate the fact that simple rules can work well. However, many more protein engineering experiments of this type are required to elucidate fully the role of loops, surface, and core regions of globular proteins in general.

Redesigning protein topology

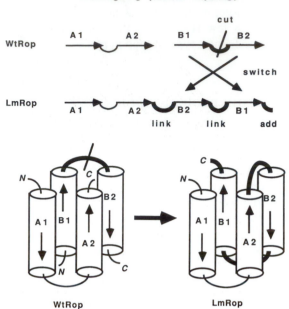

Figure 2. Re-engineering the topology of loop connections. Design steps from wild-type four-helix Rop dimer to the LmRop monomer involved deletion of one loop, addition of two new loops, and a C-terminal extension. Top: The orientation in space and packing of the four helices remains intact. Bottom: Two helical segments are switched at the gene level. In detail: wild-type Rop has two identical subunits, say A and B, and a total of four helices, say A1, A2 and B1, B2 (cylinders/arrows). To achieve the desired topology of a left-handed four-helix bundle which starts with helix A1, delete the loop between helices B1 and B2; switch the order (in the chain sense) of these helices from B1,B2 to B2,B1; connect by a new loop helices A2, B2 and by another loop helices B2, B1; complete helix B1 by addition of a few residues. Loops involved in these changes are emphasized (thick lines).

3. Designing proteins from scratch: example, a β-sheet sandwich

Designing proteins from scratch, rather than modifying naturally evolved ones, is a fascinating and challenging key problem of molecular engineering. Given a desired function and/or structure, the task is to design or calculate a suitable amino acid sequence that fulfils the design specifications. The goal in *de novo* protein design is to gain total control over the design process and to overcome the constraint of having to use existing amino acid sequences.

3.1 Build backbone model

The methods and tools used in the design process are rapidly evolving, so only an outline of some of the current techniques can be given here, illustrated

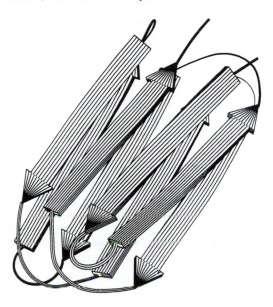

Figure 3. Ribbon cartoon of Shpilka, an antiparallel sandwich of two four-stranded β-sheets designed *de novo* as a scaffold on which to explore variations in the topology of loop connections (4, 5). The 96 residues of its amino acid sequence are: G {IKMTITLEL} GT {TKHSIPIET} GSPGG {IRMTITLEL} GT {TRHSVPVEG} GT {KHSVPVDT} GGGGG {VKWTVTMDL} GT {TRHSTPVDT} SGSPN {VRMTVTVD} LG, with β-strands enclosed in curly brackets. This sequence was designed in 2 weeks as an exercise. The design has recently been improved by A. V. Finkelstein (personal communication). The atomic coordinates of the corresponding model can be obtained from the electronic mail file server *netserv@embl-heidelberg.de* by sending the message *send proteindata:shpilka.brk*.

with a simple example (4, 5), a hypothetical protein called Shpilka (*Figure 3*). Suppose the design specification is for a globular protein consisting of a single-chain β-sheet sandwich with loops that can be varied flexibly. First, one makes a rough sketch of the structure: how many strands and which approximate strand lengths are to be used? Are the β-sheets to cross at approximately right angles or to run essentially parallel to each other? How strongly are the sheets twisted? Then, using molecular modelling software (15, 16, 17, 18) one builds a backbone model, best achieved by picking appropriate fragments from the data-base of known 3D structures so as not to deviate too strongly from well-tested structural units.

3.2 Choose sequence to fit structure

In the crucial next step, one chooses a sequence to fit the desired structure (see also Sections 6 and 7). This step is not yet automated, but instead is performed interactively, using empirical criteria based on physics, statistics, or data-base analysis (3, 4, 5, 19, 20). For choosing individual residues, statistical tables of residue preferences are available (21, 22, 23, 24). A very useful parameter set gives preferences for specific positions in secondary structure segments (see *Figures 5a, 6, 7*), e.g. near the beginning, middle, and end of strands or helices (21, 25, 26, 27, 28), turns (29, 30, 31), or loops (32). An additional useful distinction is between solvent-exposed and interior positions on helices and strands (see *Figures 4b, 5b*) (28, 33). Statistical parameters of a new type presented here are preferences for interfaces (see *Figure 11*), e.g. between two β-strands, between two β-sheets, between a β-sheet and an α-helix, or between a β-sheet and solvent (34). Specific examples of parameter usage are shown in *Figures 8* and *12*.

Particular physical properties of individual residues can also be exploited directly, without recourse to statistics. For example, Pro can be used to lock in local backbone conformation (the dihedral ϕ angle) and to exclude the formation of a backbone hydrogen bond (to nitrogen); Gly can be used for maximum backbone flexibility; and Trp may be chosen as a spectroscopic probe of folding. The full set of all types of rules for choosing amino acid types for a desired structural purpose cannot be given here.

For choosing pairs and clusters of residues, criteria include local complementarity of shape, size, and polarity, optimization of packing, hydrogen bonding, and charge–charge interactions. In this part of the design process frequent comparison of the current model with similar 3D constellations in the data-base of known structures is extremely useful (ref. 17 and K. Robson and C. Sander, unpublished). To assure good solvation properties, the overall amino acid composition of the apolar cores and solvent-exposed surfaces of globular proteins can be used as a guide.

3.3 Example: choosing a sequence for Shpilka

In the example of Shpilka (*Figure 3*), the following criteria were considered particularly important for the formation of the desired eight-stranded β-sandwich:

- no possibility to form α-helices
- eight regions with the capability to form β-strands
- stronger hydrophobicity in internal strands compared to edge strands
- well-defined non-polar and polar surfaces in all β-sheet regions
- edge strands that are restricted in backbone H-bond formation on one side
- close packing of side-chains in the interior

The designed sequence of Shpilka (*Figure 3*) has eight regions enriched in β-forming residues, separated by β-breaking connections. Non-polar and polar residues alternate to form well-defined inner and outer surfaces for β-strands, not suitable for α-helices. The interior face of the strands contains Val, Ile, and Met residues, the best β-formers. The interior face of edge strands contains β-forming threonines, which are partly polar, making contact with the hydrophobic core as well as with solvent. The solvent-accessible outer surfaces of all strands include charged groups that can form salt bridges. Prolines, which have no backbone NH group, are incorporated into the edge strands so that only one side of an edge strand can form a continuous hydrogen-bond net. Turns and loops are made of β-breaking residues; to help ensure unambiguous folding, they have minimal length, just enough to connect the strands in the designed structure. A Trp was included on the interior face of the sheet to serve as an experimental marker for correct folding. At the interior face of internal strands bulky side-chains alternate with small alanines in order to form complementary surfaces for close packing.

3.4 Avoid alternate folds

Simply adapting a sequence to a desired static scaffold is, however, not sufficient. How does one guarantee that the sequence actually allows a sufficient number of folding pathways from an unfolded to the final fold? That it is stable with respect to conformational fluctuations? That it will not equally well, or better, fold up into another globular shape? The current theories of protein folding do not yet provide satisfactory answers to these questions. However, empirical rules are being developed for some of these aspects. For example, in the design of the four-helix bundle, 'Felix', the number of residues in a loop was chosen so as to disfavour formation of an uninterrupted helix of excessive length (1). Also, in a helix bundle design, DeGrado *et al.* (35) attempted to make use of repulsive ion-pair interactions to force loop formation and prevent incorrect helix association, by insertion of a Pro–Arg–Arg sequence in a loop.

3.5 Refine three-dimensional model and sequence

After a first sequence has been chosen, one fully optimizes the 3D model by Monte Carlo or molecular dynamics exploration, followed by energy minimization. In our experience, Monte Carlo methods are vastly preferable for this purpose (36). Next, one evaluates the overall quality of the designed protein according to general criteria that were perhaps not used in the construction process, e.g. estimates of solvation energy (37), transfer energy (38), packing rules (39), or distribution of contacts around residue types (47). Such analysis will invariably lead to suggestions for point mutations in the model. Before proceeding to experiment, it is advisable to subject the sequence and 3D model to several or many cycles of sequence change, conformational optimization, and analysis.

3.6 Evaluate by structural experiments

With a designed sequence in hand, the protein or corresponding gene can be synthesized (see other chapters in this volume). For *in vivo* expression of synthetic genes, the main problem in some cases has been insufficient recovery of overexpressed product and purification into a soluble fraction. *In vitro* expression systems side-step some of these problems (O. Ptitsyn, personal communication), but suffer from low yields. Finally, spectroscopic methods, such as circular dichroism, Trp fluorescence, and 1D NMR are used as first tests of structure. Alternatively, in the case of a designed function that is strongly dependent on a specific 3D structure, a functional assay is used. The free energy of unfolding can be estimated by following a spectral signal as a function of the concentration of denaturant. A full determination of correct 3D structure requires X-ray crystallography or higher-dimensional (at least 2D) NMR spectroscopy, but so far the crystal structure of only one *de novo* designed protein has been completed (40). In general, several or many cycles of design, verification, and redesign are required to evolve a correctly folded protein.

4. Limitations

4.1 Difficulty in designing well-packed protein core

In spite of some successes in *de novo* protein design (1, 2), recent experiments show that currently used rules of protein design are incomplete or insufficient. None of the experimentally tested *de novo* designs has yielded a crystal or NMR structure. There is one exception (40), but in that case the 3D arrangement of α-helices in the crystal structure did not conform with the designers' intentions. It appears that so far none of the *de novo* designed proteins has achieved a unique fold comparable in quality to that of natural proteins, in spite of considerable stability against unfolding in some cases (2, 41). Both the physical theory of protein folding and the statistical and data-derived rules need to be improved.

4.2 Need for more powerful modelling computer tools

What tools are required in the near future? For *de novo* design, there is a clear need for more powerful constructive tools for going from a raw sketch to a first model. Example: given a sketch of a four-on-four β-sandwich, build a first backbone model in which the strands have typical twist angles and the intersheet angle is approximately as specified. Then there is a need for globally modifying structures. Example: increase the overall twist of a β-sheet while maintaining optimal packing and good loop geometry.

4.3 Need for improved energy calculations

There are two key problems in energy calculations. With current potentials, even in simulations of molecular dynamics including water, there is in many cases little correlation between correctness of structure and low energy; and it is still extremely time consuming to explore a significant fraction of conformational space. So we urgently need more precise energetics that can reliably distinguish between correct and incorrect structures, and more efficient simulation tools for exploring many alternate protein conformations.

4.4 Novel protein structure motifs? Enzyme design?

A principle limitation of current work is the fact that protein design tends to stay within the limits of the structural motifs discovered in natural proteins. For example, all *de novo* designed proteins so far have been helical bundles, β-sheet sandwiches, or regular types of β–α proteins (1, 2). Protein engineers are generally not yet bold (or foolish) enough to attempt designs that deviate significantly from the architecture of natural proteins. Finally, the design principles for functional interfaces and for enzymatically active sites are still in their infancy (2).

5. Outlook

5.1 More rapid protein design cycle

Protein design methods are evolving rapidly. In our opinion, key advances will be: the development of more efficient cycles of design and verification, with the current bottleneck being, in many cases, protein expression and purification; and the development of better methods for the evaluation of models by empirical or energetic criteria, with the current bottleneck being proper treatment of solvation effects, coupled to electrostatics.

5.2 *De novo* design of structures

It appears that *de novo* protein design is much more difficult than the re-engineering of natural proteins. However, there is one way in which *de novo* design may turn out to be very simple. The sequences of natural proteins contain a superposition of information about structure, function, protein transport, average lifetime before degradation, etc., and, in addition, have been randomized within the bounds of selective pressure by mutational events. Sequences of *de novo* designed proteins, however, can be very simple, if the design goal is limited solely to the attainment of a stable structure. So the simplest application of the type of rules described here may soon lead to the successful design of properly folded and well-packed protein, once the major deficiency of current designs, i.e. insufficiently optimized residue–residue interactions in the protein core and excessive flexibility of alternate packings, have been overcome.

5.3 Learn from natural evolution and evolution experiments

The two goals of protein design are to achieve a complete understanding of the folding of natural proteins and to develop design principles to achieve any physically possible target structure, such as enzyme scaffolds, self-assemblying structures, catalytically active sites, and structural regulators of activity. On the way to achieving these goals, both the study of natural protein families (9) and the planning, execution, and analysis of evolution experiments in the laboratory complement activities in protein design. All these ways need to be pursued, with evolution experiments being the most underdeveloped. We are in the early stages of protein design, in which skills are actively being developed at a high rate. 'The future is bright, but the road is tortuous.'

Acknowledgements

We thank François Colonna-Cesari for his important contributions several years ago to the definition and analysis of contact interfaces and John Priestle for his ribbon drawing program. Research was supported in part by the BRIDGE program of the Commission of the European Communities and the Human Frontiers Science Program.

Appendix

6. Tool: choosing amino acid types in helices, strands, and loops

6.1 Analysis of known protein structures

The data-base of known protein structures (6) is the basis for many empirical rules of protein design. In order to derive such rules, the very complicated set of 3D coordinates of a protein is first reduced to a simplified representation in terms of structurally characteristic regions. Statistical analysis can then be performed and preference rules can be derived for which types of amino acids to use in which regions. We now introduce two different ways of representing protein structure. The first (this section) refers to particular positions in secondary structure segments, the second (Section 7) refers to interactions, or contacts, between such segments and with water.

6.2 Representing protein structure: positions in secondary structure segments

The classical representation of 3D structure is in terms of α-helices, β-strands, hydrogen-bonded turns, and loops (42). A more refined description, presented here, distinguishes between various positions on these elements, e.g. positions near the ends or in the middle of a helix or strand, or positions on the solvent-exposed or interior face of a segment (*Figures 4–8*) (28). In an even more refined statistical analysis (not presented here), preferences for pairs, triplets, tetra-peptides, and pentapeptides in certain structural states can be derived (28).

6.3 Definition of sequence–structure preference parameters

How are preference parameters defined? In a given structural state (S), the preference parameter for a residue type (R) is calculated from the number of observed cases of R in S, $N(R,S)$, as

$$\mathrm{pref}(R,S) = \mathrm{ld}\left(\frac{N(R,S)\ N}{N(R)\ N(S)}\right)$$

where

$$N(R) = \sum_{S} N(R,S), \ N(S) = \sum_{R} N(R,S), \ N = \sum_{R,S} N(R,S)$$

and ld is the logarithm base 2, so that $\mathrm{pref}(R,S)$ is in information units called bits. The expression in parentheses is the ratio of the number of observed

(a)

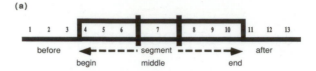

(b)

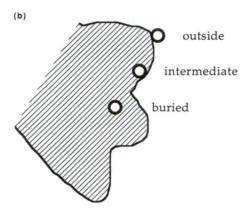

Figure 4. Definitions of positions on the secondary structure segment used in *Figures 5–8*. (a) Definition of linear positions along the secondary structure segment. Three positions each are distinguished on either side of the segment boundaries; positions in the middle of the segment are lumped together in one class. Positions are numbered 1–13 for convenience. Note that position 4 is at the beginning (N-terminus) of a segment and position 10 is at the segment end (C-terminus). (b) Definition of outside/inside/buried positions on the secondary structure segment. The solvent-accessible surface area is used as a basis for the classification. Residues with large accessibility are classified as outside, those with intermediate values as inside, and the rest as buried. The notation used is: 'exposed to solvent' (O, for outside); 'intermediate' (I, for intermediate or inside); 're-moved from solvent' (B, for buried). Residue accessibility is defined as the ratio of the actual accessible surface area and its maximal value, in per cent, as tabulated by Baumann *et al.* (37). The cut-offs used in the definition vary with secondary structure type and were determined on the basis of residue accessibility histograms (28). In terms of the percentage of maximal surface exposed to solvent, the cut-offs are as follows, for all residue types:

	B	I	O
β-strand	E 0% – buried – 5%	– interm.– 15%–	outside– 100%
loop	L 0% – buried – 35%–	interm.– 60%–	outside– 100%
α-helix	H 0% – buried – 25%–	interm.– 50%–	outside– 100%

These cut-off values reflect the fact that the average exposure is largest for loops, followed by helices and strands (28). Such cut-offs are, of course, somewhat arbitrary. Note that solvent exposure can be quantified in terms of interatomic contacts without cut-offs (see *Figure 9*).

(a) preference parameter of a single amino acid for a secondary structure type

S	V	L	I	M	F	W	Y	G	A	P	S	T	C	H	R	K	Q	E	N	D
E	0.6	0.4	0.6	0.5	0.5	0.1	0.3	-0.5	-0.1	-0.6	-0.1	0.2	0.0	-0.2	-0.1	-0.4	-0.1	-0.4	-0.6	-0.7
L	-0.4	-0.4	-0.5	-0.5	-0.3	0.0	-0.1	0.3	-0.1	0.4	0.1	0.0	-0.1	0.1	0.1	0.1	0.0	0.0	0.3	0.2
H	0.0	0.1	0.0	0.2	0.0	-0.2	-0.3	-0.4	0.3	-0.5	-0.2	-0.2	0.2	0.0	-0.1	0.2	0.1	0.3	-0.1	0.0

(b) preference parameter of a single amino acid for an inside/outside position in a secondary structure type

S X	V	L	I	M	F	W	Y	G	A	P	S	T	C	H	R	K	Q	E	N	D
E B	1.0	0.9	1.2	0.9	0.9	0.4	0.2	-0.2	0.2	-0.9	-0.4	-0.1	0.6	-1.0	-1.4	-2.2	-1.0	-1.7	-1.6	-1.8
E I	0.6	0.5	0.7	0.6	0.8	0.3	0.8	-0.4	-0.4	-0.8	-0.3	-0.1	0.0	0.3	-0.1	-0.6	-0.3	-0.7	-0.7	-1.1
E O	0.0	-0.2	0.0	0.1	-0.1	-0.3	0.2	-0.8	-0.3	-0.3	0.1	0.3	-1.0	0.1	0.4	0.3	0.3	0.1	-0.1	-0.2
L B	0.0	0.0	0.0	-0.1	0.2	0.3	0.3	0.3	-0.1	0.2	0.1	-0.1	0.5	0.2	-0.2	-0.6	-0.2	-0.4	0.0	0.0
L I	-0.5	-0.6	-0.8	-0.7	-0.8	-0.4	-0.3	0.2	-0.2	0.5	0.1	0.2	-0.3	0.0	0.3	0.3	0.1	0.2	0.4	0.3
L O	-1.0	-0.9	-1.2	-0.9	-0.9	-0.1	-0.5	0.5	-0.1	0.6	0.3	0.0	-1.4	0.0	0.2	0.4	0.1	0.3	0.5	0.4
H B	0.6	0.8	0.7	0.8	0.6	0.5	0.1	-0.5	0.4	-1.2	-0.4	-0.2	0.9	-0.2	-0.8	-0.8	-0.6	-0.6	-0.7	-1.0
H I	-0.4	-0.4	-0.6	-0.3	-0.6	-0.6	-0.1	-0.5	0.2	-0.4	-0.3	-0.3	-0.6	0.1	0.4	0.6	0.5	0.6	0.1	0.3
H O	-0.8	-1.1	-1.4	-0.7	-1.1	-1.3	-1.1	-0.3	0.3	0.0	0.1	0.0	-1.2	0.1	0.1	0.6	0.4	0.8	0.4	0.6

Figure 5. Preference parameters for amino acid types in various structural states, going beyond the simple description in terms of helices, strands, and loops. These parameters are useful for choosing residues for particular structural positions in protein design. They can also be used in protein structure prediction from sequence (28). The parameters are based on a non-redundant set of 38 selected high-resolution protein structures (list not shown). In order to improve the effective data-base size, for each of the 38 proteins all clearly homologous (10 percentage points above the threshold for structural homology (9), i.e. more than 35% identical residues for alignment lengths longer than 80 residues) protein sequences are also used, counting each distinct sequence–structure pair (R, S) only once at each sequence position, for a total of 52 426 unique residue occurrences in the 38 protein families. In this way, refined distinctions of structural states become amenable to statistical analysis. (a) Secondary structure preference parameters (set S3). Secondary structure can be S = E, H, L where E = β-strand; L = loop; H = α-helix (see *Figure 4b*). Parameters are information values in bits. For example, Val in β-strands has a preference of 0.6 bits, i.e. the odds are $2^{0.6}$ = 1.52 to 1 to observe Val in β-strands. (b) Combined accessibility and secondary structure preference parameters (set SX9). Secondary structure in S = E, L, H as above. The three accessibility states are: 'exposed to solvent' (X = O for outside), 'intermediate' (X = I for intermediate or inside), and 'removed from solvent' (X = B for buried). Numerical example: pref(Met,H,B) = 0.8, but pref(Met,H,O) = 0.7, showing that Met is preferentially found at the solvent-protected side of helices rather than on the surface. Note that these preferences are clearer than the average preference for the helical state, pref(Met,H) = 0.2. Averaging obviously results in loss of information and more refined distinctions of structural states enhance the information content of preference parameters. Notation is, for example, EB, β-strand, buried; LO, loop, outside; HI, helix, intermediate.

103

position preference parameter of a single amino acid in a secondary structure segment

β-strand

S	P	V	L	I	M	F	W	Y	G	A	P	S	T	C	H	R	K	Q	E	N	D
E	1	-0.5	-0.4	-0.5	-1.0	-0.3	-0.1	0.1	0.2	0.0	0.3	0.2	-0.2	-0.6	0.3	0.3	0.1	-0.1	0.0	0.4	0.3
E	2	-0.6	-0.6	-0.9	-0.5	-0.4	0.1	-0.7	0.6	-0.2	0.4	0.1	-0.2	-0.7	0.1	0.0	0.0	0.2	0.2	0.4	0.6
E	3	-0.5	-0.4	-0.4	0.1	0.1	0.4	0.2	0.4	-0.3	0.3	0.1	-0.2	-0.3	0.2	0.0	0.1	0.1	-0.2	0.2	0.1
E	4	0.2	0.1	0.5	0.6	0.4	0.0	0.4	-0.7	-0.2	-0.5	0.0	0.2	0.0	-0.1	0.0	-0.1	0.3	-0.3	-0.7	-0.9
E	5	0.9	0.4	0.9	0.7	0.5	0.7	0.2	-0.9	-0.1	-0.9	-0.4	0.1	0.1	-0.5	-0.3	-0.5	-0.4	-0.5	-0.6	-1.0
E	6	0.7	0.6	0.6	0.4	0.6	0.2	0.7	-0.8	0.0	-1.1	-0.2	0.1	0.3	0.2	0.3	-0.7	-0.3	-0.7	-1.1	-1.6
E	7	0.4	0.3	0.6	0.4	0.2	-0.2	-0.1	-0.1	0.0	-0.7	0.0	0.2	-0.4	-0.2	-0.3	-0.3	0.0	-0.3	-0.5	-0.7
E	8	0.7	0.6	0.9	0.0	0.7	-0.7	0.2	-0.2	0.2	-1.0	-0.3	0.3	0.0	-0.6	-0.5	-0.7	-0.5	-0.7	-0.8	-0.9
E	9	0.7	0.5	0.7	0.6	0.5	-0.1	0.2	-0.9	-0.1	-0.8	-0.2	0.0	-0.3	-0.2	0.0	-0.4	-0.1	-0.5	-0.7	-0.9
E	10	0.3	0.6	0.5	0.2	0.2	0.0	0.2	-0.4	-0.5	0.0	-0.1	0.0	0.0	-0.2	0.0	-0.4	-0.4	-0.3	-0.3	0.0
E	11	0.0	-0.2	-0.3	-0.7	-0.3	0.0	-0.1	0.3	0.0	0.2	0.2	0.2	-0.2	0.1	-0.1	-0.3	-0.4	-0.1	0.3	0.1
E	12	-0.7	-0.7	-0.6	-0.9	-0.5	0.1	-0.3	0.5	-0.1	0.5	0.3	0.2	-0.8	-0.1	0.1	0.1	0.0	0.1	0.1	0.3
E	13	-0.4	-0.6	-0.5	-0.5	-0.2	0.2	0.2	0.2	-0.2	0.3	0.1	0.0	-0.1	0.2	0.1	0.1	-0.1	0.2	0.0	0.4

loop

S	P	V	L	I	M	F	W	Y	G	A	P	S	T	C	H	R	K	Q	E	N	D
L	1	0.4	0.4	0.5	0.1	0.3	0.3	0.1	-0.3	0.1	-0.4	-0.2	0.0	-0.2	0.0	-0.1	-0.2	-0.3	-0.2	-0.5	-0.6
L	2	0.4	0.2	0.4	0.5	0.2	-0.3	0.0	-0.7	0.0	-1.0	-0.1	0.0	0.4	-0.2	0.0	0.0	0.1	0.0	-0.3	-0.5
L	3	0.1	0.4	0.2	0.0	0.2	-0.1	0.3	-0.3	-0.1	-0.7	-0.1	-0.1	0.3	0.0	-0.1	-0.1	0.0	-0.2	-0.1	-0.1
L	4	-0.1	-0.1	-0.4	-0.1	-0.1	-0.1	-0.1	0.5	0.0	-0.2	0.1	0.0	0.1	0.1	0.0	-0.2	-0.3	-0.1	0.3	0.0
L	5	-0.5	-0.5	-0.5	-0.9	-0.4	-0.2	-0.3	0.5	0.0	0.6	0.1	0.1	-1.4	0.0	0.2	0.2	0.2	0.0	0.1	0.2
L	6	-0.5	-0.3	-0.6	-0.4	-0.6	0.2	0.0	0.3	-0.2	0.4	-0.1	0.0	-0.2	0.1	0.3	0.3	0.0	0.1	0.2	0.3
L	7	-0.3	-0.4	-0.3	-0.6	-0.4	0.1	0.1	0.1	-0.1	0.4	0.0	0.0	0.2	-0.1	0.2	0.0	0.1	0.0	0.1	0.2
L	8	-0.4	-0.3	-0.5	-1.5	-0.3	-0.4	-0.1	0.1	-0.2	0.5	0.1	-0.4	0.0	0.3	0.4	0.2	0.0	0.1	0.4	0.4
L	9	-0.4	-0.3	-0.4	-0.3	-0.3	0.3	-0.4	0.4	-0.2	0.4	0.0	-0.1	-0.1	0.0	0.1	0.1	0.0	0.0	0.2	0.4
L	10	-0.9	-0.5	-0.7	-0.3	-0.1	0.0	-0.1	0.3	-0.4	0.4	0.4	0.1	-0.1	0.2	-0.1	0.1	0.0	-0.1	0.5	0.3
L	11	0.1	0.0	0.2	0.4	0.2	0.0	0.2	-0.5	0.0	0.4	0.0	0.1	0.0	-0.2	-0.1	0.0	0.2	-0.1	-0.6	-0.5
L	12	0.4	-0.1	0.3	0.2	0.0	0.2	-0.2	-0.4	0.1	-0.2	-0.1	0.0	0.1	-0.3	-0.3	-0.2	-0.1	0.3	-0.2	0.1
L	13	0.4	0.3	0.3	0.3	0.4	0.1	0.4	-0.5	0.0	-0.8	-0.1	-0.1	0.2	0.0	-0.2	-0.4	-0.1	0.0	-0.2	-0.1

α-helix

S	P	V	L	I	M	F	W	Y	G	A	P	S	T	C	H	R	K	Q	E	N	D
H	1	0.0	-0.2	-0.1	0.0	0.1	-1.5	0.1	0.3	0.1	0.0	-0.1	-0.1	0.3	-0.1	-0.1	0.3	-0.1	-0.2	0.0	-0.3
H	2	0.2	0.4	0.3	0.3	0.1	-0.4	-0.2	-0.1	0.2	0.2	0.0	0.2	-0.1	-0.2	-0.2	-0.2	-0.4	-0.7	-0.4	-0.1
H	3	-1.7	-0.8	-1.7	-0.9	-0.9	-1.4	-0.7	0.2	-0.3	0.7	0.7	0.4	0.4	0.2	-0.2	-0.2	-0.4	-0.1	0.7	0.7
H	4	0.0	0.0	-0.1	-0.3	-0.2	0.1	-0.2	-0.3	0.2	1.2	0.0	-0.1	0.0	-0.5	-0.4	0.1	-0.2	0.2	-0.4	-0.1
H	5	-0.6	-1.3	-1.0	-0.5	-0.8	-0.1	-1.0	0.1	0.4	0.1	0.1	0.0	0.3	-0.1	-0.4	0.0	0.1	0.8	0.2	0.7
H	6	0.0	-0.3	-0.4	0.4	-0.1	-0.5	0.0	-1.2	-0.1	-0.3	-0.4	-0.2	-0.7	0.0	-0.7	-0.2	0.6	1.0	0.1	0.8
H	7	0.2	0.4	0.3	0.5	0.1	-0.2	-0.4	-0.5	0.5	-1.8	-0.4	-0.2	0.3	-0.1	0.0	0.2	0.0	0.2	-0.2	-0.4
H	8	0.2	0.6	0.5	0.5	0.0	0.2	-0.8	-0.6	0.3	-1.9	-0.5	-0.3	-0.8	0.4	0.2	0.4	-0.1	0.0	-0.2	-0.5
H	9	0.1	0.3	0.2	0.6	0.1	-0.6	-0.4	-0.8	0.2	-1.0	-0.2	-0.2	0.0	0.0	-0.1	0.4	0.3	0.2	0.0	-0.4
H	10	-0.3	0.1	-0.6	-0.2	0.1	-0.1	0.5	-0.3	0.3	-2.2	-0.2	-0.3	0.6	0.2	0.1	0.3	0.3	0.0	0.2	-0.1
H	11	-0.4	0.2	-0.4	0.2	0.3	-0.2	0.2	0.6	-0.2	-0.9	0.0	-0.3	0.4	0.1	0.1	0.0	-0.1	-0.2	0.4	-0.1
H	12	-0.6	-0.3	-0.5	-0.5	-0.3	-1.1	-0.3	0.6	0.1	0.5	0.0	-0.1	-0.4	0.0	-0.1	0.4	0.1	-0.3	0.3	0.0
H	13	-0.1	-0.1	-0.5	0.0	-0.6	-0.2	-0.3	0.0	-0.2	0.5	-0.2	0.1	0.0	0.1	0.0	0.4	0.1	0.0	0.2	0.2

Figure 6. Combined positional and secondary structure preference parameters (set SP39). Positions P = 1–13 are defined in *Figure 4a*, secondary structure types S = E, L, H in *Figure 5a*. Notation is, for example, E1, third residue position before the beginning of a β-strand; L6, third residue position in a loop; H12, second residue after end of a helix.

position preference parameter of a single amino acid in an inside/outside position of a secondary structure segment

β-strand

S X P	V	L	I	M	F	W	Y	G	A	P	S	T	C	H	R	K	Q	E	N	D
E B 1	0.2	0.7	0.5	0.0	0.8	1.0	0.2	-0.1	0.3	-2.2	0.3	-0.5	1.4	0.1	-0.2	-2.3	-0.4	-1.5	-0.7	-1.5
E B 2	0.5	0.2	0.3	0.8	0.8	0.3	-0.8	0.8	0.2	0.0	-0.1	-0.6	0.1	-2.0	-1.4	-1.5	-3.6	-1.4	0.4	0.4
E B 3	0.0	0.6	0.5	0.8	0.5	1.8	0.4	0.1	-0.3	0.7	0.1	-0.3	0.4	-0.1	-1.9	-1.8	-0.8	-1.4	-1.0	-0.5
E B 4	0.2	0.5	1.2	1.0	0.9	0.5	0.3	-0.2	-0.1	-0.4	0.0	-0.1	0.9	-0.9	-1.4	-1.3	0.1	-1.3	-1.4	-2.2
E B 5	1.3	0.9	1.3	1.1	0.9	0.9	0.1	-0.8	0.2	-1.8	-0.9	-0.4	0.7	-0.6	-1.8	-1.9	-1.1	-1.7	-1.7	-2.4
E B 6	1.1	1.0	0.9	0.7	1.0	0.4	0.7	-0.3	0.4	-1.2	-0.5	-0.5	0.5	-0.6	-0.7	-2.4	-1.4	-2.6	-2.0	-2.5
E B 7	1.0	0.9	1.1	0.7	0.8	0.3	0.0	0.2	0.2	-1.9	-0.4	0.2	0.4	-0.9	-1.3	-2.5	-0.8	-1.6	-2.4	-2.4
E B 8	1.1	1.0	1.3	0.1	0.8	-1.0	-0.2	0.0	0.4	-0.7	-0.9	0.2	-0.2	-0.8	-1.1	-3.1	-1.8	-1.4	-1.4	-1.6
E B 9	1.2	0.9	1.1	0.9	0.7	0.4	0.2	-0.5	0.3	-0.7	-0.4	0.1	0.4	-2.0	-1.8	-2.8	-1.0	-2.1	-1.5	-2.0
E B 0	0.9	1.1	1.1	0.7	0.5	0.1	0.0	-0.2	-0.4	-0.4	-0.3	-0.1	0.9	-1.6	-1.2	-1.9	-1.5	-1.4	-1.2	-0.5
E B 1	0.7	0.6	0.5	0.0	0.5	1.1	0.2	0.8	0.3	-1.0	0.1	0.2	0.8	-0.5	-1.7	-2.4	-2.1	-1.6	-2.1	-0.9
E B 2	0.3	-0.2	-0.4	-0.9	0.2	-1.5	-0.9	1.1	0.5	0.1	0.4	0.3	0.1	-1.1	-0.9	-0.4	-1.3	-1.5	-0.7	-0.1
E B 3	0.5	0.2	0.8	0.1	1.0	1.7	0.8	0.2	0.0	-0.1	-0.8	0.1	1.1	-0.7	-1.6	-2.0	-1.3	-1.2	-1.5	-0.3
E I 1	0.0	0.6	0.6	-0.4	0.2	0.0	0.4	-0.3	0.0	-0.1	0.2	-0.7	0.4	0.6	-0.1	-0.8	-0.3	-0.8	-0.3	-0.5
E I 2	0.4	0.4	0.5	-1.1	0.1	0.0	0.4	0.6	0.3	-1.5	-0.1	0.0	1.1	-1.3	-1.2	-0.8	0.3	-2.0	-0.5	-0.2
E I 3	-0.1	0.1	-0.3	0.2	0.6	0.8	0.7	0.4	-0.1	-0.7	-0.3	-0.4	0.0	-0.7	-0.4	-0.5	0.3	-0.6	-0.1	-0.5
E I 4	0.4	0.5	0.5	0.6	0.9	0.3	1.0	-0.5	-0.4	-0.1	-0.6	0.0	-0.1	0.4	-0.1	-0.8	-0.2	-0.7	-1.7	-0.7
E I 5	0.7	0.3	0.4	0.7	1.0	0.6	0.5	-0.6	-0.5	-0.3	-0.2	-0.4	0.8	-0.6	-0.1	-0.8	-0.2	-0.8	-0.1	-1.0
E I 6	0.3	0.2	0.6	0.9	0.1	0.0	0.8	-1.2	-0.9	-1.5	-0.3	0.4	0.8	0.8	0.9	0.0	-0.4	-0.5	-1.3	-3.6
E I 7	0.3	0.3	0.3	0.3	0.2	-0.2	-0.1	-0.5	-0.4	-0.4	0.3	0.2	-0.7	-0.2	-0.6	-0.6	0.5	0.1	-0.5	-0.9
E I 8	0.4	0.4	0.4	0.0	1.4	1.1	0.7	-0.1	-0.2	-1.9	-0.7	-0.5	0.7	0.6	-0.2	-0.6	-0.7	-1.5	-0.2	-0.9
E I 9	0.9	0.8	1.0	0.9	0.9	-1.1	1.0	-1.3	-0.2	-2.5	-0.6	-0.3	-0.9	0.1	0.0	-0.7	-1.1	-1.1	-0.5	-1.5
E I 0	0.2	0.6	0.7	0.3	0.6	0.9	0.8	0.2	-0.4	-0.8	-0.4	-0.5	0.0	0.3	-0.5	-0.7	-0.4	-0.4	-0.7	-0.7
E I 1	0.4	0.2	0.2	-0.1	0.2	0.7	0.8	0.2	0.1	0.3	-0.1	-0.1	1.1	0.1	-0.6	-1.7	-1.6	-0.7	-0.4	-0.3
E I 2	-0.4	-0.8	0.3	-1.3	-0.5	-0.9	0.4	0.2	0.3	0.7	-0.5	0.2	0.3	0.6	0.0	-0.3	0.1	-0.2	-0.4	0.0
E I 3	-0.1	0.3	0.4	-0.8	0.4	-0.1	0.7	-0.4	-0.4	0.0	-0.3	-0.3	0.5	0.6	-0.3	-0.5	0.0	0.0	-0.2	0.1
E O 1	-0.7	-0.8	-1.0	-1.3	-0.7	-0.3	-0.1	0.3	-0.1	0.4	0.2	-0.2	-1.4	0.2	0.3	0.2	0.0	0.1	0.5	0.4
E O 2	-0.9	-0.8	-1.2	-0.8	-0.7	0.1	-0.8	0.6	-0.3	0.4	0.1	-0.2	-1.1	0.3	0.1	0.1	0.3	0.4	0.4	0.6
E O 3	-0.8	-0.9	-0.7	-0.2	-0.2	-0.6	0.0	0.4	-0.4	0.3	0.1	-0.1	-0.7	0.2	0.2	0.3	0.1	0.0	0.4	0.2
E O 4	0.1	-0.4	-0.1	0.4	-0.2	-0.4	0.2	-1.0	-0.2	-0.7	0.1	0.4	-0.8	0.0	0.3	0.4	0.5	0.1	-0.4	-0.7
E O 5	0.3	-0.2	0.2	0.1	-0.4	0.2	0.2	-1.2	-0.4	-0.5	-0.1	0.6	-1.6	-0.3	0.4	0.2	-0.1	0.2	-0.1	-0.3
E O 6	0.0	0.0	-0.2	-0.9	0.1	0.0	0.7	-1.3	-0.3	-0.8	0.0	0.3	-0.4	0.4	0.7	0.0	0.5	0.1	-0.3	-0.6
E O 7	-0.4	-0.5	0.0	0.0	-0.6	-0.5	-0.2	-0.3	-0.2	-0.2	0.1	0.2	-1.7	0.1	0.3	0.4	0.3	0.2	0.2	-0.1
E O 8	0.1	-0.1	0.3	-0.1	0.1	-0.9	0.3	-0.4	-0.2	-0.9	0.1	0.5	0.1	-0.7	-0.1	0.2	0.2	-0.1	-0.4	-0.4
E O 9	-0.1	-0.1	-0.1	0.0	0.0	-0.3	-0.2	-1.2	-0.5	-0.5	0.1	0.0	-0.9	0.4	0.7	0.4	0.5	0.3	-0.3	-0.3
E O 0	-0.1	0.1	-0.1	-0.2	-0.2	-0.8	0.0	-0.8	-0.6	0.3	0.0	0.1	-0.9	0.1	0.4	0.2	0.0	0.1	0.0	0.3
E O 1	-0.4	-0.8	-0.8	-1.3	-0.9	-0.8	-0.6	0.2	-0.2	0.4	0.3	0.2	-1.5	0.2	0.2	0.1	-0.1	0.2	0.7	0.3
E O 2	-0.9	-0.7	-0.8	-0.9	-0.6	0.2	-0.4	0.5	-0.2	0.5	0.3	0.2	-1.1	-0.1	0.1	0.1	0.0	0.2	0.1	0.4
E O 3	-0.6	-0.9	-1.0	-0.5	-0.5	-0.1	0.0	0.2	-0.2	0.4	0.2	0.1	-0.4	0.2	0.2	0.2	0.0	0.3	0.1	0.4

loop

S X P	V	L	I	M	F	W	Y	G	A	P	S	T	C	H	R	K	Q	E	N	D
L B 1	0.8	0.7	0.9	0.5	0.7	0.5	0.2	-0.4	0.1	-0.9	-0.6	0.0	0.0	0.1	-0.5	-0.9	-0.6	-1.0	-1.1	-1.2
L B 2	0.8	0.7	0.9	0.8	0.7	-0.2	0.3	-0.7	-0.1	-1.9	-0.4	-0.1	0.8	-0.4	-0.4	-0.9	-0.4	-0.9	-0.9	-1.2
L B 3	0.4	0.8	0.6	0.2	0.5	0.4	0.5	-0.1	-0.1	-0.7	-0.2	-0.2	0.8	-0.2	-0.6	-0.9	-0.6	-0.8	-0.7	-0.5
L B 4	0.1	0.3	-0.1	0.1	0.2	0.3	0.4	0.4	-0.1	-0.4	0.1	0.0	0.6	0.1	-0.3	-1.0	-0.8	-0.5	-0.2	-0.2
L B 5	0.1	0.0	0.1	-0.7	0.1	-0.9	0.4	0.4	0.1	0.3	-0.1	0.0	-1.4	0.3	-0.3	-0.5	0.0	-0.4	-0.5	0.1
L B 6	-0.3	0.1	-0.1	0.3	0.4	0.2	0.3	-0.1	-0.6	0.0	-0.2	-0.3	0.4	0.4	0.2	-0.3	0.0	-0.1	0.0	0.3
L B 7	0.0	-0.1	0.1	-0.4	0.0	0.4	0.2	0.3	-0.1	0.1	-0.1	-0.1	0.7	0.1	0.0	-0.6	-0.1	-0.3	-0.1	0.0
L B 8	-0.2	0.1	0.0	-1.4	0.2	-0.2	0.4	-0.2	-0.2	0.5	0.0	-0.7	0.9	0.0	0.3	-0.4	0.1	-0.2	0.2	-0.1
L B 9	0.2	0.3	0.5	0.2	0.2	0.6	-0.3	0.3	-0.1	0.2	-0.2	-0.1	0.6	-0.5	-0.5	-0.7	-0.8	-0.5	-0.1	0.1
L B 0	-0.6	-0.2	-0.3	0.0	0.2	0.2	0.2	-0.3	0.4	0.3	-0.1	0.2	0.3	-0.4	-0.3	0.0	0.5	0.2	0.0	0.0
L B 1	0.2	0.3	0.6	0.5	0.5	0.2	0.4	-0.4	0.0	0.1	-0.1	0.0	0.2	-0.1	-0.4	-0.4	0.0	-0.5	-1.0	-1.0
L B 2	0.9	0.5	0.9	0.7	0.6	0.5	0.3	-0.4	0.0	-1.0	-0.4	-0.1	0.8	-0.4	-0.8	-1.3	-0.7	-0.8	-1.0	-0.9
L B 3	0.6	0.5	0.6	0.6	0.6	0.4	0.5	-0.5	0.0	-1.3	-0.3	-0.2	0.5	0.0	-0.4	-0.9	-0.4	-0.6	-0.7	-0.6
L I 1	-0.5	-0.3	-0.4	-0.9	-0.4	0.0	0.0	-0.2	-0.3	0.2	0.1	-1.0	-0.3	0.4	0.5	0.2	0.4	0.1	0.0	
L I 2	-0.1	-0.7	-0.7	0.1	-0.7	0.1	-0.1	-1.5	-0.1	-0.8	0.2	0.2	0.0	0.2	0.6	0.6	0.5	0.4	-0.1	-0.5
L I 3	-0.5	-0.2	-0.8	-0.1	-0.6	-1.5	0.0	-0.5	-0.2	-0.6	-0.2	0.1	-0.9	0.4	0.4	0.5	0.4	0.3	0.3	0.2
L I 4	-0.4	-0.6	-0.6	-0.6	-0.7	-0.7	-0.8	0.2	-0.1	0.1	0.1	0.1	-1.1	0.0	0.4	0.5	-0.1	0.1	0.5	0.1
L I 5	-1.0	-0.5	-0.9	-0.8	-0.9	0.1	-0.6	0.4	-0.3	0.7	0.1	0.1	-0.7	-0.4	0.4	0.4	0.3	0.2	0.1	0.3
L I 6	-0.5	-0.4	-0.8	-0.7	-2.1	0.2	-0.5	0.3	-0.1	0.6	-0.4	-0.2	-0.3	-0.2	0.5	0.5	0.1	0.1	0.2	0.2
L I 7	-0.2	-0.5	-0.5	-0.6	-0.8	-0.9	0.0	-0.1	-0.2	0.4	0.0	0.1	0.1	-0.1	0.3	0.2	0.2	0.2	0.2	0.1
L I 8	-0.2	-0.5	-0.8	-1.5	-0.3	-0.2	-0.9	0.0	-0.2	0.0	0.1	-0.2	-1.4	0.5	0.6	0.2	-0.2	0.1	0.4	0.6
L I 9	-0.6	-0.6	-1.1	-0.4	-0.5	0.1	-0.3	0.4	-0.4	0.2	0.0	-0.2	0.2	0.0	0.4	0.3	0.4	0.2	0.2	0.3
L I 0	-1.3	-1.1	-1.6	-1.1	-0.9	-0.6	-0.4	0.1	-0.6	0.6	0.4	-0.5	0.0	0.0	0.4	-0.1	0.1	0.7	0.7	
L I 1	-0.1	-0.4	-0.6	0.2	-0.1	-0.2	-0.1	-0.9	0.0	0.0	0.2	0.2	0.1	-0.5	0.3	0.4	0.3	0.3	-0.3	-0.4
L I 2	-0.4	-0.9	-0.3	-0.9	0.1	-0.8	-0.6	0.0	0.0	0.0	0.1	0.1	-1.4	-0.1	0.3	0.5	0.4	0.6	0.2	0.3
L I 3	-0.5	-0.5	-1.1	-0.9	-0.6	-1.5	0.0	-0.8	-0.3	-0.4	0.0	0.2	-1.6	0.0	0.1	0.4	0.5	0.9	0.5	0.5
L O 1	-0.6	-1.3	-0.9	-2.2	-1.9	-1.7	-1.8	-0.1	0.2	0.5	0.2	-0.1	-0.8	-0.4	0.2	0.4	0.2	1.1	0.4	0.4
L O 2	-0.7	-0.7	-1.7	0.0	-0.8	-1.6	-1.4	-0.2	0.2	-0.1	0.2	-0.2	-1.7	-0.1	0.0	0.6	0.4	0.7	0.5	0.4
L O 3	-0.7	-0.7	-1.2	-1.1	-0.6	-1.0	-0.4	-0.8	0.2	-0.8	0.1	-0.1	-1.5	-0.5	0.5	0.7	0.5	0.5	0.6	0.4
L O 4	-0.9	-0.9	-1.5	-0.3	-0.9	-1.6	-1.5	0.7	0.2	-0.2	0.2	-0.2	-0.8	0.1	0.0	0.2	0.3	0.3	0.8	0.3
L O 5	-1.1	-1.1	-1.1	-1.1	-0.9	0.0	-1.0	0.6	0.0	0.7	0.2	0.0	-2.2	-0.2	0.2	0.5	0.2	0.2	0.4	0.3
L O 6	-0.8	-0.9	-1.0	-1.1	-1.0	0.1	0.1	0.5	-0.1	0.4	0.3	0.0	-0.8	-0.1	-0.1	0.5	-0.2	0.2	0.3	0.3
L O 7	-0.8	-0.7	-0.9	-0.9	-0.9	0.2	-0.3	0.1	-0.1	0.7	0.2	-0.1	-1.1	-0.3	0.3	0.4	0.1	0.2	0.2	0.4
L O 8	-1.4	-1.1	-1.7	-1.3	-1.5	-1.0	-0.4	0.3	-0.1	0.7	0.0	-0.2	-1.5	0.5	0.1	0.7	0.1	0.2	0.6	0.6
L O 9	-1.2	-1.0	-1.5	-0.9	-0.9	0.0	-0.7	0.4	-0.2	0.6	0.1	-0.1	-2.5	0.4	0.2	0.4	0.2	0.3	0.4	0.6
L O 0	-1.6	-0.9	-1.3	-0.4	-0.6	-1.7	-1.1	0.8	-0.4	0.0	0.5	0.0	-1.1	-0.5	0.2	0.3	0.0	0.3	0.7	0.4
L O 1	-0.5	-0.8	-0.7	-0.3	-1.0	-0.8	-1.1	-0.4	0.1	1.4	0.2	-0.1	-1.6	-0.3	-0.2	0.4	0.4	0.5	-0.2	0.2
L O 2	-0.7	-1.4	-1.6	-0.6	-1.3	-0.9	-1.7	-0.3	0.4	0.4	0.0	0.1	-1.2	-0.2	-0.3	0.0	0.2	1.1	0.3	0.9
L O 3	-1.1	-1.8	-1.6	-1.0	-0.9	-1.0	-0.4	0.2	0.3	0.6	0.3	-0.1	-0.8	-0.2	-0.4	0.3	-0.2	0.5	0.5	0.8

α–helix

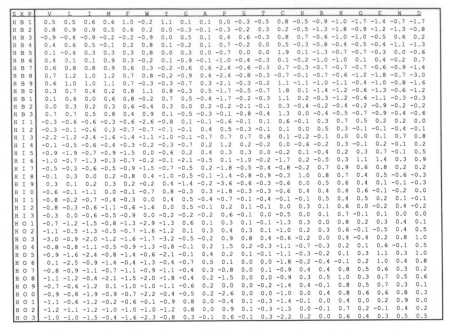

S X P	V	L	I	M	F	W	Y	G	A	P	S	T	C	H	R	K	Q	E	N	D
H B 1	0.5	0.5	0.6	0.6	1.0	-0.2	1.1	0.1	0.1	0.0	-0.3	-0.5	0.8	-0.5	-0.9	-1.0	-1.7	-1.4	-0.7	-1.7
H B 2	0.8	0.9	0.9	0.5	0.6	0.2	0.0	-0.3	-0.1	-0.3	-0.2	0.3	0.2	-0.5	-1.3	-1.8	-0.9	-1.2	-1.3	-0.8
H B 3	-0.9	-0.4	-0.9	-0.2	-0.2	-0.9	0.0	0.5	0.1	0.4	0.6	-0.3	0.8	0.7	-0.6	-1.0	-1.0	-0.5	0.6	0.2
H B 4	0.4	0.6	0.5	-0.1	0.2	0.8	0.1	-0.2	0.1	0.7	-0.2	0.0	0.5	-0.3	-0.8	-0.4	-0.5	-0.4	-1.1	-1.3
H B 5	0.1	-0.4	0.3	0.3	0.3	0.8	0.0	0.3	0.0	-0.7	0.0	0.0	1.9	0.1	-1.3	-0.7	-0.7	-0.3	0.0	-0.6
H B 6	0.4	0.1	0.1	0.9	0.3	-0.2	0.1	-0.9	-0.1	-1.0	-0.4	-0.3	0.1	-0.2	-1.0	-1.0	0.1	0.4	-0.2	0.7
H B 7	0.6	0.8	0.8	0.9	0.6	0.3	-0.2	-0.6	0.6	-2.4	-0.6	-0.3	0.7	-0.3	-0.7	-0.7	-0.7	-0.6	-0.9	-1.4
H B 8	0.7	1.2	1.0	1.2	0.7	0.8	-0.2	-0.9	0.4	-2.4	-0.8	-0.3	-0.7	-0.1	-0.7	-0.6	-1.2	-1.8	-0.7	-3.0
H B 9	0.6	1.0	1.0	1.1	0.7	-0.3	-0.3	-0.7	0.3	-2.1	-0.3	-0.2	1.1	-1.1	-1.0	-1.1	-0.4	-1.0	-0.8	-1.6
H B 0	0.3	0.7	0.4	0.2	0.8	1.1	0.8	-0.3	0.5	-1.7	-0.5	-0.7	1.8	0.1	-1.4	-1.2	-0.6	-1.3	-0.6	-1.2
H B 1	0.1	0.6	0.0	0.6	0.8	-0.2	0.7	0.5	-0.4	-1.7	-0.2	-0.3	1.1	0.2	-0.3	-1.2	-0.6	-1.1	-0.3	-0.3
H B 2	0.0	0.3	0.2	0.3	0.6	-0.4	0.3	0.0	0.2	-0.2	-0.1	-0.1	0.3	-0.4	-0.2	-0.4	-0.2	-0.9	-0.2	-0.2
H B 3	0.7	0.7	0.5	0.8	0.4	0.9	0.1	-0.5	-0.3	-0.1	-0.8	-0.4	1.3	0.0	-0.4	-0.5	-0.7	-0.9	-0.4	-0.6
H I 1	-0.3	-0.6	-0.6	-0.3	-0.6	-2.6	-0.8	0.1	-0.1	-0.6	-0.1	0.1	0.6	-0.1	0.3	0.7	0.5	0.2	0.2	0.0
H I 2	-0.3	-0.1	-0.6	0.3	-0.7	-0.7	-0.1	-0.1	0.4	0.5	-0.3	-0.1	0.1	0.0	0.5	0.3	-0.1	-0.1	-0.4	-0.1
H I 3	-2.2	-1.2	-2.4	-1.6	-1.4	-1.1	-1.0	-0.1	-0.7	0.7	0.7	0.8	0.1	-0.2	-0.1	0.0	0.0	0.1	0.7	0.8
H I 4	-0.1	-0.5	-0.6	-0.4	-0.3	-0.2	-0.3	-0.7	0.2	1.2	0.2	-0.2	0.0	-0.6	-0.2	0.3	-0.1	0.2	-0.1	0.2
H I 5	-0.9	-1.9	-0.7	-0.9	-1.5	0.0	-0.6	0.2	0.4	0.3	0.3	0.0	-0.2	0.1	-0.4	0.2	0.3	0.7	-0.1	0.5
H I 6	-1.0	-0.7	-1.3	-0.3	-0.7	-0.2	-0.1	-2.1	-0.5	0.1	-1.0	-0.2	-1.7	0.2	-0.5	0.3	1.1	1.4	0.3	0.9
H I 7	-0.5	-0.3	-0.6	-0.5	-0.9	-1.5	-0.7	-0.5	0.2	-1.8	-0.5	-0.4	-0.8	-0.2	0.7	0.9	0.6	0.8	0.2	0.2
H I 8	-0.1	0.3	0.0	0.2	-0.8	0.4	-1.0	-0.5	-0.1	-1.4	-0.8	-0.9	-0.3	1.0	0.8	0.7	0.4	0.5	-0.6	-0.3
H I 9	0.3	0.1	0.2	0.3	0.2	-0.2	0.4	-1.4	-0.2	-3.6	-0.6	-0.3	-0.6	0.0	0.5	0.6	0.4	0.5	0.2	-0.1
H I 0	-0.6	-0.1	-1.1	0.0	-0.1	-0.7	0.8	-0.3	0.3	-1.8	-0.3	-0.3	-0.6	0.4	0.4	0.4	0.6	-0.1	-0.2	0.0
H I 1	-0.8	-0.2	-0.7	-0.4	-0.3	0.0	0.4	0.5	-0.4	-0.7	-0.1	-0.4	-0.1	-0.1	0.5	0.4	0.5	0.2	0.1	-0.1
H I 2	-0.8	-0.3	-0.6	-1.1	-0.6	-1.6	0.0	0.5	-0.1	0.2	0.1	-0.1	0.0	0.3	0.1	0.6	0.0	-0.2	0.4	-0.2
H I 3	-0.3	0.0	-0.6	-0.5	-0.9	0.0	-0.2	-0.2	-0.2	0.6	-0.1	0.0	-0.5	0.0	0.1	0.7	-0.1	0.1	0.0	0.0
H O 1	-0.7	-1.2	-1.5	-0.8	-1.3	-2.9	-1.3	0.6	0.1	0.3	0.1	-0.1	-1.3	0.3	0.0	0.8	0.2	0.3	0.4	0.1
H O 2	-1.1	-0.5	-1.3	-0.5	-0.7	-1.6	-1.2	0.1	0.3	0.4	0.3	0.1	-1.0	0.2	0.3	0.6	-0.1	-0.5	0.4	0.5
H O 3	-3.0	-0.9	-2.0	-1.2	-1.6	-1.7	-3.2	-0.5	-0.2	0.9	0.8	0.4	-0.6	0.0	0.4	-0.4	0.2	0.8	1.0	
H O 4	-0.8	-0.8	-1.1	-0.5	-0.9	-1.3	-0.8	-0.1	0.2	1.5	0.2	-0.3	-1.1	-0.7	-0.3	0.2	0.1	0.6	-0.1	0.5
H O 5	-0.9	-1.6	-2.4	-0.8	-1.4	-0.6	-2.1	-0.1	0.4	0.2	0.1	-0.1	-1.1	-0.3	-0.2	0.1	0.3	1.1	0.3	1.0
H O 6	0.1	-2.5	-0.9	-1.4	-0.4	-1.3	-0.4	-0.7	0.5	0.1	0.0	0.0	-1.8	-0.2	-0.4	-0.1	0.2	1.0	0.4	0.8
H O 7	-0.8	-0.9	-1.1	-0.7	-1.1	-0.9	-1.1	-0.4	0.3	-0.8	0.0	0.0	-0.9	0.4	0.4	0.8	0.5	0.6	0.3	0.2
H O 8	-1.1	-1.2	-0.4	-2.1	-1.5	-2.0	-1.8	-0.4	0.2	-1.5	0.0	0.0	-0.9	0.3	0.5	1.0	0.3	0.7	0.5	0.6
H O 9	-0.7	-0.6	-1.2	0.1	-1.0	-1.0	-1.1	-0.6	0.2	0.0	0.0	-0.2	-1.4	0.4	-0.1	0.8	0.5	0.7	0.3	0.1
H O 0	-0.9	-0.8	-1.9	-0.9	-0.7	-2.2	-0.4	-0.5	0.2	-2.6	0.0	0.0	-1.0	0.0	0.4	0.8	0.6	0.6	0.8	0.3
H O 1	-1.1	-0.6	-1.2	-0.2	-0.6	-0.1	-0.9	0.8	0.0	-0.4	0.1	-0.3	-1.4	-0.1	0.0	0.4	0.0	0.2	0.9	0.0
H O 2	-1.2	-1.1	-1.2	-1.0	-1.0	-1.0	-1.2	0.8	0.0	0.9	0.1	-0.3	-1.5	0.0	-0.1	0.7	0.2	-0.1	0.4	0.2
H O 3	-1.0	-1.0	-1.5	-0.4	-1.6	-2.3	-0.8	0.3	-0.1	0.6	-0.1	0.3	-2.2	0.2	0.0	0.6	0.4	0.3	0.5	0.5

Figure 7. Combined positional, accessibility, and secondary structure, preference parameters (set SXP117). This is the most sophisticated set of parameters reported here, using the finest distinction in structural states. Positions 10, 11, 12, 13 are labelled 0, 1, 2, 3 (bottom of lists). The many interesting and strong preferences and antipreferences have not yet been fully interpreted. Note, for example, that the tendency for Phe in solvent-protected positions on a helix increases from pref(Phe,H,B,4) = +0.2 at the N-terminus to pref(Phe,H,B,10) = +0.8 at the C-terminus; or, that the tendency for Pro in solvent-exposed positions on a helix decreases from pref(Pro,H,O,4) = +1.5 at the N-terminus to pref(Pro,H,O,10) = −2.6. Comparison with the collapsed tables (*Figures 5* and *6*) is an aid in interpreting the underlying effects. Notation is, for example, EB1, third residue position before the beginning of a β-strand, buried; LO6, third residue position in a loop, outside; HI12, second residue position after the end of a helix, intermediate solvent exposure.

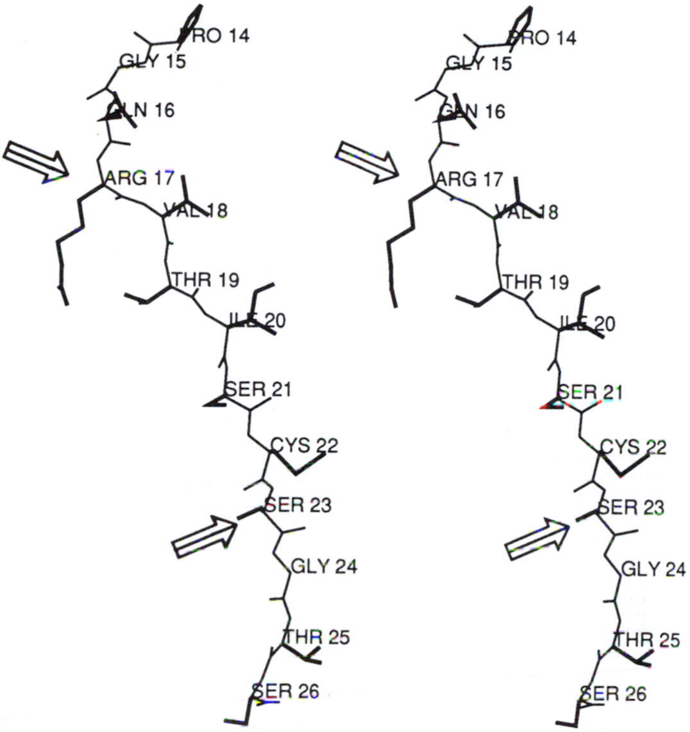

Figure 8. Stereo view. Example of how single residue preference parameters are combined to evaluate an amino acid sequence in a given structural context, using position (P) dependent and accessibility (X) dependent secondary structure (S) preference parameters from *Figures 5–7*. This 13-residue fragment from a known structure, contains a seven-residue β-strand (between arrows) and has the sequence PGQRVTISCSGTS (β-strand residues underlined). To evaluate the suitability of the sequence, look up the single residue preferences in, say, *Figure 7* and add them up along the fragment. For example, the first position in the fragment (labelled Pro14) has the structural state, as derived from the atomic coordinates, of S=E, X=O, P=1 (segment is β-strand; high solvent accessibility; position -3 before the segment) and Pro has a preference of +0.4 bits in this state; the fourth position in the fragment (labelled Arg17) is in the state S=E, X=O, P=4, with pref(Arg,E,O,4) = +0.3 bits; similarly pref(Val,E,I,5) = +0.7, pref(Ile,I,B,7) = +1.1, pref(Cys,E,O,9) = −0.9 (negative value—not preferred), and so on. Using 'PXS' preference parameters, the sum over the 13 residue is +3.3 bits, a relatively high value, corresponding to a probability ratio of observed/expected of $2^{3.3}$ = 9.85, i.e. the odds are 10 to 1 to find this sequence in this structural state. The fragment is residues 14–26 from an immunoglobulin Fab fragment, data-set 2FB4 in the Protein Data Bank. Cys22 is involved in an intramolecular disulphide bond. The entire strand has state S = (LLLEEEEEEELLL), X = (OOOOIOBOOOOOO), and P = (1–13).

cases, $N(R,S)$, to the number of expected cases, $E(R,S) = N(R)N(S)/N$, assuming a random model. A ratio of 2.0 (pref = 1.0) means that the observation is made twice as frequently as the random model would suggest. A value of pref = -1.0 indicates that the number of observed cases is half as large as expected from random. A numerical example: N(Pro, helix, first residue) = 82, N(Pro) = 2037, N(helix, first residue) = 893, N(all) = 52 426 gives pref(Pro, helix, first residue) = $+1.2$.

7. Tool: choosing amino acid types in interfaces

7.1 Representing protein structure: interface states

Another description of protein structure, more sophisticated than the standard helix–strand–turn–loop classification, is in terms of contact interfaces. The contact view was developed in collaboration with F. Colonna-Cesari in 1984–85 and is published here for the first time. This type of description is motivated by observations that residue preferences vary strongly on different sides of secondary structure segments (24). For example, the amino acid composition of solvent-exposed faces of helices is very different compared to that of helix–sheet interfaces (*Figure 11a*). The interface definition described here is in terms of the residue type and secondary structure of a central residue and the secondary structure and chain distance of its contact partners (*Figure 9*). As protein–solvent interaction appears to be the physically dominant effect in the folding of globular proteins, water contacts are explicitly taken into account.

7.2 Contacts between protein atoms

Contacts are best represented in terms of atom–atom interactions. A very simple form, fairly insensitive to errors in atomic coordinates, is that of a 'linear-square well'. A single contact event is counted with a strength of 1.0 when two atoms are 'touching' (interatomic distance less or equal to the sum of the van der Waals radii). With increasing interatomic distance the strength decreases linearly and reaches 0.0 when a water molecule just fits between two atoms. No contacts are counted when the interatomic distance is larger than the sum of the van der Waals radii plus the diameter of a water molecule. The contact strength of a residue is the sum of the contact strengths over all of its atoms.

7.3 Contacts of protein atoms with water molecules

In order to account for protein–water contacts, one is faced with the difficulty that the positions of (most) water molecules are not known and, in any event, vary strongly with time. In a reasonable hydration shell model the number of water molecules in contact with the protein is proportional to the volume of a single hydration shell. In terms of the classical definition of solvent-

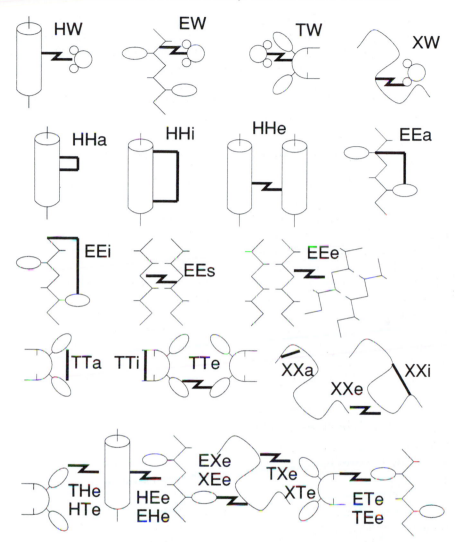

Figure 9. Schematic representation of the interface types used for the derivation of contact preferences given in *Figure 11*. Cylinders are helices (H), zig-zags are β-strands (E), semi-circles are hydrogen-bonded turns (T), smooth curves are loops (X), two small and one big circle is a water molecule (W), and ellipses are amino acid side-chains. Contacts are shown as bold face lines (squashed Z shapes or rectangular brackets). Interfaces are, for example, HHe for helix–helix, HEe for sheet–sheet. HW for helix–water, etc. For details of notation, see *Figure 11*.

#PID	C	SIZ	RES	%H	%B	%BP	%BA	SID	ORIGIN	PROTEIN_NAME
351C		82	1.6	50	4		0	100	C551$PSEAE	PSEUDOMONAS AERUGINOSA	CYTOCHROME C 551
256B	Ā	106	1.4	79	0		0		C562$ECOLI	ESCHERICHIA COLI	CYTOCHROME B 562
8ADH		374	2.4	28	24	45		55	ADHE$HORSE	EQUUS CABALLUS	ALCOHOL DEHYDROGENASE
8ATC	Ā	310	2.5	40	15	100		0	PYRB$ECOLI	ESCHERICHIA COLI	ASPARTATE CARBAMOYLTRANSFERASE (ASPARTATE TRANSCARBAMYLASE)
8ATC	B	146	2.5	15	34	1		98	PYRI$ECOLI	ESCHERICHIA COLI	ASPARTATE CARBAMOYLTRANSFERASE (ASPARTATE TRANSCARBAMYLASE)
2AZA	A	129	1.8	16	35	36		63	AZUR$ALCDE	ALCALIGENES DENITRIFICANS	AZURIN
3B5C		85	1.5	31	23	25		75	CYB5$BOVIN	BOS TAURUS	CYTOCHROME B 5
3BLM		257	2.0	42	17		0	100	BLAC$STAAU	STAPHYLOCOCCUS AUREUS	BETA-LACTAMASE
2CA2		256	1.9	16	30	23		76	CAH2$HUMAN	HOMO SAPIENS	CARBONIC ANHYDRASE II (CARBONATE DEHYDRATASE)
1CCR		111	1.5	42	1		0	100	CYC$ORYSA	ORYZA SATIVA	CYTOCHROME C
2CCY	Ā	127	1.7	74	1		0	100	CYCP$RHOMO	RHODOSPIRILLUM MOLISCHIANUM	CYTOCHROME C'
1CD4		173	2.3	5	41	11		88	CD4$HUMAN	HOMO SAPIENS, recombinant	T-CELL SURFACE GLYCOPROTEIN CD4 (N-TERMINAL FRAGMENT)
3CLA		213	1.8	29	28	23		76	CAT3$ECOLI	ESCHERICHIA COLI, engineered	CHLORAMPHENICOL ACETYLTRANSFERASE TYPE III
5CPA		307	1.5	38	16	63		36	CBPA$BOVIN	BOS TAURUS	CARBOXYPEPTIDASE A
2CPP		405	1.6	51	10	11		88	CPXA$PSEPU	PSEUDOMONAS PUTIDA	CYTOCHROME P450CAM (CAMPHOR MONOOXYGENASE)
4CPV		108	1.5	56	1		0	100	PRVB$CYPCA	CYPRINUS CARPIO	CALCIUM-BINDING PARVALBUMIN
1CSE	Ē	274	1.2	30	20	73		26	SUBT$BACLI	BACILLUS SUBTILIS	SUBTILISIN
1CSE	I	63	1.2	22	33	44		55	ICIC$HIRME	HIRUDO MEDICINALIS	EGLIN-C
1CTF		68	1.7	55	26		0	100	RL7$ECOLI	ESCHERICHIA COLI	50S RIBOSOMAL PROTEIN L7/L12 (C-TERMINAL DOMAIN)
2CYP		293	1.7	50	7	8		91	CCPR$YEAST	SACCHAROMYCES CEREVISIAE	CYTOCHROME C PEROXIDASE
8DFR		186	1.7	23	33	57		42	DYR$CHICK	GALLUS GALLUS	DIHYDROFOLATE REDUCTASE
1ECN		136	1.4	75	0		0		GLB3$CHITH	CHIRONOMOUS THUMMI)	HEMOGLOBIN (ERYTHROCRUORIN) (FRACTION III)
2ER7	Ē	330	1.6	11	45	13		86	CARP$CRYPA	ENDOTHIA PARASITICA	ASPARTIC PROTEINASE (ENDOTHIAPEPSIN)
4FD1		106	1.9	33	14		0	100	FER1$AZOVI	AZOTOBACTER VINELANDII	FERREDOXIN
4FXN		138	1.8	36	22	95		4	FLAV$CLOSP	CLOSTRIDIUM MP	FLAVODOXIN
3GAP	Ā	208	2.5	30	14		0	100	CRP$ECOLI	ESCHERICHIA COLI	CATABOLITE GENE ACTIVATOR PROTEIN
2GBP		309	1.9	43	19	90		10	DGAL$ECOLI	ESCHERICHIA COLI	D-GALACTOSE/D-GLUCOSE BINDING PROTEIN
1GCR		174	1.6	7	46		0	100	CRGB$BOVIN	BOS TAURUS	CRYSTALLIN GAMMA-II
1GD1	Ō	334	1.8	29	29	52		47	G3P$BACST	BACILLUS STEAROTHERMOPHILUS	D-GLYCERALDEHYDE-3-PHOSPHATE DEHYDROGENASE
1GOX		350	2.0	44	13	78		21	2HAO$SPIOL	SPINACIA OLERACEA	GLYCOLATE OXIDASE
1GP1	Ā	183	2.0	32	18	47		52	GSHP$BOVIN	BOS TAURUS	GLUTATHIONE PEROXIDASE
2HLA	B	99	2.6	0	49		0	100	HA1H$HUMAN	HOMO SAPIENS	HISTOCOMPATIBILITY CLASS I ANTIGEN
1HOE		74	2.0	0	48		0	100	IAA$STRTE	STREPTOMYCES TENDAE	ALPHA-AMYLASE INHIBITOR
1I1B		151	2.0	5	47		0	100	IL1B$HUMAN	HOMO SAPIENS, recombinant	INTERLEUKIN-1 BETA
4ICD		414	2.5	39	18	52		47	IDH$ECOLI	ESCHERICHIA COLI	ISOCITRATE DEHYDROGENASE
1IL8	Ā	71	NMR	26	25		0	100	IL8$HUMAN	HOMO SAPIENS, recombinant	INTERLEUKIN 8
1L13		164	1.7	64	9		0	100	LYCV$BPT4	BACTERIOPHAGE T4, mutant	LYSOZYME
6LDH		329	2.0	43	17	51		48	LDHM$SQUAC	SQUALUS ACANTHIAS	LACTATE DEHYDROGENASE
2LIV		344	2.4	44	19	73		26	LIVJ$ECOLI	ESCHERICHIA COLI	LEU/ILE/VAL-BINDING PROTEIN
2LTN	Ā	181	1.7	1	43		0	100	LEC$PEA	PISUM SATIVUM, recombinant	LECTIN
2LTN	B	47	1.7	8	63		0	100	LEC$PEA	PISUM SATIVUM, recombinant	LECTIN
1LZ1		130	1.5	39	12	11		88	LYC$HUMAN	HOMO SAPIENS	LYSOZYME
1MBD		153	1.4	77	0		0		MYG$PHYCA	PHYSETER CATODON	MYOGLOBIN
2MHR		118	1.7	70	0		0		HEMM$THEZO	THEMISTE ZOSTERICOLA	MYOHEMERYTHRIN
2PAB	Ā	114	1.8	7	51	16		83	TTHY$HUMAN	HOMO SAPIENS	PREALBUMIN
1PAZ		120	1.6	16	37	35		64	AZUP$ALCFA	ALCALIGENES FAECALIS	PSEUDOAZURIN
4PTP		223	1.3	10	34	2		97	TRYP$BOVIN	BOS TAURUS	BETA TRYPSIN
1R69		63	2.0	63	0		0		RPC1$BP434	PHAGE 434	434 REPRESSOR (N-TERMINAL DOMAIN)
1RHD		293	2.5	29	13	87		12	THTR$BOVIN	BOS TAURUS	RHODANESE
7RSA		124	1.3	20	35	3		96	RNP$BOVIN	BOS TAURUS	RIBONUCLEASE A
2RSP	Ā	115	2.0	5	41	17		82	GAG$RSVP	ROUS SARCOMA VIRUS	RSV PROTEASE
5RXN		54	1.2	16	22		0	100	RUBR$CLOPA	CLOSTRIDIUM PASTEURIANUM	RUBREDOXIN
2SGA		181	1.5	9	55	6		93	PRTA$STRGR	STREPTOMYCES GRISEUS	PROTEINASE A
4SGB	Ī	51	2.1	0	29	11		88	IPR2$SOLTU	SOLANUM TUBEROSUM	SERINE PROTEINASE B INHIBITOR PCI-I
2SNS		141	1.5	20	22	15		85	NUC$STAAU	STAPHYLOCOCCUS AUREUS	STAPHYLOCOCCAL NUCLEASE
2SOD	Ō	151	2.0	1	42	2		97	SODC$BOVIN	BOS TAURUS	CU,ZN SUPEROXIDE DISMUTASE
2SSI		107	2.6	15	28	5		95	ISUB$STRAO	STREPTOMYCES ALBOGRISEOLUS	SUBTILISIN INHIBITOR
2STV		184	2.5	11	47	1		98	COAT$STNV	SATELLITE TOBACCO NECROSIS VIRUS	COAT PROTEIN
2TMN	Ē	316	1.6	40	17	26		73	THER$BACTH	BACILLUS THERMOPROTEOLYTICUS	THERMOLYSIN
1TNF	A	152	2.6	1	44		0	100	TNFA$HUMAN	HOMO SAPIENS, recombinant	TUMOR NECROSIS FACTOR-ALPHA
2TS1		317	2.3	54	10	85		14	SYY$BACST	BACILLUS STEAROTHERMOPHILUS	TYROSYL-tRNA SYNTHETASE
1UBQ		76	1.8	23	34	25		75	UBIQ$HUMAN	HOMO SAPIENS	UBIQUITIN
1UTG		70	1.3	75	0		0		UTER$RABIT	ORYCTOLAGUS CUNICULUS	UTEROGLOBIN
2WRP	R̄	104	1.6	78	0		0		TRPR$ECOLI	ESCHERICHIA COLI	TRP REPRESSOR
1WSY	A	248	2.5	50	13	100		0	TRPB$SALTY	SALMONELLA TYPHIMURIUM	TRYPTOPHAN SYNTHASE
4XIA	A	393	2.3	47	10	85		14	XYLA$ARTS7	ARTHROBACTER SP	D-XYLOSE ISOMERASE
1YPI	A	247	1.9	43	17	96		3	TPIS$YEAST	SACCHAROMYCES CEREVISIAE	TRIOSE PHOSPHATE ISOMERASE

Figure 10. List of 67 protein chains used to derive the contact parameters given in *Figure 11*, with a total of 12 460 residues. The list was produced with the aid of an algorithm that selects a representative set of proteins from a protein data-bank, assuring that no two proteins have higher than 30% sequence identity after optimal alignment (46). Membrane proteins, very small proteins, proteins with a large number of disulphide bonds, proteins with many heteroatoms in the data-set, with a low percentage of secondary structure, or with nominal resolution not better than 2.7 Å were excluded. PID is the Protein Data Bank identifier (6); C, the chain index; SIZ, the number of residues in the chain; RES, the nominal crystallographic resolution (or 'NMR'); %H, the number of hydrogen bonds involved in $(i,i+4)$ and $(i,i+3)$ type H-bonds (helices and turns) per 100 residues (42); %B, the same in β-structure (parallel and antiparallel bridges); %BP and %BA, the percentage thereof in parallel or antiparallel bridges (%BA+%BP = 100 if there is any β-structure). SID is the corresponding SWISS-PROT sequence identifier (8) and ORIGIN the species origin.

(a) Contact Preferences AInt29:

	V	L	I	M	F	W	Y	G	A	P	S	T	C	H	R	K	Q	E	N	D
HW	-1.48	-0.74	-1.56	-0.49	-1.60	-1.10	-1.06	-1.25	0.20	-0.28	-0.41	-0.66	-2.59	-0.55	0.73	1.14	0.83	1.24	-0.14	0.55
HHa	0.01	0.27	-0.02	0.17	-0.30	-0.36	-0.46	-0.55	0.89	-0.10	-0.15	-0.20	-0.43	-0.35	-0.24	-0.05	0.16	0.41	-0.37	0.05
HHi	-0.02	0.47	0.17	0.50	-0.27	-0.34	-0.62	-0.80	0.82	-1.59	-0.49	-0.36	-0.83	-0.42	0.24	0.16	0.46	0.36	-0.53	-0.28
HHe	-0.01	0.74	0.50	0.52	0.79	0.68	0.43	-1.40	0.25	-1.15	-1.01	-0.68	-0.85	-0.07	0.21	-0.66	-0.20	-0.36	-1.10	-1.03
HEe	0.13	0.54	0.39	1.07	0.65	1.02	0.37	-1.81	0.26	-1.33	-0.85	0.12	0.87	0.27	-0.71	-0.92	-0.32	-0.54	-0.72	-0.98
HTe	-0.38	0.19	-0.42	0.56	0.12	0.84	0.19	-1.19	0.10	-0.91	-0.84	-0.79	-0.44	0.04	0.97	-0.06	0.18	0.22	-0.38	-0.30
HXe	-0.34	0.11	-0.37	0.37	0.18	0.61	0.38	-1.23	-0.12	-1.03	-0.37	-0.49	-0.42	0.02	0.44	-0.54	0.53	0.61	-0.33	-0.05
EW	-0.44	-0.67	-0.48	-0.73	-0.98	-1.68	0.02	-0.59	-1.62	-0.74	0.23	0.67	-2.41	-0.20	0.79	1.02	0.64	0.46	0.08	-0.54
EEa	1.02	0.30	0.81	-0.02	0.25	-0.49	0.10	-0.04	-0.29	-0.95	0.11	0.47	0.32	-0.13	-0.76	-0.66	-0.52	-0.81	-0.56	-0.98
EEi	0.93	0.52	0.71	0.26	0.84	-0.68	0.71	-1.33	-1.02	-1.66	-0.75	0.09	-0.08	0.51	0.32	-1.07	-0.55	-0.92	-0.97	-1.07
EEs	0.89	0.34	0.84	0.12	0.65	0.36	0.50	-0.65	-0.61	-2.05	-0.20	0.23	0.24	-0.03	-0.54	-0.85	-0.47	-0.76	-0.80	-1.38
EEe	0.75	0.78	0.87	0.61	1.12	1.04	0.43	-0.87	-1.19	-1.88	-1.06	-0.15	0.21	-0.27	-1.05	-1.14	-0.59	-1.29	-1.25	-1.62
EHe	1.04	0.51	1.15	0.34	1.19	0.64	0.67	-1.37	-1.13	-1.14	-0.70	-0.73	0.16	-0.24	-0.71	-1.35	-1.33	-2.15	-1.15	-1.60
ETe	0.24	-0.33	0.21	-0.46	-0.05	-0.36	0.80	-0.99	-1.20	-1.11	0.36	0.23	0.40	0.09	0.58	-0.12	-0.58	-0.88	0.32	0.09
EXe	0.59	0.15	0.55	-0.07	-0.45	0.44	0.74	-1.15	-0.86	-1.12	-0.24	0.00	0.15	0.21	0.19	-0.63	-0.42	-0.53	-0.48	-0.73
TW	-2.79	-2.53	-2.36	-1.42	-1.95	-2.60	-0.63	1.72	-0.32	0.81	0.44	-0.49	-1.34	-0.52	-0.18	1.07	0.43	0.74	1.06	0.68
TTa	-1.34	-1.19	-1.45	-1.09	-0.81	-1.29	-0.33	1.70	0.16	1.25	0.49	-0.07	0.03	-0.22	-0.81	0.26	-0.05	0.24	0.95	0.52
TTi	-0.93	-0.34	-1.17	-0.52	-0.24	-0.73	-0.63	0.34	0.20	0.01	0.07	0.51	0.07	-0.09	-0.12	0.76	-0.05	0.34	0.51	-0.14
TTe	-1.96	-0.46	-0.59	-0.55	-0.58	-1.59	0.46	0.57	-0.40	1.00	-0.06	0.07	0.09	0.24	0.54	0.11	-0.85	-0.17	1.15	0.12
THe	-0.57	-0.04	-0.93	-0.39	0.31	-0.49	0.09	0.99	0.04	-0.11	-0.30	-0.78	0.11	0.12	-0.39	0.45	0.31	0.14	0.63	-0.09
TEe	-0.78	-0.79	-1.39	-0.97	-0.13	-0.86	0.83	1.32	-0.06	0.50	0.46	0.11	0.51	-0.05	-0.61	-0.35	0.13	-0.27	0.94	0.15
TXe	-1.48	-1.06	-1.19	-0.91	-0.28	-0.42	0.43	0.63	-0.29	0.67	0.16	-0.03	0.82	-0.38	0.17	0.09	-0.02	-0.02	1.08	0.60
XW	-0.85	-1.21	-1.50	-0.44	-1.63	-1.46	-0.82	0.62	-0.71	0.62	0.46	0.44	-1.19	-0.11	0.41	1.01	0.38	0.26	0.41	0.56
XXa	-0.16	-0.49	-0.40	-0.60	-0.57	-0.53	-0.58	0.79	-0.03	1.13	0.64	0.51	0.48	0.16	-0.53	-0.11	-0.35	-0.48	0.32	0.44
XXi	-0.24	-0.47	-0.53	-0.59	-0.06	0.19	0.09	0.02	-0.61	0.70	0.29	0.25	0.74	0.19	0.11	-0.18	0.27	-0.47	0.46	0.19
XXe	-0.42	-0.47	-0.40	-0.16	-0.28	0.43	0.21	0.22	-0.44	0.33	0.16	0.20	0.74	0.55	0.12	-0.30	-0.19	-0.31	0.50	0.39
XHe	-0.34	0.17	-0.15	0.23	0.35	0.45	-0.03	-0.15	-0.52	0.59	0.38	0.01	-0.42	-0.38	-0.10	-0.40	-1.19	-0.93	0.14	0.42
XEe	0.17	-0.07	0.41	-0.30	0.26	0.87	-0.07	-0.21	-0.11	0.64	0.21	-0.07	0.86	0.29	-0.53	-0.83	-0.93	-0.48	-0.16	0.13
XTe	-0.54	-0.62	-0.48	-0.24	-0.51	0.40	0.03	-0.18	-0.69	0.34	0.12	-0.18	0.87	0.88	0.38	-0.02	-0.17	-0.53	0.53	0.76

(b) Contact Preferences AS5:

	V	L	I	M	F	W	Y	G	A	P	S	T	C	H	R	K	Q	E	N	D
H	0.01	0.39	0.18	0.32	0.21	0.04	-0.13	-0.42	0.47	-0.60	-0.23	-0.29	-0.42	-0.10	-0.04	-0.13	-0.01	0.01	-0.29	-0.13
E	0.71	0.29	0.65	0.15	0.50	0.31	0.29	-0.24	-0.30	-0.62	-0.09	0.23	0.48	0.06	-0.62	-0.80	-0.55	-0.71	-0.47	-0.75
T	-0.49	-0.39	-0.54	-0.09	-0.31	0.04	0.08	0.42	-0.12	0.08	0.14	-0.14	0.40	0.36	0.25	0.06	-0.03	-0.11	0.47	0.43
X	-0.17	-0.28	-0.24	-0.27	-0.13	0.16	0.17	0.17	-0.27	0.63	0.17	0.14	0.33	0.10	0.03	-0.26	-0.02	-0.10	0.22	0.13
Wat	-1.19	-1.09	-1.45	-0.62	-1.57	-1.48	-0.73	0.42	-0.33	0.27	0.17	0.03	-1.72	-0.33	0.50	1.07	0.59	0.59	0.35	0.48

(c) Contact Preferences AM2:

	V	L	I	M	F	W	Y	G	A	P	S	T	C	H	R	K	Q	E	N	D
Wat	-1.19	-1.09	-1.45	-0.62	-1.57	-1.48	-0.73	0.42	-0.33	0.27	0.17	0.03	-1.72	-0.33	0.50	1.07	0.59	0.77	0.35	0.48
Prot	0.13	0.12	0.14	0.08	0.15	0.14	0.09	-0.08	0.05	-0.05	-0.03	-0.00	0.16	0.05	-0.10	-0.29	-0.13	-0.18	-0.07	-0.10

(d) Contact Preferences AA21:

	V	L	I	M	F	W	Y	G	A	P	S	T	C	H	R	K	Q	E	N	D
V	0.39	0.23	0.33	0.20	0.26	0.20	-0.02	-0.04	0.18	-0.08	-0.14	-0.07	0.12	-0.06	-0.36	-0.36	-0.34	-0.41	-0.20	-0.35
L	0.22	0.39	0.32	0.21	0.28	0.09	-0.03	-0.22	0.09	-0.18	-0.19	-0.09	-0.16	-0.19	-0.09	-0.37	-0.16	-0.27	-0.20	-0.40
I	0.33	0.33	0.40	0.18	0.23	0.09	0.01	-0.05	0.06	-0.12	-0.12	-0.07	0.01	-0.25	-0.37	-0.30	-0.18	-0.33	-0.30	-0.35
M	0.20	0.23	0.13	0.61	0.42	0.40	-0.05	-0.23	0.07	-0.08	-0.34	-0.23	0.12	0.05	-0.30	-0.36	-0.18	-0.15	-0.14	-0.42
F	0.20	0.27	0.23	0.35	0.55	0.28	0.06	-0.25	-0.08	-0.21	-0.22	-0.06	0.10	0.14	-0.09	-0.54	-0.30	-0.38	-0.27	-0.48
W	0.19	0.10	0.11	0.40	0.31	0.08	0.14	-0.01	-0.16	0.01	-0.17	-0.18	0.15	0.21	0.06	-0.24	-0.19	-0.41	-0.22	-0.18
Y	0.00	0.00	0.09	0.03	0.14	0.18	0.12	-0.03	-0.11	0.15	0.10	-0.15	0.01	0.04	-0.02	-0.04	0.02	-0.31	-0.07	-0.10
G	0.12	-0.08	0.08	-0.25	-0.06	0.09	0.09	-0.02	0.00	0.02	0.20	0.18	0.14	0.09	-0.04	-0.24	-0.03	-0.32	0.02	0.06
A	0.23	0.16	0.16	0.08	-0.04	-0.14	-0.07	-0.02	0.35	-0.11	-0.02	0.05	-0.03	-0.24	-0.34	-0.11	-0.04	-0.15	-0.04	-0.11
P	0.03	-0.10	0.01	0.04	-0.08	0.19	0.25	-0.01	-0.02	-0.26	0.06	0.29	0.32	-0.06	0.23	0.12	0.21	-0.09	0.13	0.05
S	-0.03	-0.13	-0.03	-0.24	-0.08	0.03	0.21	0.16	-0.00	0.06	0.20	0.19	0.02	-0.12	-0.21	-0.09	-0.13	0.03	0.09	0.09
T	-0.00	0.01	0.01	-0.13	0.07	-0.06	-0.04	0.08	0.07	-0.03	0.17	0.19	-0.18	0.24	-0.08	-0.33	-0.11	-0.13	0.03	0.12
C	0.04	-0.18	0.01	0.06	0.13	0.15	-0.00	0.09	-0.09	0.19	0.10	-0.31	1.84	0.29	0.00	-0.61	-0.14	-0.34	0.09	-0.50
H	-0.03	-0.15	-0.16	0.07	0.21	0.31	0.04	0.02	-0.24	0.27	-0.10	0.13	0.29	0.35	-0.09	-0.31	-0.44	0.12	-0.05	0.06
R	-0.16	0.05	-0.11	-0.20	0.12	0.23	0.14	-0.09	-0.24	-0.11	-0.12	-0.06	0.10	0.04	0.04	-0.89	0.18	0.33	-0.18	0.40
K	-0.08	-0.09	-0.01	-0.18	-0.25	0.04	0.25	-0.12	0.05	-0.07	-0.07	-0.13	-0.26	-0.06	-0.63	-0.30	-0.07	0.50	0.08	0.41
Q	-0.11	0.00	-0.01	-0.05	-0.10	0.05	0.12	0.05	0.04	0.00	-0.09	-0.04	0.18	-0.27	0.20	-0.17	-0.03	-0.07	0.24	0.07
E	-0.19	-0.05	0.13	0.04	-0.13	-0.22	-0.11	-0.19	-0.02	0.08	-0.03	-0.05	-0.14	0.18	0.35	0.43	-0.05	-0.17	-0.01	-0.16
N	-0.08	-0.11	-0.17	0.05	-0.10	0.02	0.09	0.04	0.05	0.08	0.05	0.04	0.25	-0.00	-0.17	-0.03	0.17	-0.12	0.12	0.16
D	-0.17	-0.23	-0.17	-0.28	-0.29	0.09	0.03	0.04	0.00	0.03	0.15	0.16	-0.34	0.12	0.41	0.26	0.05	-0.23	0.16	-0.29
Wat	-0.81	-0.73	-0.98	-0.45	-1.06	-1.00	-0.49	0.25	-0.22	0.18	0.12	0.02	-1.15	-0.24	0.35	0.75	0.40	0.54	0.24	0.33

Figure 11. Preference parameters for amino acid types in 29 different contact interfaces. Contact preference parameters extracted from the non-redundant data-set of 67 protein chains. The parameters can be used to evaluate how well a particular sequence fits into a particular 3D structure, e.g. in sequence/structure alignment. For example, line 'HHe' contains the single-residue preferences for helix–helix interfaces, line 'HEe' those for helix–sheet interfaces. The 20 standard amino acids are given in one-letter code. Second-ary structure notation is: 'H', helix; 'E', extended or β-sheet; 'T', hydrogen-bonded turn; 'X', everything else, called loop (42). Contacts with water are labelled by a 'W'. Chain distance (proximity) of two contacting residues is: 'a', adjacent—the two residues are adjacent in sequence; 'i', internal—the two residues are on the same element of second-ary structure, i.e. on the same helix or strand, but not adjacent; 's', strand–strand—the two residues are on adjacent strands in the same β-sheet; 'e', external—the two residues are on two different elements of secondary structure (for strands, in different sheets). (a) Preferences of amino acid side-chains for 29 interface types, e.g. helix–helix, helix–sheet (parameter set AInt29). These 'structure–structure' interface types are classified according to: the secondary structure of a specific residue type, S_1 = H,E,T, or X; the secondary structure of the contacting residue(s) of any type S_2 = H,E,T, or X (or S_2 = W for contacts with water); and the chain distance (proximity) of any two contacting residues, p_{12} = a,i,s,e. Notation for a contact type is $S_1S_2p_{12}$, e.g. HHa, HHi, HHe, HEe, etc. For example, Pro has a strong preference for TTa (Pro located in a turn makes contacts with other residue(s) in the same turn); Lys has clear preference for HW, EW, TW, and XW (Lys

111

Figure 11. *Continued*

located on any element of secondary structure make contacts with water); the strongest preferences for EHe are expressed by Ile and Phe (Ile, Phe in a β-strand make strand–helix contacts). Writing 'strand' for β-strand, the notation is, for example, HHa, intra-helix contact between residues *i* and *i*+1; XXi, contact between two residues in the same loop; EEi, intra-strand contact; EEs, strand–strand contact; EEe, sheet–sheet contact; HEe, helix–sheet contact, first residue in helix; EHe, helix–sheet contact, first residue in strand; EXe, contact between strand and loop; TW, contact between an H-bonded turn and water. (b) Preferences of amino acid side-chains for four interface types, e.g. contact with a helix, contact with a sheet (parameter set AS5). These simpler 'anything–structure' interface types are derived from the 29 interface types in (a) by summing over the secondary structure state 'S$_1$' of the first residue. So the four 'contact-with' interface types are: 'H', residue contact of the first residue (in any secondary structure state) with a helix residue; 'E', contact with a β-sheet residue; 'T', contact with a turn residue; 'X', contact with a loop residue; 'W', contact with water. For example, Val and Ile have a clear preference to be in contact with β-sheet residues; Ala, for making contacts with helix residues; Lys, for making contact with water; and so on. These preferences are dominated by the secondary structure state of the first residue, as residues in a helix are likely to make contacts with other helix residues. Notation is, for example, H, residue (e.g. Ala) makes contact with a helix; E, residue (e.g. Leu) makes contact with a strand; Wat, residue (e.g. Lys) makes contact with water. (c) Preferences of amino acid side-chains for two interface types, i.e. contact with protein atoms ('Prot') or contact with water molecules ('Wat') (parameter set AM2). These very simple 'protein–water' interface types are derived from the parameters in (a) by summing protein–protein contacts over the secondary structure states of both participating residues, or from those in (b) by summing over the secondary structure state 'S$_2$' of the second residue. The only remaining distinction is that between contacts of a protein atom with other protein atoms or with solvent atoms. These parameters resemble a hydrophobicity scale, e.g. Lys has the strongest preference for water contacts while Ile, Phe, Trp, and Cys have the weakest preference for contacts with protein atoms. The apparently weaker contrast in line 'Prot' is a numerical effect, due to the fact that, in the data-base used, the total number of contacts with protein atoms exceeds by far that with water atoms. Notation is: Wat, residue (e.g. Lys) makes contact with water; Prot, residue (e.g. Leu) makes contact with other protein atoms. (d) Preferences of amino acid side-chains for contacting another side-chain or water, irrespective of secondary structure (parameter set AA21). These parameters are analogous to effective residue–residue potential energy parameters for a polypeptide chain, but also include a residue–water contact term. Water contacts are in line 'Wat'. When these parameters are summed over one of the contact partners, the parameters in (c) result, for contacts of a side-chain with 'protein' or with 'water'. Notation is (column/row), for example, V/I, Val and Ile make contact (in any structural state); K/D, Lys and Asp make contact (in any structural state); K/Wat, Lys makes contact with water.

accessible surface area, this translates to 0.31 atom–water contact for each Å^2 of surface area (43). So water–protein contacts can be included in the definition of contact interfaces by using this simple equivalence.

7.4 Statistical evaluation of contacts

For the statistical evaluation, contacts are labelled with the residue type of the central residue and the interface type. The contact strengths of all residues in

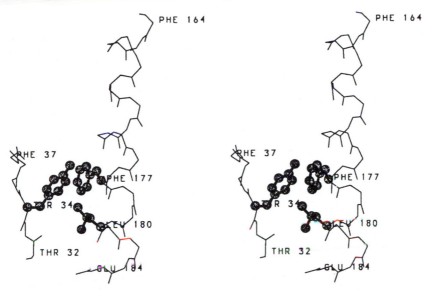

Figure 12. Stereo view. Example of how single-residue contact preference parameters are used to evaluate amino acid residues in a given structural context, using interface-dependent parameters from *Figure 11a*. The two contacting segments (backbone trace), a β-strand (left) and an α-helix (right) make a central contact involving Phe37, Phe177, and Leu180 (side-chains highlighted as balls and sticks). To evaluate the suitability of these residues in this interface sequence, look up the single residue preferences in *Figure 10*. Phe37 participates in the interface of type 'EHe' (residue in β-strand (=E) in contact with an α-helix (=H) in two different secondary structure segments (=e), with preference pref(Phe,EHe) = +1.19; similarly for Phe177 we have pref(Phe,HEe) = +0.65 and for Leu180 pref(Leu,HEe) = +0.54. So the combination has a combined preference of +2.38 bits or an observed/expected ratio of $2^{2.38}$ = 5.21. Note that residues in general participate in more than one interface so that all other contacts of each residue have to be evaluated in order to judge the entire structure. The fragments are residues 32–37 and 164–184 from tyrosyl tRNA synthase, Protein Data Bank entry 3TS1.

a data-base of selected high-resolution proteins (*Figure 10*) are summed in each category and are written as C(Ri,I), where R is the amino acid type and I the type of the contact interface. An example is C(Ala, HHe), i.e. the total contact strength of alanine in helix–helix contact interfaces. Preference parameters, pref(R,I), that express the over- or under-representation of certain residue types in certain interface types can then be calculated in a way completely analogous to those for residue occurrences in secondary structures, by simply replacing the index S by I in the equation for pref(R,S) above (see Section 6.3). However, the information contained in the contact interface parameters is much richer and reflects directly the important hydrophobic effect by including water–protein contacts in a natural way.

References

1. Richardson, J. and Richardson, D. C. (1989). *TIBS*, **13**, 304–9.
2. Sander, C. (1991). *Current Opinion in Structural Biology*, **1**, 630–8.
3. *Protein design exercises*, EMBL BIOcomputing Technical Document 1. European Molecular Biology Laboratory, Heidelberg.
4. Sander, C. and Vriend, G. (ed.) (1991). *ProDes90*, EMBL BIOcomputing Technical Document 6. European Molecular Biology Laboratory, Heidelberg.
5. Sander, C., *et al.* (1991). *Proteins*, **12**, 105–10.
6. Bernstein, F. C., *et al.* (1977). *J. Mol. Biol.*, **112**, 535–42.
7. Barker, W. C., George, D. G., and Hunt, L. T. (1990). In *Methods in enzymology* (ed. R. F. Doolittle), Vol. 183, pp. 31–49. Academic Press, San Diego.
8. Bairoch, A. and Boeckmann, B. (1991). *Nucl. Acids Res.*, **19**, 2247–50.
9. Schneider, R. and Sander, C. (1991). *Proteins*, **9**, 56–68.
10. Karpusas, M., Baase, W. A., Matsumura, M., and Matthews, B. W. (1989). *Proc. Natl Acad. Sci. USA*, **86**, 8237–41.
11. Zhang, X. J., Baase, W. A., and Matthews, B. W. (1991). *Biochemistry*, **30**, 2012–17.
12. Castagnoli, L., Scarpa, M., Kokkinidis, M., Banner, D. W., Tsernoglou, D., and Cesareni, G. (1989). *EMBO J.*, **8**, 621–9.
13. Sander, C. (1990). *Biochem. Soc. Symp.*, **57**, 25–33.
14. Luger, K., Hommel, U., Herold, M., Hofsteenge, J., and Kirschner, K. (1989). *Science*, **243**, 206–10.
15. Jones, T. A. (1978). *J. Appl. Cryst.*, **11**, 268–72.
16. Dayringer, H. E., Tramontano, A., Sprang, S. R., and Fletterick, R. J. (1986). *J. Mol. Graphics*, **4**, 82–7.
17. Vriend, G. (1990). *J. Mol. Graphics*, **8**, 52–6.
18. Vriend, G. and Sander, C. (1991). *Proteins*, **11**, 52–8.
19. Finkelstein, A. V. and Ptitsyn, O. B. (1987). *Prog. Biophys. Mol. Biol.*, **50**, 171–90.
20. Finkelstein, A. V. and Reva, B. A. (1990). *Biofisika (USSR)*, **35**, 401–6.
21. Chou, P. Y. and Fasman, G. D. (1974). *Biochemistry*, **13**, 211.
22. Robson, B. and Suzuki, E. (1976). *J. Mol. Biol.*, **107**, 327–56.
23. Levitt, M. (1978). *Biochemistry*, **17**, 4277–85.
24. Lifson, S. and Sander, C. (1979). *Nature*, **282**, 109–11.
25. Oefner, C. (1981). Diplomarbeit (Master thesis), Universität Heidelberg.
26. Argos, P. and Palau, J. (1982). *Int. J. Pep. Prot. Res.*, **19**, 380.
27. Richardson, J. S. and Richardson, D. C. (1988). *Science*, **240**, 1648–52.
28. Schneider, R. (1989). Diplomarbeit (Master thesis), Universität Heidelberg.
29. Chou, P. Y. and Fasman, G. D. (1979). *Biophys. J.*, **26**, 367–73.
30. Isogai, Y., Nemethy, G., Rackovsky, S., Leach, S. J., and Scheraga, H. A. (1980). *Biopolymers*, **19**, 1183–210.
31. Wilmot, C. M. and Thornton, J. M. (1988). *J. Mol. Biol.*, **203**, 221–32.
32. Leszczynski, J. F. and Rose, G. D. (1986). *Science*, **234**, 849–55.
33. Ptitsyn, O. B. and Finkelstein, A. V. (1983). *Biopolymers*, **22**, 15–25.
34. Scharf, M. (1989). Diplomarbeit (Master thesis), Universität Heidelberg.
35. Regan, L. and DeGrado, W. F. (1988). *Science*, **241**, 976–8.
36. Holm, L. and Sander, C. (1991). *J. Mol. Biol.*, **218**, 183–94.

37. Baumann, G., Froemmel, C., and Sander, C. (1990). *Prot. Engng,* **2,** 329–34.
38. Eisenberg, D. and McLachlan, A. D. (1986). *Nature,* **319,** 199–203.
39. Gregoret, L. M. and Cohen, F. E. (1990). *J. Mol. Biol.,* **211,** 959–74.
40. Hill, C. P., Anderson, D. H., Wesson, L., DeGrado, W. F., and Eisenberg, D. (1990). *Science,* **249,** 543–5.
41. DeGrado, W. F., Wassermann, Z. R., and Lear, J. D. (1989). *Science,* **243,** 622–8.
42. Kabsch, W. and Sander, C. (1983). *Biopolymers,* **22,** 2577–637.
43. Colonna-Cesari, F. and Sander, C. (1990). *Biophys. J.,* **57,** 1103–7.
44. Banner, D. W., Kokkinidis, M., and Tsernoglou, D. (1987). *J. Mol. Biol.,* **196,** 657–75.
45. Eberle, W., Pastore, A., Sander, C., and Roesch, P. (1991). *Biomol. NMR,* **1,** 71–82.
46. Hobohm, U., Scharf, M., Schneider, R., and Sander, C. (1992). *Protein Science,* **1,** 409–17.
47. Vriend, G. and Sander, C. (1992). *Acta. Cryst.,* in press.

5

From protein sequence to structure

PAUL A. BATES and MICHAEL J. E. STERNBERG

1. Introduction

Knowledge of the three-dimensional structure of a protein is central to the systematic engineering of the function and stability of the molecule. The structural information can define the residues forming the binding site for a ligand or identify residues to change in order to alter thermal stability by the introduction of disulphide bridges or salt bridges.

The most accurate method of obtaining protein structure is X-ray crystallography but, despite improvements in methodology (see Chapter 1), the technique remains dependent on obtaining crystals suitable for analysis. Recently, NMR (see Chapter 2) has been used successfully to determine the structure of proteins of up to 200 residues. However, at the present time, fewer than 100 new protein structures will be revealed in a year by these experimental methods, whereas the sequences of at least 1000 proteins are obtained from their cloned gene sequences. Accordingly, computer algorithms are being used increasingly to obtain structural information for a protein from its sequence. In addition, these algorithms can provide guidelines for modifications to experimentally determined structures and are essential for the design of novel proteins.

This chapter will describe the following analyses of any newly determined sequence:

- hydrophobicity plots
- searches for homologous proteins in a data-base
- multiple sequence alignments
- searches for known local sequence motifs
- searches for profiles of residues
- secondary structure prediction
- modelling by homology

2. ICAM-1

The analysis of the sequence of human intracellular adhesion molecule 1 (ICAM-1) will be used in this chapter to illustrate the algorithms. This molecule binds to the lymphocyte function associated antigen (LFA-1), promoting cell adhesion in immune and inflammatory reactions. ICAM-1 has also been identified as a cell-surface receptor for rhinovirus and malaria (1, 2).

These analyses will show that ICAM-1 consists of five extracellular domains, a single-spanning transmembrane region, and an intracellular region. Each of the five extracellular domains has sequence similarities with an immunoglobulin fold and this enables a three-dimensional model to be predicted based on the crystal structure of an antibody.

3. Hydrophobicity plots

The simplest analysis of a sequence is a plot of the local hydrophobicity (or hydrophilicity) along the chain. There are a variety of different implementations of this notion, differing in the numerical values of the hydrophobicity assigned to the residues and the method of smoothing. Two of the most popular algorithms are by Kyte and Doolittle (3) and Hopp and Woods (4).

There are two major applications of these plots:

- the location of membrane-spanning regions
- the prediction of exposed loops, including B-cell epitopes

3.1 Location of membrane-spanning regions

Typically, membrane-spanning regions are 20 or more, mainly hydrophobic, residues that span the non-polar lipid bilayer. Accordingly, the smoothing of residue hydrophobicity typically is over a segment of length 15 residues. These plots generally are successful in locating a single transmembrane region in a protein but where the chain crosses the membrane several times the plot can be harder to interpret unambiguously. *Figure 1a* shows a hydrophobic plot for ICAM-1 averaged over 15 residues. The transmembrane region can clearly be seen.

Several other algorithms have been developed to search for membrane-spanning regions (for a review see ref. 5). We have implemented the approach of Rao and Argos (6) based on the frequencies of residues within manually assigned non-polar transmembrane regions combined with the frequencies of residues just outside the non-polar region.

3.2 The prediction of exposed loops, including B-cell epitopes

The B-cell epitopes tend to be exposed loops in proteins that are hydrophilic, and the same hydrophobicity plot when averaged over a shorter segment

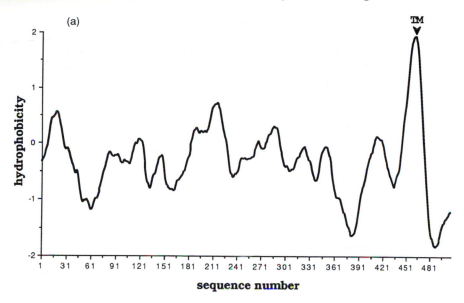

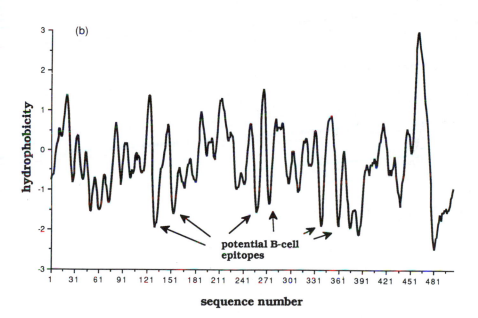

Figure 1. Hydrophobicity plots for ICAM-1 using the scale of Kyte and Doolittle (3). (a) The average over 15 residues is given and the predicted transmembrane region is shown by TM; (b) the average over seven residues is given and the minima represent potential B-cell epitopes.

length (say seven residues) is a more sensitive approach to find these minima of hydrophobicity (i.e. maxima of hydrophilicity). The highest peaks of hydrophilicity define peptides that are often successful in raising antibodies that can cross-react with the native protein (see *Figure 1b*). The regions of high local hydrophibicity tend to be the partially or fully buried secondary structures.

4. Homology searches

The most powerful method of obtaining information about the structure and function of a newly determined sequence is to identify sequences that are homologous. The major choices for a data-base scan for homologous sequences are:

- the scoring scheme relating amino acids
- the protein data-base scanned
- the type of search procedure
- the evaluation of the significance of the results

4.1 Scoring schemes

The simplest scoring scheme is to assign 1 to identical residues (i.e. the identity matrix). But this fails to include the knowledge obtained from sequences of evolutionarily related proteins that mutations occur preferentially between particular pairs of residues, especially when they have similar chemical properties. A widely used approach is based on the analysis of Dayhoff and colleagues who evaluated the frequencies of observed substitutions in families of evolutionarily related proteins. The scoring scheme is called a PAM matrix (point acceptable mutations) (7). An estimate of evolutionary distance can be included, so a PAM100 matrix is more appropriate for closely related proteins and a PAM250 for more distant relationships. The weights for identities compared to conservative changes is higher in the PAM100 than in the PAM250 matrices.

4.2 Which data-base?

There are a variety of available protein sequence data-bases (see Chapter 3). We use the OWL data-base developed at Leeds University (8) which is a non-redundant, composite of several protein data-bases combined with translations of DNA data-bases.

4.3 Search procedure

There are several fast methods to scan sequence data-bases, such as FASTP (9) and BLAST (10). However, at present the availability of computing power for a given price is increasing faster than the number of sequences to scan, and the emphasis should be on sensitive methods to detect sequence similarities.

Most sensitive schemes are based on the dynamic programming approach for sequence alignment (11).

Protocol 1. Sequence homology searching by dynamic programming

1. Decide on the choice of whether the algorithm finds the best global or local alignment. Increasingly, sequence relationships are between parts of proteins and, accordingly, local searches such as the Smith–Waterman (12) algorithm provide a better approach.
2. Decide on the choice of the gap penalty that is added whenever a gap must be introduced to optimize the alignment.
3. Decide on the scoring matrix for residue similarity.
4. For each sequence in the data-base:
   (a) First, generate a weight matrix where the element $W(i,j)$ holds the score for the residue at position i in one sequence against the residue at position j in the other.
   (b) Then, find the optimal path through the matrix to maximize the similarity between the two sequences. (In some approaches the algorithm minimizes the difference between the two sequences, but this chapter will only consider maximizing similarity.)
5. Keep the scores for matching the new sequence against each data-base protein and examine the highest scoring matches as potential homologies.

4.4 Evaluation of significance

A major problem in data-base scanning is evaluating the significance of a potential homology. One approach is to take the equivalence between two proteins (or protein regions) and compare the observed score to the distribution of scores obtained when one sequence is randomized. Generally, a normal distribution of random scores is assumed and the number of standard deviations that the observed score is above the mean of random scores provides an evaluation of significance. However, protein sequences are not a random string of residues but have periodicities of local hydrophobicities and hydrophilicities, as observed in nearly all hydrophobic plots. An alternative method of evaluating the significance of a match has been introduced (13). The lowest 97% of all observed scores from the data-base scan is taken to represent random matches and from this distribution the likelihood of observing a particular high score is evaluated.

One method used to assess true and false matches is provided by scanning the sequence of every protein of known structure (as given in the Brookhaven

From protein sequence to structure

```
          *+  +   +*   *+**  *  *  ++          *   *+ *++

1FB4   92 WNSSDNSYVFGTGT-KVTVLGQPKANPTVTLFPPSS 126
4APE  195 WEWTSTGYAVGSGTFKSTSIDGIADTGTTLLYLPAT 230

1FB4      EE   EEEEEEE   -EEEEE        EEEEE  HH
4APE           EEEEEEEEE    EEE EEEEEE      EEEHHH
```

Figure 2. A seemingly significant sequence alignment between parts of two unrelated proteins. IFB4 is the light chain of the Fab fragment from human IgG Kol and 4APE is endothiapepsin. *Denotes identity and '+' a conservative change (from ref. 16).

data-base (14)) against every other sequence in Brookhaven. As the structures are known, most true homologies can be identified from sequence and structural similarities. When this procedure was followed, several seemingly significant, but actually false, sequence similarities were detected (15, 16).

Figure 2 shows an example of a 36-residue alignment with 10 identities (i.e. 28%), nine conservative changes (25%), and one gap between two evolutionarily and unrelated proteins—an immunoglobulin Fab fragment and an acid proteinase. Both structures have a rough equivalence of the β-sheets but, as shown in *Figure 3*, the three-dimensional conformations are quite different. This example is typical of the false matches that can be located with this level of sequence homology. Thus, one requires a compelling biological explanation to propose an evolutionary or a structural similarity between different proteins given this level of sequence similarity.

5. Multiple sequence alignments

Frequently, homology is found between more than two proteins and additional information can be obtained if a multiple alignment of all sequences is obtained. In particular, identification of conserved residues can suggest a local region that contains a motif that specifies a particular structure or function (see Section 6). The rigorous dynamic programming algorithm can only be applied to a few sequences because of memory storage requirements and computing time. Accordingly, approximate methods have been developed to align numerous sequences (17, 18, 19).

One algorithm was developed by Barton and Sternberg (18):

Protocol 2. Multiple sequence alignment

1. Obtain a rigorous dynamic programming alignment between each pair of sequences and note each alignment score.
2. Sort the scores so that the sequences are in order of maximum similarity. Thus proteins 1 and 2 are the closest, then protein 3 is closest to either protein 1 or 2, etc.

3. Obtain the multiple alignment by sequential use of the dynamic programming method:

   (a) Sequences 1 and 2 are aligned.

   (b) Sequence 3 is aligned against the 1+2 alignment with the score in the weight matrix element (i,j) being the average of 3 v. 1 + 3 v. 2.

   (c) Sequences 4 to n are similarly aligned.

   (d) Sequence 1 is then aligned against the alignment of 2, 3, 4 ... n.

   (e) Sequence 2 is then aligned against the alignment of 1, 3, 4 ... n.

   (f) Sequences 3 to n are similarly aligned.

   (g) Steps d–f can be repeated.

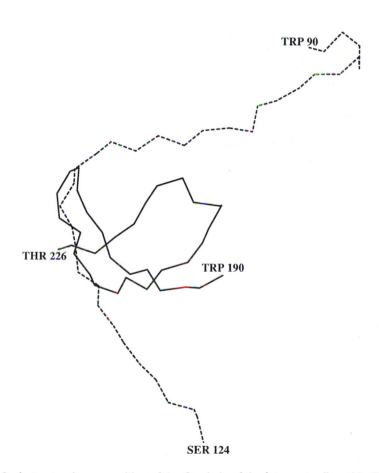

Figure 3. A structural superposition of the Cα-chain of the fragments aligned in *Figure 2*, showing that they have quite different conformations.

Trials with the sequences of proteins whose structures are known allow the accuracy of the method to be bench-marked and the multiple alignment can lead to an improvement in accuracy of any pair of sequences.

6. Sequence motifs

The problem with the standard homology searches is that each position along a sequence is given the same emphasis. However, increasing sequence relationships due to a functional and/or structural similarity between proteins are being found, based on the common occurrence of a local sequence motif. Recently Bairoch (20) has established a computer library of such motifs (called PROSITE). Several computer programs have been developed to scan sequences against this library, including MacPattern for the Macintosh (21) and PROMOT in FORTRAN (22).

6.1 PROSITE motif library

Version 7 of PROSITE describes over 500 motifs which specify a range of features, such as protein modification, binding sites, and enzyme active sites and structural folds. A typical motif is for an ATP/GTP binding site and is coded as:

$$[AG] - x(4) - G - K - [ST]$$

where [] denotes that any of the list of residues is allowed and ×(4) that any four residues are allowed.

Each motif must match exactly a protein sequence with the feature. The motifs were carefully defined so that the exact match will locate most, or preferably all, protein regions that have the particular feature. In addition, most motifs are sufficiently specific for there to be, at most, a few false matches when the motif is scanned against a sequence data-base.

6.2 Scanning sequences using PROMOT

PROMOT is a FORTRAN program to scan a newly determined sequence against the PROSITE library. PROMOT allows the following simple extension to the PROSITE motif definitions:

- a maximum specified number of mismatches to the motif is permitted
- certain positions must match
- regions before and after a motif can be output

The program is run interactively and the user simply answers a few questions. The output is a list of sequence positions where there is a match between the PROSITE motif and the test sequence. *Figure 4* shows a typical output for ICAM-1. The PROSITE data-base identifies potential *N*-glycosylation sites. In addition there are only two local regions that are identified with one mismatch with an immunoglobulin signature. The presence of the immunoglobulin fold

```
        protein pat   nos  seq  sequence
          code   id   miss nos  of motif
----------------------------------------------

001= N-glycosylation site.
P1;ICA1$HUMAN   001     0   130  NLTL
P1;ICA1$HUMAN   001     0   145  NLTV
P1;ICA1$HUMAN   001     0   183  NFSC
P1;ICA1$HUMAN   001     0   202  NTSA
P1;ICA1$HUMAN   001     0   267  NDSF
P1;ICA1$HUMAN   001     0   296  NQSQ
P1;ICA1$HUMAN   001     0   385  NQTR
P1;ICA1$HUMAN   001     0   406  NWTW

476= Immunoglobulins and major histocompatibility complex proteins signature
P1;ICA1$HUMAN   476     1   369  FSCSATL
P1;ICA1$HUMAN   476     1   455  YLCRARS


STATISTICS OF MATCHES

pat probability expected no observed no  probability expected no observed no
 no exact match exact match exact match  mismatch 1  mismatch 1 mismatch 1

001 0.5112E-02  0.2704E+01  0.8000E+01   0.0000E+00  0.0000E+00 0.0000E+00
476 0.4233E-05  0.2226E-02  0.0000E+00   0.4908E-03  0.2582E+00 0.2000E+01
```

Figure 4. Output from PROMOT. The sequence of ICAM-1 was scanned against version 7 of the PROSITE data-base. Some of the matches are given.

in the other five domains is not even suggested by the matches. These results show that the formalism of specifying motifs used in PROSITE is well suited to locate functional features, although structural features such as the presence of the immunoglobulin fold are less well characterized.

6.3 Statistics for motif matching

PROMOT evaluates for each motif the number of matches in the scan. To evaluate the significance of a match, it is helpful to know the number of matches expected by chance. The probability of matching a motif can simply be calculated from the frequencies of residues in a standard data-base. In conventional homology searches, the calculated significance of a particular alignment score is known to be unreliable (see Section 4.4). This raises the question of whether the evaluation of chance matches in PROMOT is accurate.

For many motifs in PROSITE, the observed number of false matches in the SWISS-PROT data-base is given. The number of false matches expected by chance can be evaluated from the residue frequencies. It is found that there is good agreement between the observed and expected number of false matches (23). Thus the expected number of false matches provides a reliable guide to evaluate the significance of motif matches.

7. Sequence profiles based on weights

The definition of sequence motifs used in PROSITE is based on exact matches to a list of residues. For many functional and some structural features, this is a powerful and sensitive approach to locate true matches without too many false matches. However, some features in proteins cannot be well characterized by this approach. An alternative method is to weight matches of the target sequence against a profile.

There are several implementations of profile scanning (24, 25) and here we describe the approach of Barton and Sternberg (26). The profile is constructed from a multiple alignment of several (m) sequences, each of length l. $F(i,j)$ is the residue in protein i at position j. For each alignment against the target sequence, T, the score is evaluated by:

$$\text{Score} = \sum_{i=1}^{l} \sum_{j=1}^{m} M[F(i,j),\ T(j)]$$

where M is a matrix relating all pairs of amino acids, such as the identity matrix or a PAM matrix. The profile is compared against each possible alignment in the target sequence(s) and the highest scoring matches kept as possible matches. The significance of a score can be evaluated from the distribution of scores against randomized sequences, as in the standard homology search. The profile search can be extended to consider two or more fixed-length sections separated by gaps of defined ranges.

8. Secondary structure predictions

Secondary structure predictions aim to identify α-helices, β-sheets/strands, tight turns (e.g. a β-turn), and coil regions of a globular protein from its sequence.

8.1 Early methods

In the 1970s several methods were developed for secondary structure predictions. Most, such as the methods of Chou and Fasman (27) and of Garnier *et al*. (28), were based on empirically derived residue propensities for different conformational states. An alternative approach was developed by Lim (29) based on heuristically derived patterns of residues based on forming hydrophobic, hydrophilic, or amphipathic surfaces on α-helices and β-strands. Tests (30) of these types of methods showed that they were around 60% accurate for a three-state (α-, β-, or coil) prediction.

Eliopoulos *et al*. (31) developed one approach for improving the accuracy, whereby the results of several methods were combined and averaged. An alternative approach for improvement was bench-marked by Zvelebil *et al*. (32) who used the information from a multiple alignment of a family of related sequences. The empirical parameters of Garnier *et al*. (28) are averaged over the aligned residues. In addition, regions of high sequence variability are likely to be exposed loops and, accordingly, can be assigned lower values for potential α- and β- predictions. This method can yield an improvement of about 10% in a three-state prediction.

8.2 Complex pattern recognition

Recently, several groups have applied neural networks to predict secondary structure. For example, Kneller *et al.* (33) achieved improvements in accuracies above the earlier methods when the approach was applied to proteins of a defined structural class (α/α, β/β, and α/β).

An alternative approach to obtaining patterns is based on machine learning. This has the potential advantage over neural networks that the rules are expressed as intelligible patterns of residues rather than as a series of weights. We are collaborating with the artificial intelligence group at the Turing Institute, Glasgow to develop such predictive patterns. The first attempt yielded algorithms comparable to most other approaches (34), and recent work suggests that these initial results can be improved to yield about 80% accuracy for the α/α class of proteins.

All these approaches simply incorporate local sequence information; clearly this is insufficient to specify a local structure that is stabilized by long-range tertiary interactions. Cohen and co-workers (35) are developing algorithms that incorporate longer-range periodicities of exposed loops and regular secondary structures, together with heuristically derived patterns of residues.

9. Tertiary structure prediction by homology model building

The most powerful method of obtaining structural information occurs when the new sequence is homologous to a protein of known conformation. Examinations of the structures of homologous proteins show that most of the α and β secondary structures are arranged in a similar three-dimensional fold. The major variations tend to occur in the loop regions connecting the secondary structures.

The steps in modelling by homology are:

Protocol 3. Modelling by homology

1. Find the best parent structure.
2. Accurately align the new sequence to that of the parent.
3. Model the main-chain conformation.
4. Model the side-chain conformations.
5. Energy refinement.
6. Evaluate the accuracy of the models.

The steps necessary to model the first N-terminal domain, ICAM-1(D1), have been outlined. Experimental evidence suggests that this domain contributes most to both LFA-1 and rhinovirus binding (2).

9.1 Find the best parent structure

Often there is more than one parent or template structure from which the unknown could be modelled. The approach we follow is to use the single structure that is closest in sequence to that of the unknown, based on sequence alignments with each of these possible templates. This approach is based on the general observation that conformational similarity correlates with sequence similarity within a family of proteins. For ICAM-1(D1), the best sequence match occurs with REI, a Bence–Jones dimer (36).

An alternative approach (e.g. refs 37, 38) is to generate an average three-dimensional template from superposition of the structures of all related proteins. In either approach, the structural alignment indicates conserved three-dimensional features within a fold with which the unknown sequence should comply (39).

9.2 Accurate alignment of the new sequence to that of the parent

Generally, the actual pairwise equivalence of residues obtained from a sequence alignment program is not always correct in terms of the structural equivalence (40). Accordingly, manual intervention is often used, incorporating the general principles that insertions and deletions occur predominantly in the loop regions and that buried hydrophobic residues are conserved. It is most important that any key markers of a particular fold, such as conserved disulphide bonds, are also equivalenced in the alignment.

Figure 5a shows a manual alignment between ICAM-1(D1) and REI that incorporates these principles. However, several subjective judgements are made and so the alignment of a secondary structure element can be misplaced by several residues. Thus alternative alignments (e.g. *Figure 5b*), which could be rejected later, should also be considered in the early stages of modelling. Towards this aim, we have developed (41) an automatic and rapid approach to generate suboptimal alignments from the standard dynamic programming method.

ICAM-1(D1) has four cysteine residues, two of which are expected to form the conserved disulphide link of the immunoglobulin fold. Our favoured alignment places the other two cysteines in the same region of space, to form a second disulphide link. An alternative alignment presents the possibility of the fourth cysteine being part of the conserved disulphide bond, leading to a rearrangement of residues in strands E and F of the connecting loop. This alternative alignment would not give the possibility of a second stabilizing disulphide link. However, there is a compensating factor in that the buried tyrosine residue at the base of the E strand is conserved. In either alignment, *Figure 5* shows that the edge strand, D, of REI is missing in ICAM-1(D1).

Figure 5. Sequence alignment between ICAM-1(1D) and REI; (a) is considered to be the most probable alignment based on conserved residues within the immunoglobulin fold, but (b) represents an alternative that should also be considered in the early stages of the modelling. In the IDENTITY lines *denotes identical residues in ICAM-1(D1) and REI. In the BURIED lines, *denotes those residues in the template that are greater than 95% buried. The STRANDS lines show each β-strand of the template, and these are labelled from A to G. The conserved disulphide bond of the fold lies between the two cysteine residues in strands B and F.

9.3 Model the main-chain conformation

9.3.1 Identify structurally variable regions to be modelled

The key principle of modelling the main-chain is to identify the structurally conserved regions (SCR) between homologous proteins and those sections that are structurally variable (VR) (42, 43). The simplest assignment of SCRs is the regular secondary structure elements. Indeed, if only one parent structure has a determined structure, this represents the simplest approach. However, if several possible parent structures are known, then a structural superposition will identify the SCRs more accurately. The definition of the SCRs from the known structure then has to be modified from the sequence alignment of the unknown structure against the known conformations. Sometimes sections of secondary structure elements also need to be replaced, as for example when a proline or a glycine residue is in an unfavourable conformation relative to the residue replacing it.

9.3.2 Data-base search for connecting fragments

The next step is to predict the main-chain conformation of the VRs. We have developed a program called 3D-JIGSAW to generate a plausible model, or models, from sequence alignments. An overview of the program is given in *Figure 6*. The present implementation of the program centres around a fragment data-base search (44).

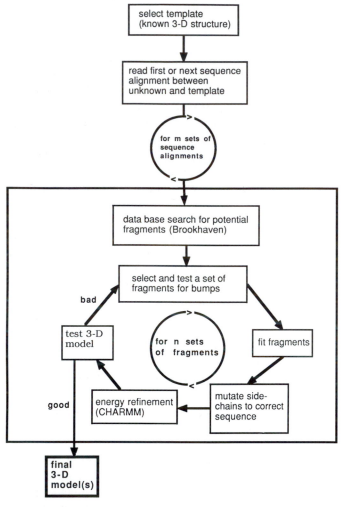

Figure 6. An overview of the steps carried by our program 3D-JIGSAW to generate a three-dimensional model from a set of homologous sequence alignments.

For each region to be modelled, the two fixed ends in the parent structure are defined (*Figure 7*). The object is to build a region in the new structure that has a similar conformation for the fixed region and the correct number of residues in the connection. The approach is to find fragments in the data-base of X-ray structures that meet these requirements. Thus the search is for all structures equal to or below a user-defined resolution cut-off (typically ⩽2.0 to ⩽3.0 Å) in the Brookhaven data-base (14). Each search is based on the root mean square (r.m.s.) deviation between the four C^α positions, two either side of the fragment to be replaced. For computational speed, the r.m.s. is estimated (45) from the interatomic distances (RMS(D)). The data-base search finds all fragments that have RMS(D) below a chosen cut-off.

All fragments from the data-base that satisfy this constraint are then fitted on to the fixed ends of the template by a least-squares superposition of the four C^α atoms of the fragment and the parent (*Figure 8*). The r.m.s. deviation between the position of the four C^α atoms of the parent and the fragment calculated after spatial superposition (RMS(C^α)) is not exactly proportional to the r.m.s. calculated from the interatomic distances (RMS(D)). We therefore impose a second constraint that RMS(C^α) should also lie below a cut-off value.

These distance constraints ensure that the C^α positions of the connections of the fragment match the ends of the template. However, the dihedral angle between planes of the equivalent peptide units may be considerably different. We have therefore imposed a further selection procedure where the r.m.s. of the angle, RMS(CO), between equivalenced $C = O$ vectors should be below a certain threshold.

The choice of the number of C^α atoms to use in the fixed region can be varied. Two C^α positions each side of the gap are needed to ensure that fragments are oriented correctly, i.e. loops are not least-squared into an upside-down position. The use of more C^α positions and hence more distance constraints is possible, but we find this over-restrictive, giving fewer fragments as candidates for replacement.

In 3D-JIGSAW, each fragment search produces an output file. This is illustrated in *Figure 9* which shows the fragments found for a data-base search on the loop between the F and G strands for the preferred alignment. Distance constraints were calculated between C^α positions 89, 90, 97, and 98 of the template. Fragments listed were below cut-offs for RMS(D), RMS(C^α), and RMS(CO).

These constraints are insufficient to limit the number of possible replacement fragments. Accordingly we are exploring the use of other filters.

(a) The compatibility of the main-chain ϕ, ψ angles of the data-base loop with the sequence of the region to be modelled are checked. If there is a proline in the region to be modelled, then the range of allowed ϕ, ψ is restricted. Similarly, if the data-base loop has a glycine then this residue may be in a conformation disallowed for the region to be modelled.

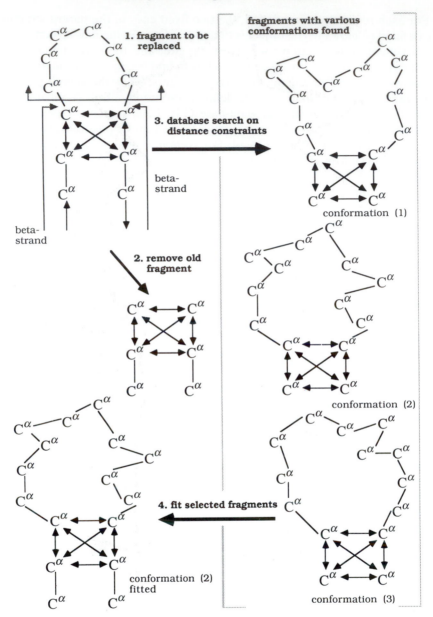

Figure 7. The process of searching for a new connecting fragment in the Brookhaven data-base. This example shows the replacement of a loop connecting two strands, where the loop is to be expanded from six residues in the parent to 10 residues. For simplicity, only the C$^\alpha$ backbone atoms are shown. Double-headed arrows indicate the distance constraints between C$^\alpha$ atoms. Only fragments that have similar distances between the four C$^\alpha$ positions at the ends of each fragment are selected for the next stage.

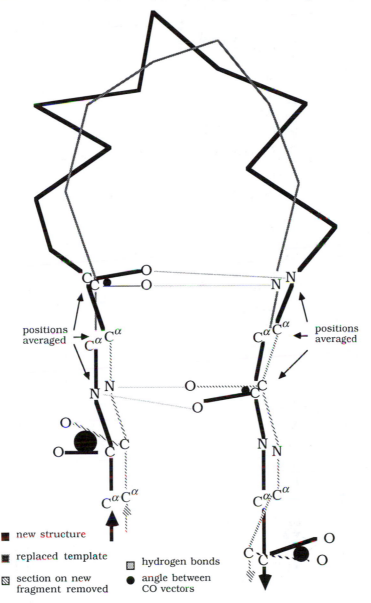

positions averaged

positions averaged

- ■ new structure
- ▨ replaced template
- ▨ section on new fragment removed
- ▨ hydrogen bonds
- ● angle between CO vectors

Figure 8. A fragment being slotted into position. This example is based on the modelling described in *Figure 7*. The fragment has been selected because it has passed the distance and C^α position constraints. Also, the angles between the C = O vectors between the base of the template and the fragment are sufficiently small so that perturbations in the final structure are not too great when the backbone atoms of the template and the fragment in the fitting region are averaged. This is particularly important for the top two C = O angles denoted by smaller solid circles.

CI=A   Ca(1)=89, Ca(2)=90, Ca(3)=97, Ca(4)=98   NR=14
(chain identifier)   (fixed Cα positions for distance constraints)   (total number of residues in new fragment)

MAX_RMS(D)=1.0   MAX_RMS(Ca)=1.5   MAX_RMS(CO)=60   SORT=DS
(maximum rms for distance constraints)   (maximum rms for Cα constraints)   (maximum rms for CO angle constraints)   (sorted on Dayhoff score)

No.	RMS(D)	RMS(Ca)	RMS(CO)	DS	PROTEIN	SEQID	TURN	RMS(F)	DA	(sequence of new fragment) YSNCPDGQSTAKTF
1	0.50	0.48	19	16	2FB4	L 61		2.20	27	FSGSKSGASASLAI
2	0.97	0.85	28	13	2PLV	3 55		2.26	42	FDLSATKKNTMEMY
3	0.55	1.21	28	10	2PLV	92		3.02	86	VDNPASTNKDKLF
4	0.61	0.74	28	5	1CA2	57		2.41	25	LRILNNGHAFNVEF
5	0.43	0.46	48	5	2MCP	H 49		2.24	28	AASRNKGNKYTTEY
6	0.36	0.48	21	3	2MEV	14		2.17	38	VSQDTAGNTVTNTQ
7	0.91	0.90	58	3	2FB4	L 121		2.84	14	PPSSEELQANKATL
8	0.50	0.84	51	2	4RHV	3 54		6.24	30	PMNNTHTKDEVNSY
9	0.46	0.56	27	2	2CAB	57		2.32	21	KEIINVGHSFHVNF
10	0.48	0.43	15	1	2TBV	C 293		1.97	34	LVLTRTPTVLTHTF
..	....	....	....	.	....	. ..		....	..	··············
..	....	....	....	.	....	. ..		....	..	··············

(sequence unknown)

(sequences of sorted fragments)

Figure 9. A typical output file of fragments for a data-base search on the loop between the F and G strands. PROTEIN indicates from which protein the fragment was found, Brookhaven code, and SEQID the starting position for the fragment within that protein. TURN is a useful flag when turned on by the user as it indicates whether a particular fragment has a characteristic β-turn conformation. As this particular loop is not expected to be a β-turn this facility has not been used here.

(b) The sequence similarity between the fragment to be modelled and each data-base loop is evaluated. This is quantified by the score (DS) based on the PAM250 matrix of residue similarity introduced by Dayhoff and co-workers (7). Thus, fragments with sequences similar to that of the region to be modelled can be chosen.

(c) The overall geometry of the loop region is quantified. A parameter, RMS(F) measures how far each C^α atom of each residue in the section of template to be replaced and its equivalent C^α in the new fragment are apart. Since C^α atoms of the template and data-base fragments can be difficult to assign equivalence to, due to different conformations and often different numbers of residues involved, a dynamic program algorithm was used to find the closest fit. RMS(F) is then a measure of the r.m.s. difference between equivalenced pairs of C^α atoms. This quantifies how similar the conformation of each fragment is to the conformation of the template in that region. A second parameter used is the dihedral angle (DA) between least-squares planes through each fragment and the template fragment. *Figure 10* is a scatter plot of RMS(F) and DA which shows clustering of similar conformations. Thus fragments can be selected to span the major possible loop connections by inserting fragments that lie in different regions in this scatter plot.

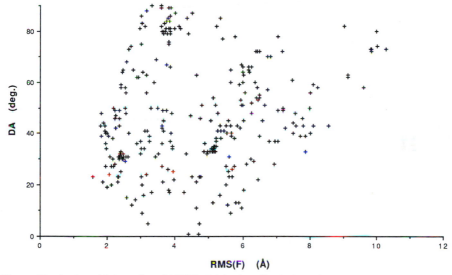

Figure 10. A plot of DA against RMS(F). DA is the dihedral angle between each fragment and the template fragment that is to be replaced. RMS(F) is the r.m.s. of the difference in distances between equivalenced C^α positions between the replaced template and new fragment. Dihedral angles are measured between least-squares planes to the C^α positions of the replaced template and each fragment in turn. The plot clusters similar conformations, since points close to each other have the same relationship with the template fragment and thus with each other.

9.3.3 Generation of the tertiary fold

The next stage is for the user to select a set of data-base fragments to represent each of the VRs modelled. Each set is then placed into the program by order of preference, for example:

SET1:	1	6	13	9	3	1	1
SET2:	1	6	13	9	4	1	1
SET3:	2	6	11	8	1	1	7

This means that for SET1 fragment 1 from fragment output file 1 is taken for the first replacement region, fragment 6 from output file 2 for the second, etc. For SET2, only the fifth fragment would be different.

For each SET, the data-base fragments are fitted to the template by simply deleting a residue at each end of the fragment, and averaging the key back-bone atoms from the next residue in from each end and appropriate backbone atoms from ends of the template gap (see *Figure 8*). The generated structure is then tested for bumps with the unchanged sections of the protein, plus possible bumps between the individual fragments themselves. Bumps are only calculated between main-chain atoms and, once again, these depend on user-defined distance criteria. If a set of fragments pass the bumps filter, the resulting structure is considered further (see below).

If the above procedure fails to identify a possible model, then the user can change the cut-offs used. However, if a less stringent cut-off still cannot yield an acceptable structure, then the particular sequence alignment needs to be examined carefully with a view to using an alternative.

9.3.4 Principles of this approach

This approach to model the main-chain incorporates several different principles. The first is that if there is a family of homologous proteins, then in each VR, the best model for the unknown structure can be from a different member of the family, e.g. ref. 42. The best connection will have the same length and a similar sequence to that of the unknown. As the coordinates of several members of a homologous family can be included in the data-base, then the program can automatically identify the most suitable fragment.

The immunoglobulin hypervariable regions are a special application of inserting loops from the homologous family. Chothia and Lesk (46) have shown that for several complementary determining loops there are a few structural families, known as canonical forms. Identifying the unknown loop as a member of a particular family provides a powerful guide for its likely conformation. However, canonical forms can be adopted by loops of different length and so the fragment data-base search may not identify the appropriate immunoglobulin data-base fragment.

The second principle is that segments from non-homologous proteins can adopt similar three-dimensional structures and thus can provide a good model. This is particularly true for short loops connecting secondary structures as there are certain well-defined families (47, 48). Examination of the sequence of the unknown connection and of the data-base fragments in terms of these structural principles can help to identify the most probable choice of data-base fragment.

The alternative approach for loop modelling is to sample all allowed conformation for the loop, followed by an evaluation of energy for that conformation (49, 50, 51). For further details see Chapter 6. This has been explored particularly for the immunoglobulin hypervariable loops, and good agreement with X-ray structure is being reported (49, 52).

It is, however, most important to recognize that the above approach is simply the best suggestion for any connection. Particularly for longer connections, the algorithm simply shows that the two ends can be tethered in an allowed conformation.

9.4 Model the side-chain conformations

The next step is to replace the side-chains of the parent and of the data-base fragments to those of the new sequence. Then the appropriate side-chain conformation is selected. In a family of closely homologous sequences, side-chains of similar size and electrostatic character tend to retain the same conformation (53). Therefore in the program these side-chains are modelled into a similar conformation to that of the residue being replaced. Side-chains that are larger than the side-chains to be replaced have more degrees of freedom, and additional information must be used to predict their conformations. The present implementation of this program simply takes the more likely conformations of side-chains in various secondary structure elements (54).

The alternative approach is to sample possible conformations and evaluate their energy. Global searching remains computationally intractable (49) but, recently, simulated annealing (55) has been used to sample probable side-chain conformations, and good agreement with the X-ray conformation has been obtained (see also Chapter 6).

9.5 Energy refinement

In order that steric clashes between side-chains and bad geometries around the fragment fitting areas can be alleviated, each three-dimensional test model generated by the program must be energy minimized. In this laboratory the energy minimization package CHARMM (56) is used. Generally, 100–200 cycles of steepest descent minimization is required for convergence. However, energy minimization does not have the radius of convergence to alter the conformation of an incorrect loop or make major shifts to the relative positions of secondary structures. Thus, the main use of minimization is simply to remove the most drastic, local steric clashes.

9.6 Evaluate accuracy of models

Many test models may be generated by the combinatorial selection of fragments as well as the different starting alignments. These test models need to be filtered and the best model, or possibly models, suggested. There are two approaches for filtering.

The first is based on energetic criteria of the overall fold. The absolute value of the minimum potential energy is not a sufficient filter on its own to select possible models. Indeed, Novotny *et al.* (57) have shown that a totally wrongly folded structure can have as low a minimum energy as the correct structure after minimization. However, a consideration of the exposure of non-polar regions, together with identification of buried ionizable groups, is a useful guide to the correctness of the overall fold (58). Similar concepts are encoded in a program by Baumann (59) that is available from the EMBL laboratory at Heidelberg.

The second approach is to use experimental data such as information from site-directed mutagenesis (2). For example, for the first domain of ICAM-1, Staunton *et al.* (2) have identified mutants that interfere with LFA-1 and rhinovirus binding. These mutants can be mapped on to different test models

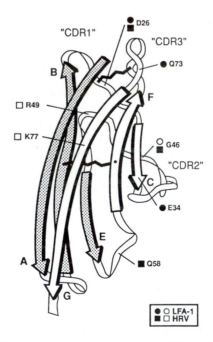

Figure 11. A ribbon diagram of the first domain of ICAM-1. Mutants that affect LFA-1 binding are indicated by circles and those that affect rhinovirus binding, by squares. Filled symbols indicate stronger inhibition than unfilled symbols. Disulphide bonds are denoted by dark zig-zag lines.

and the degree of correctness of that model evaluated empirically. Thus mutants that block rhinovirus binding but not LFA-1 binding should not be in the same region. These concepts were incorporated (60) to yield our resultant predicted model for ICAM-1(D1) shown in *Figure 11*. Based on this model, further experiments can be designed, such the design of peptides to block a particular function.

Acknowledgements

This review incorporates work from several of our colleagues at the ICRF working on modelling (Drs G. Barton, S. Islam, M. Saqi, and M. Zvelebil). The model for ICAM-1 incorporated experimental results from Drs A. Berendt, A. Craig, K. Marsh, C. Newbold at the Institute of Molecular Medicine, Oxford, and Drs A. McDowall and N. Hogg at the ICRF, London.

References

1. Berendt, A. R., Simmons, D. L., Tansey, J., Newbold, C. I., and Marsh, K. (1989). *Nature,* **341,** 57–9.
2. Staunton, D. E., Dustin, M. L., Erickson, H. P., and Springer, T. A. (1990). *Cell,* **61,** 243–54.
3. Kyte, J. and Doolittle, R. F. (1982). *J. Mol. Biol.,* **157,** 105–32.
4. Hopp, T. P. and Woods, K. R. (1981). *Proc. Natl Acad. Sci., USA,* **78,** 3824–8.
5. Fasman, G. D. (1989). *TIBS,* **14,** 295–9.
6. Argos, P., Rao, J. K., and Hargrave, P. A. (1982). *Eur. J. Biochem.,* **128,** 565–75.
7. Dayhoff, M. (1978). *Atlas of protein sequence and structure.* National Biomedical Research Foundation, Silver Spring, Md.
8. Bleasby, A. and Wootton, J. C. (1990). *Prot. Engng,* **3,** 153–9.
9. Lipman, D. J. and Pearson, W. R. (1985). *Science,* **227,** 1435–41.
10. Altschul, S. F., Gish, W., Miller, W., Myers, E. W., and Lipman, D. J. (1990). *J. Mol. Biol.,* **215,** 403–10.
11. Needleman, S. B. and Wunsch, C. D. (1970). *J. Mol. Biol.,* **48,** 443–53.
12. Smith, T. F. and Waterman, M. S. (1981). *J. Mol. Biol.,* **147,** 195–7.
13. Collins, J. F., Coulson, A., and Lyall, A. (1988). *CABIOS,* **4,** 67–71.
14. Bernstein, F. C., *et al.* (1977). *J. Mol. Biol.,* **112,** 535–42.
15. Sander, C. and Schneider, R. (1991). *Proteins,* **9,** 56–68.
16. Sternberg, M. J. E. and Islam, S. A. (1990). *Prot. Engng,* **4,** 125–31.
17. Bacon, D. J. and Anderson, W. F. (1990). In *Methods in enzymology* (ed. R. F. D. Doolittle), Vol. 183, pp. 438–46. Academic Press, San Diego.
18. Barton, G. J. and Sternberg, M. J. E. (1987). *J. Mol. Biol.,* **198,** 327–37.
19. Taylor, W. R. (1987). *Comput. Appl. Biosci.,* **3,** 81–7.
20. Bairoch, A. (1991). *Nucl. Acids Res.,* **19,** 2241–5.
21. Fuchs, R. (1990). *CABIOS,* **7,** 105–6.
22. Sternberg, M. J. E. (1991). *CABIOS,* **7,** 257–60.

23. Sternberg, M. J. E. (1991). *Nature, 349,* 111.
24. Bowie, J. U., Lüthy, R., and Eisenberg, D. (1991). *Science,* **253,** 164–70.
25. Gribskov, M., McLachlan, A. D., and Eisenberg, D. (1987). *Proc. Natl Acad. Sci. USA,* **84,** 4355–8.
26. Barton, G. J. and Sternberg, M. J. E. (1990). *J. Mol. Biol.,* **212,** 389–402.
27. Chou, P. Y. and Fasman, G. D. (1978). *Adv. Enzymol.,* **47,** 45–148.
28. Garnier, J., Osguthorpe, D. J., and Robson, B. (1978). *J. Mol. Biol.,* **120,** 97–120.
29. Lim, V. I. (1974). *J. Mol. Biol.,* **88,** 873–94.
30. Kabsch, W. and Sander, C. (1983). *FEBS Lett.,* **155,** 179–82.
31. Eliopoulos, E. F., Geddes, A. J., Brett, M., Pappin, D. J. C., and Findlay, J. B. C. (1982). *Int. J. Biol. Macromol.,* **4,** 263.
32. Zvelebil, M. J. J. M., Barton, G. J., Taylor, W. R., and Sternberg, M. J. E. (1987). *J. Mol. Biol.,* **195,** 957–61.
33. Kneller, D. G., Cohen, F. E., and Langridge, R. (1990). *J. Mol. Biol.,* **214,** 171–82.
34. King, R. D. and Sternberg, M. J. E. (1990). *J. Mol. Biol.,* **216,** 441–57.
35. Cohen, F. E., Abarbanel, R. M., Kuntz, I. D., and Fletterick, R. J. (1986). *Biochemistry,* **25,** 266–75.
36. Epp, O., Lattman, E. E., Schiffer, M., Huber, R., and Palm, W. (1975). *Biochemistry,* **14,** 4943–52.
37. Sutcliffe, M. J., Hayes, F., and Blundell, T. L. (1987). *Prot. Engng,* **1,** 385–92.
38. Sutcliffe, M. J., Haneef, I., Carney, D., and Blundell, T. L. (1987). *Prot. Engng,* **1,** 377–84.
39. Sali, A., Overington, J. P., Johnson, M. S., and Blundell, T. L. (1990). *TIBS,* **15,** 235–40.
40. Barton, G. J. and Sternberg, M. J. E. (1987). *Prot. Engng,* **1,** 89–94.
41. Saqi, M. A. S. and Sternberg, M. J. E. (1991). *J. Mol. Biol.,* **219,** 727–32.
42. Greer, J. (1990). *Proteins,* **7,** 317–34.
43. Blundell, T. L., Sibanda, B. L., Sternberg, M. J. E., and Thornton, J. M. (1987). *Nature,* **326,** 347–52.
44. Jones, T. A. and Thirup, S. (1986). *EMBO J.,* **5,** 819–22.
45. Cohen, F. E. and Sternberg, M. J. E. (1980). *J. Mol. Biol.,* **138,** 321–33.
46. Chothia, C., *et al.* (1989). *Nature,* **342,** 877–83.
47. Wilmot, C. M. and Thornton, J. M. (1988). *J. Mol. Biol.,* **203,** 221–32.
48. Sibanda, B. L., Blundell, T. L., and Thornton, J. M. (1989). *J. Mol. Biol.,* **206,** 759–77.
49. Bruccoleri, R. E., Haber, E., and Novotny, J. (1988). *Nature,* **335,** 564–8.
50. Schiffer, C. A., Caldwell, J. W., Kollman, P. A., and Stroud, R. M. (1990). *Proteins: Structure, Function, and Genetics,* **8,** 30–43.
51. Moult, J. and James, M. N. G. (1986). *Proteins,* **1,** 146–63.
52. Martin, A. C. R., Cheetham, J. C., and Rees, A. R. (1989). *Proc. Natl Acad. Sci., USA,* **86,** 9268–72.
53. Summers, N. L., Carlson, W. D., and Karplus, M. (1987). *J. Mol. Biol.,* **196,** 175–98.
54. McGregor, M. J., Islam, S. A., and Sternberg, M. J. E. (1987). *J. Mol. Biol.,* **198,** 295–310.
55. Lee, C. and Subbiah, S. (1991). *J. Mol. Biol.,* **217,** 373–88.

56. Brooks, B. R., Bruccoleri, R. E., Olafson, B. D., States, D. J., Swaminathan, S., and Karplus, M. (1983). *J. Comput. Chem.,* **4,** 187–217.
57. Novotny, J., Bruccoleri, R., and Karplus, M. (1984). *J. Mol. Biol.,* **177,** 787–818.
58. Novotny, J., Rashin, A. A., and Bruccoleri, R. E. (1988). *Proteins Struc. Func. Genet.,* **4,** 19–30.
59. Baumann, G., Frommel, C., and Sander, C. (1989). *Prot. Engng,* **2,** 329–34.
60. Berendt, A. R., McDowall, A., Craig, A. G., Bates, P. A., Sternberg, M. J. E., Marsh, K., Newbold, C. I., and Hogg, N. (1992). *Cell,* **68,** 71–81.

6

The use of theoretical methods in protein engineering

VALERIE DAGGETT and PETER A. KOLLMAN

1. Introduction

The ability to modify selectively proteins has increased our understanding of a variety of important, biologically relevant processes, including protein stability, catalysis, and substrate binding. These processes often defy experimental characterization at the molecular level; hence, there is a need for theoretical approaches to better understand these properties. With the advent of fast and affordable computers, it is now feasible to study complex macromolecular systems with computer simulation methods.

Before studying mutant proteins, however, these methods must be able to simulate known properties of wild-type proteins. This goal has now been realized in a number of cases. Often, though, it is difficult in a simulation of a naive protein to stay near the crystal structure; root-mean-square deviations of 2 Å for the α-carbons are common. Deviations of this magnitude can put any conformational changes that accompany mutation at the level of the noise. There is a further complication, in that proteins can adopt different structures in solution and in the crystalline environment. This appears to be the case with tendamistat (1, 2) and with a recently characterized mutant of bovine pancreatic trypsin inhibitor (C. Woodward, A. Wlodawer, and co-workers, in preparation). These results illustrate that one must be careful in choosing appropriate controls, and that comparison to the crystal structure may not always be appropriate. Examples like these also provide good testing grounds for simulations aimed at exploring the effect of the environment, and for fine tuning potential functions to deliver realistic simulations of proteins. In any case, with reliable simulation methods, one can then attempt to explain the effects of known mutations on proteins and, hopefully, successfully predict the effect of potential mutations. And, even though simulation methods are evolving and improvements can still be made, one can gain interesting insights into important biological processes by employing the types of methods discussed in this review. To be most effective, theoretical and experimental studies should be linked. Our focus in this chapter is to survey the theoretical methodologies that have been used, or can easily be adapted, to study single-site mutations in proteins.

A variety of methods have been developed to predict the effect of single-site mutations in proteins for which reasonable starting structures are available. All of the theoretical methods are best, and almost exclusively, suited for cases where the mutant protein closely resembles the parent molecule, with any structural changes localized to the region of the mutation. Fortunately, there are now a variety of protein systems showing that single-site mutations generally yield these localized conformational alterations with little effect on the overall structure of the protein (3–5). The theoretical approaches used to study the effects of mutations fall into three broad groups:

- force-field methods
- conformational search procedures
- knowledge-based approaches relying on structural data-bases

No one of these methods is vastly superior to the others, such that the choice of method depends on the system of interest, the computational resources, and the goal of the work. A description, computational details, and some applications of each of these methods are discussed briefly in turn.

2. Force-field methods

2.1 General issues

In principle, one could calculate the atomic positions of the mutated residue using *ab initio* quantum mechanics. In the case of proteins, however, one must use classical approaches because of the large number of atoms that must be considered. Therefore, an empirical force field, like the one shown below, is generally employed (Equation 1).

$$V(x) = \Sigma\, k_r(r\text{-}r_{eq})^2 + \Sigma\, k_\theta(\theta - \theta_{eq})^2 + \Sigma\, V_n/2\, [1+\cos(n\phi\text{-}\gamma)]$$

$$\sum_{i \neq j}\sum [(A_{i,j}/r_{i,j}{}^{12}) - (B_{i,j}/r_{i,j}{}^{6}) + (q_i q_j/\varepsilon r_{i,j})] \tag{1}$$

This function describes the potential energy of the system as a function of the positions of the atoms.

(a) The first three terms in the potential energy function describe the bonded interactions:

    i. the first term represents covalent bond stretching; it is harmonic, with the minimum energy bond length given by r_{eq} and the force constant by k_r

    ii. the second term represents angle bending and takes the same form as the first

    iii. the third term is a Fourier series and describes the conformational energy as a function of the dihedral angles

Valerie Daggett and Peter A. Kollman

(b) The final term represents the non-bonded contributions (van der Waals and electrostatic interactions, respectively). This is the most important term for proteins:

   i. van der Waals interactions are represented by a Lennard–Jones potential with a steep term for repulsive interactions (A/r^{12}) to avoid steric overlap and a longer-range term ($-B/r^6$) for attractive interactions

   ii. a Coulombic potential is used for the electrostatic terms, where all atoms have partial charges (q) and r is the distance between the interacting atoms.

Electrostatic contributions to the total energy are generally large and are the source of the biggest errors in simulating the properties of proteins. The parameters for these equations are derived primarily from the results of *ab initio* quantum mechanics, spectroscopic data, and crystallographic data. The force field is calibrated by fine tuning of these parameters to reproduce structures and energy trends in various model systems. (For a thorough discussion of the evolution of parameterization of force fields see ref. 6.) With both a calibrated force field and a well-defined starting structure in hand, one can easily construct an approximate model for a mutant protein using energy minimization. Central to the use of any energy method for evaluating structures is the idea that the biologically relevant conformations are of low energy.

2.1.1 Energy minimization

Energy minimization is accomplished by adjusting the positions of the atoms such that the change in the potential energy with respect to position is zero (i.e. $\partial V/\partial x = 0$). There are a number of different methods for performing minimization and the reader is referred to Allinger (7) for a more thorough discussion. Unfortunately, these minimization routines search only the local configurational space of a molecule because the system moves downhill over the energy surface and stops at the first minimum energy conformation. This is the so-called local minimum problem. This idea is shown schematically in *Figure 1*, where energy minimization from the starting position would leave the system in the energy well labelled A as opposed to finding other wells of lower energy. Molecular dynamics can be employed to search conformational space further, for it allows the system to surmount energy barriers on order of kT. Equation 1 is also used for this purpose.

2.1.2 Molecular dynamics

In molecular dynamics, a trajectory of a constantly changing molecule moving through phase space is generated:

(a) The atoms move due to the force of their own kinetic energy and the forces exerted upon them by all other atoms.

145

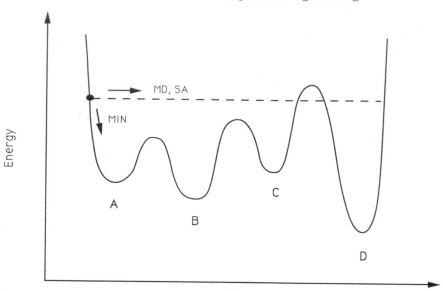

Figure 1. A simple schematic representation of a potential energy surface. From the starting point, energy minimization (MIN) would leave the system in the energy well labelled A. With molecular dynamics (MD) at some temperature, the system is able to traverse the smaller energy barriers, making regions A, B, and C accessible by quenching along the trajectory. All of the energy levels shown here would be accessible by simulated annealing (SA) because the systems can, in effect 'tunnel' through energy barriers using this method.

(b) All atoms are initially assigned velocities corresponding to a specified temperature. The temperature is determined from the mean kinetic energy for the system ($1/2 \, \Sigma \, m \, <v^2> = 3/2 \, Nk_\mathrm{B}T$, where k_B is the Boltzmann constant, m is the mass, v is the velocity, and the sum is over all atoms).

(c) New positions are determined by taking the derivative of the potential energy with respect to the position of every atom, which gives the force on every atom ($F = -\partial V/\partial x$). Using Newton's second law ($F = ma$) and known masses, one can solve for the acceleration of every atom. In effect, this is done by integrating the equations of motion by specifying a time step (Δt), knowing the initial velocity, and solving for the velocity and position at the new time point (e.g. $a = \Delta v/\Delta t$, $v = \Delta x/\Delta t$).

Figure 1 shows how molecular dynamics can arrive at energy wells that are inaccessible by energy minimization alone. However, not all energy barriers are easily traversed at moderate temperatures, so that lower energy states, such as D, may not be accessible. During a molecular dynamics simulation one can monitor specific interactions both geometrically and energetically to investigate structural transitions and the mechanism by which they occur.

2.1.3 Monte Carlo and simulated annealing

Monte Carlo methods provide another way to achieve controlled uphill steps along the energy surface and to investigate the dynamic behaviour of molecules. The Metropolis method is generally employed for these calculations (8). Using this approach:

(a) Random changes are made to the structure of the molecule.

(b) The energy of the resulting structure is then compared to the previous configuration.

(c) If the new structure is of lower energy than the previous conformation, then the structure is accepted.

(d) If, however, the energy of the new structure is higher, the new configuration is accepted with a probability that is generally determined by the Boltzmann factor (Equation 2)

$$P = \exp\left(-\Delta E / k_B T\right) \tag{2}$$

where ΔE is the energy difference between the two structures calculated from a potential energy function like that given in Equation 1, or any other appropriate function. A random number is generated between the bounds of the probability (0 and 1) and compared to the Boltzmann factor. If the random number is higher than the probability determined using Equation 2, the new, albeit high-energy, configuration of the molecule is accepted. Otherwise, the previous structure is retained and the process is repeated.

As with molecular dynamics, the degree of sampling is determined by the choice of temperature.

The method of simulated annealing is a special case of the Metropolis procedure (9). Basically, the temperature is used as a control parameter, which is not necessarily equivalent to the true physical temperature. The simulated annealing process involves:

(a) 'melting' the system at a high effective temperature (generally >1000 K)

(b) then lowering the temperature slowly in a stepwise manner (typically by 10% at each step)

The simulation is continued at each intermediate temperature until the system equilibrates. If cooling is performed too quickly, defects can remain in the structure, which again results in a local minimum problem.

This method can, in many cases, search conformational space very efficiently. Only the gross features of the conformations can be distinguished at high temperature with the details becoming more apparent as the temperature is lowered. This has obvious advantages in terms of searching potential energy surfaces because there are many opportunities to avoid local minima, but the system can sometimes become trapped, as with the other methods discussed above. Unlike molecular dynamics, energy barriers are not surmounted, rather, with simulated annealing the system 'tunnels' through barriers. This idea is demonstrated in *Figure 1*, where all of the energy wells depicted are accessible using simulated annealing.

2.1.4 Free-energy perturbation

i. Principles

Recently, the free-energy perturbation method has been used to calculate a number of properties of proteins. The free energy is generally computed as follows:

$$G(\lambda + d\lambda) - G(\lambda) = -RT \ln\, <\exp -(V(\lambda + d\lambda) - V(\lambda)/RT)>_\lambda \quad (3)$$

where V is the potential energy for the interaction between the perturbed group with its surroundings, G is the Gibbs free energy, R is the gas constant, T is the absolute temperature, and the symbol $<>_\lambda$ indicates that the ensemble average generated with molecular dynamics or Monte Carlo is taken at intermediate positions along the conversion pathway determined by the coupling parameter, λ.

As the equation is expressed above, the conversion from $\lambda = 0$ (unperturbed) $\rightarrow \lambda = 1$ (perturbed) is accomplished by considering a linear combination of the two final states, where $\Delta\lambda$ determines the number of steps, or windows, for the conversion. The free energy difference between the two states is then

$$\Delta G = \sum_{\lambda=0}^{\lambda=1} G(\lambda + \Delta\lambda) - G(\lambda)$$

ii. Calculations for mutations affecting substrate binding and
    catalysis

A free-energy diagram like the following is generally used to link calculations of protein mutations to the experimental results,

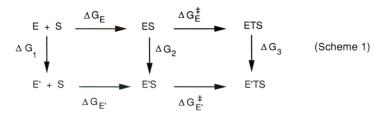

(Scheme 1)

where E represents the wild-type enzyme, E′ is the mutated form of the enzyme, S is the substrate, ES is the enzyme–substrate complex, and TS represents the enzyme transition state complex.

Transition state complexes are generally constructed by imposing a co-valent bond between the atoms of interest, doing quantum mechanical calculations to determine the new charge distribution, and changing the hybridization of the attached atom. This approach is most reliable when one has a crystal structure of a transition state inhibitor to use as a starting structure. The difference in binding free energy upon mutation is $\Delta\Delta G_{\text{bind}} = \Delta G_{E'} - \Delta G_E$ and the difference in activation free energy upon mutation is $\Delta\Delta G_{\text{cat}} = \Delta G^{\ddagger}_{E'} - \Delta G^{\ddagger}_E$. The processes shown by the horizontal arrows are measured experimentally, but they are very difficult to determine using simulation methods. Instead, the processes shown by the vertical arrows are simulated, such that wild-type enzyme is mutated in solution, in the complex with substrate, and in a model for the transition state. Since the free energy is a state function, the experimentally determined difference between the two enzymes denoted by the horizontal arrows can be related to the processes given by the vertical arrows: $\Delta\Delta G_{\text{bind}} = \Delta G_2 - \Delta G_1$ and $\Delta\Delta G_{\text{cat}} = \Delta G_3 - \Delta G_2$.

iii. Calculations for mutations affecting protein stability

Protein stability can also be evaluated using the free-energy perturbation method. Protein stability is dependent upon the difference in free energy between the native and denatured states. This phenomenon is harder to model with simulation techniques than binding and catalytic effects because of the nebulous nature of the denatured state. None the less, the following free-energy cycle can be used to estimate the relative stability of a mutant compared to the corresponding wild-type protein:

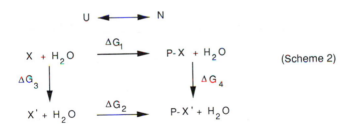

(Scheme 2)

X denotes the mutated residue and P-X and P-X′ represent the protein with the original residue and the mutated residue, respectively.

Again, the horizontal arrows represent processes followed experimentally, while the processes shown as vertical arrows are simulated. The difference in stability between the mutant and wild type is $\Delta\Delta G = \Delta G_2 - \Delta G_1 = \Delta G_4 - \Delta G_3$. The isolated residue in water may be used to model the unfolded state. This assumes that the residue is completely accessible to solvent in the unfolded state. A better model is to mutate the residue in a peptide that is free to adopt a number of configurations, some of which will have interactions between residue X and other amino acids instead of solely with solvent. One would ideally like to perform many free-energy perturbation calculations, mutating the residue of the protein starting from a number of unfolded structures and then averaging over the many possible conformations. Because of the unknown nature of the unfolded state and the large amounts of computer time necessary for these calculations, extremely simple models are used. Therefore, interpretation of these simulations must be done cautiously, except possibly where one has experimental evidence indicating that the mutation manifests its effect only in the folded state. With this general, conceptual background of force-field methods we now move to specific applications using energy minimization and molecular dynamics.

2.2 Applications

2.2.1 Applications employing predominantly energy minimization

i. Maximum overlap procedure

One of the simplest minimization protocols for modelling mutant proteins is the maximum overlap procedure (*Figure 2*). Using this method, the starting position for the new side-chain is determined by superimposing the new residue on the old, while maintaining the geometry of the original residue. In the case where the new chain extends in some direction past the parent residue, the conformation for the new residue is taken from known low-energy structures of the residue, or the residue is merely extended. At this point, the swapped residue is minimized with or without minimizing the rest of the protein.

This method has been used in building a structure of α-lactalbumin from lysozyme (10). This method was also employed to investigate a triosephosphate

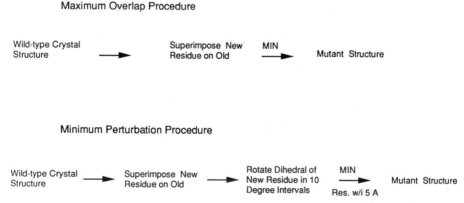

Figure 2. Minimization methods. Schematic description of the steps involved in two common minimization protocols for generating mutant protein structures. MIN, minimization.

isomerase mutant in which an active site His was mutated to Gln (11). This simulation preceded determination of the X-ray structure and suggested a possible reason for the drop in catalytic activity. The crystal structure, however, shows a similar but different interaction that may explain the kinetic results (12).

As mentioned above, the method of energy minimization suffers from the local minimum problem. The mutated residue can easily be trapped in a local energy minimum and, without human intervention, the structure may not be able to pass over the energy barrier, resulting in a better lower energy structure being overlooked. In fact, in more recent studies of the triosephosphate isomerase mutant described above, employing molecular dynamics, both the interaction observed in the previous simulation and in the crystal structure occurred during the course of the simulation (13).

Recently, a modification of the maximum overlap procedure has been used to explore inhibitor binding to the human immunodeficiency virus (HIV) protease (V. Daggett and M. Levitt, unpublished results). In this case, a new residue was built on to the substrate and minimized in the field of the protein using 'soft atoms' (14). The use of soft atoms can aid in sampling conformations by smoothing the energy surface between different configurations, thus making it easier to traverse energy barriers. Instead of stopping after minimization, varying amounts of molecular dynamics (usually 2–20 psec) were performed, allowing only the swapped residue or the entire active site to move. Then, the energy of the system was minimized using standard van der Waals potentials. Our aim in employing molecular dynamics was to circumvent the local minimum problem. Unfortunately, molecular dynamics does not always lead to improved structures and usually samples only relatively local regions of conformational space. But, in many cases, use of molecular dynamics resulted in much better structures than use of energy minimization alone. Adding the extra step involving molecular dynamics is probably at least worth trying if one is using the maximum overlap method; however, every amino acid change yields a new system. This procedure is not yet general, different minimization and molecular dynamics protocols may be needed for even related proteins. Nevertheless, the method can easily be employed to model mutant proteins and offers advantages over using energy minimization alone or even conventional molecular dynamics.

ii. Minimum perturbation procedure

Another method relying predominantly on minimization is the minimum perturbation procedure (15) (*Figure 2*). Using this approach:

(a) One begins by positioning the new residue.

(b) Then its side-chain is rotated (at 10° intervals) and the low-energy conformations are identified.

(c) The low-energy conformers are subjected to energy minimization, allowing only those residues close to the swapped residue to move (within 5–15 Å).

This method is, in principle, better than the maximum overlap procedure. The combination of systematic search and energy minimization is able to search more of conformational space than minimization alone. But, one can still fail to find the lowest energy conformation. It is especially problematic if the conformations of the residues adjacent to the swapped residue change, such that they are not accessible from the original structure with energy minimization alone. Here again, one can imagine a step involving molecular dynamics in an attempt to alleviate this problem.

Shih *et al.* (15) tested the minimum perturbation procedure on a mutant of the haemagglutinin glycoprotein of the influenza virus in which a surface Gly is replaced by Asp. They found that their procedure yielded a conformation for the Asp that matched the density in the crystal structure of the mutant. In another application of this method, Petsko and co-workers (16) attempted to rationalize the drop in activity observed when the catalytic base of triosephosphate isomerase is shortened (E165D). On the basis of the lowest energy structure found for the mutant, they argued that the drop in activity was due to orientational differences in the catalytic base and not interactions with other residues designed to stabilize the bound substrate. However, their conclusions are questionable since they did not allow the charged residues near the portion of the substrate whose charge distribution changes in the transition state to move (these residues are outside of their cut-off for evaluating groups). In contrast, in other simulations, in which all of the active-site residues were allowed to move, these interactions between the substrate and the nearby charged residues were found to be crucial for substrate stabilization (13).

iii. Caveats

In a more general vein, both of the minimization methods shown in *Figure 2* and described above base any quantitative conclusions on comparison of the potential energies. This must be done cautiously, as the total energies are generally large for both wild-type and mutant proteins while the difference between them is small. Along these same lines, the binding, catalysis, and thermodynamic stability of a protein are determined by free-energy differences and not potential energies. To complicate the matter further, these methods necessarily assume that any change upon mutation is manifested in the folded state of the protein, while differences in the unfolded form upon mutation can also yield differences in stability. Even given these caveats, these approaches are often worth trying because they are easy to implement and the calculations are fast.

2.2.2 Applications employing molecular dynamics

i. Principles of molecular dynamics

Molecular dynamics can also be used to generate mutant protein structures from the parent molecules. In addition, the process by which the wild-type

Molecular Dynamics

Free Energy Perturbation Method

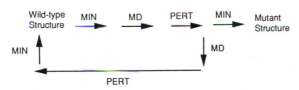

Figure 3. Schematic views of molecular dynamics procedures for generating models of mutant proteins from the native structure. The first method involves molecular dynamics to generate the model of the mutant, followed by energy minimization to relax the structure. The free-energy perturbation method involves mutation to the new residue (PERT) followed by energy minimization to relax the structure to yield the mutant protein structure. From the structure of the mutant the perturbation is performed in the opposite direction ($\lambda = 1 \rightarrow 0$) to regenerate the wild-type protein. This final step gives an estimate of the error in the calculation. Abbreviations used: MIN, energy minimization; MD, molecular dynamics; PERT, perturbation.

protein adapts to the mutation can be of interest. *Figure 3* shows one possible progression of steps for performing molecular dynamics simulations of mutant proteins.

(a) From the crystal structure, the new residue is superimposed over the wild-type residue and the structure is minimized briefly to prepare for molecular dynamics.

(b) Molecular dynamics is then performed for some period of time (at least 10 psec).

(c) Structural properties can be monitored during the run to delineate how the structure tolerates changes and, if the simulation is carried out for a sufficient period of time, to determine average properties of the mutant to compare with experiment.

(d) The final structure from molecular dynamics can be minimized to relieve any strain and compared with the original wild-type structure.

(e) If counterions and water molecules are included in the simulation, they are added to the minimized model structure. First, the counterions are added and then minimized while fixing the positions of all other atoms. Then, the water molecules are added and their energy is minimized. Following minimization, molecular dynamics is generally performed on just the water for a short period of time (1–2 psec). The water is again minimized, followed by minimization of the full system. The successive minimization and molecular dynamics steps are performed to allow re-organization of the water molecules around the protein, and relaxation to avoid bad contacts that may occur during molecular dynamics of the full system.

Realistic simulations of a protein are performed with the structure immersed in a full box of water molecules, employing periodic boundary conditions. Because of the computational requirements in doing so, other treatments of the solvent may be warranted. A number of methods have been developed to include only limited amounts of solvent or to mimic the solvation effects without actually including explicit solvent molecules. Some of these methods are listed below, in approximate order of decreasing computational requirements; there are advantages and disadvantages to all of these approaches and the reader is referred to the papers cited below for further details.

(a) Langevin dipoles can be substituted for the water molecules past the first shell to simulate the effects of bulk solvent (17).

(b) Alternatively, the solvent can be included in a limited way by adding a shell of waters around the protein, or around just the active site.

(c) Stochastic dynamics employing the Langevin equations of motion, which include a frictional term to partially account for the effects of solvent, have been employed successfully to study motion in proteins (18).

(d) Empirically determined atomic solvation parameters (19), dependent on the solvent accessibility of the residues, have been included in a force field (20).

(e) Finally, calibrated macroscopic dielectric functions may be used to screen electrostatic interactions and compensate for the lack of solvent (21).

ii. Example: triosephosphate isomerase mutants

Recently, we described a series of molecular dynamics simulations of five triosephosphate isomerase (TIM) mutants with the aim of rationalizing the experimental kinetic data (13). From the crystal structure of the wild-type protein, the residue of interest was replaced and then molecular dynamics was performed for 20 psec. For the wild-type enzyme and each mutant, we performed simulations with non-covalently bound substrate and with substrate bound covalently to the catalytic base as a model for the transition state of the rate-determining step.

154

One of these mutants involved replacement of the active site histidine (H95N), which is thought to stabilize the transition state through interactions with O1 and O2 of the substrate (*Figure 4a*). Experimentally, the relative binding affinity for substrate is approximately that of wild type for this mutant, however catalysis is severely compromised (22). A pseudorevertant (H95N, S96P) has also been characterized. This double mutant shows an improved rate of catalysis and slightly stronger binding of substrate compared to the single mutant (22). There were strong interactions between the mutants and substrate in our simulations. In the case of the single mutant, the substrate has become kinked and Lys13 interacts strongly with the phosphate portion of the substrate (*Figure 4b*). In the double mutant, the catalytically non-productive binding mode improves; the substrate is again extended and, as in wild-type TIM, Lys13 interacts with both the phosphate group and O2 to help prepare for catalysis.

Figure 5 shows TIM with covalently bound substrate following molecular dynamics. These structures serve as models for the transition state for enolization. Interactions with both O1 and O2 of the substrate were important in stabilizing the transition-state structure (*Figure 5a*). In the H95N mutant these interactions are disrupted (compare *Figures 5a* and *5b*). These interactions improve upon addition of Pro96, with Asn95 interacting strongly and exclusively with O1 of the substrate and Lys13 stabilizes O2 (*Figure 5c*). This mode of stabilization is distinct from the interactions observed in the wild-type structure.

From an evaluation of the motion of the active-site residues and the final structures obtained, we proposed a number of testable hypotheses for the behaviour of these enzymes. Molecular dynamics can be very useful for showing how structures adapt to changes in sequence as well as how an enzyme optimizes interactions important for substrate binding and catalysis. Our simulations suggest possible reasons for the changes in binding and catalysis upon mutation that could not have been predicted a priori and, as yet, provide the only structural interpretations of the effect of these mutations, as crystal structures are not available. The results using this method are only qualitative, however. In our system there was no strict correlation between the energies and either the experimental binding or catalytic properties of the enzyme. These simulations do, however, reproduce the highly co-operative nature of the interactions in the active site and suggest that this approach may be useful not only for rationalizing the effect of changes in sequence but also for identifying particularly promising sites for mutation.

iii. Principles of free-energy perturbation

The other more common and established approach for evaluating mutant proteins using molecular dynamics is the free-energy perturbation method. This method has been used to estimate differences in free energies of substrate binding, catalysis, and stability between wild-type and mutant proteins.

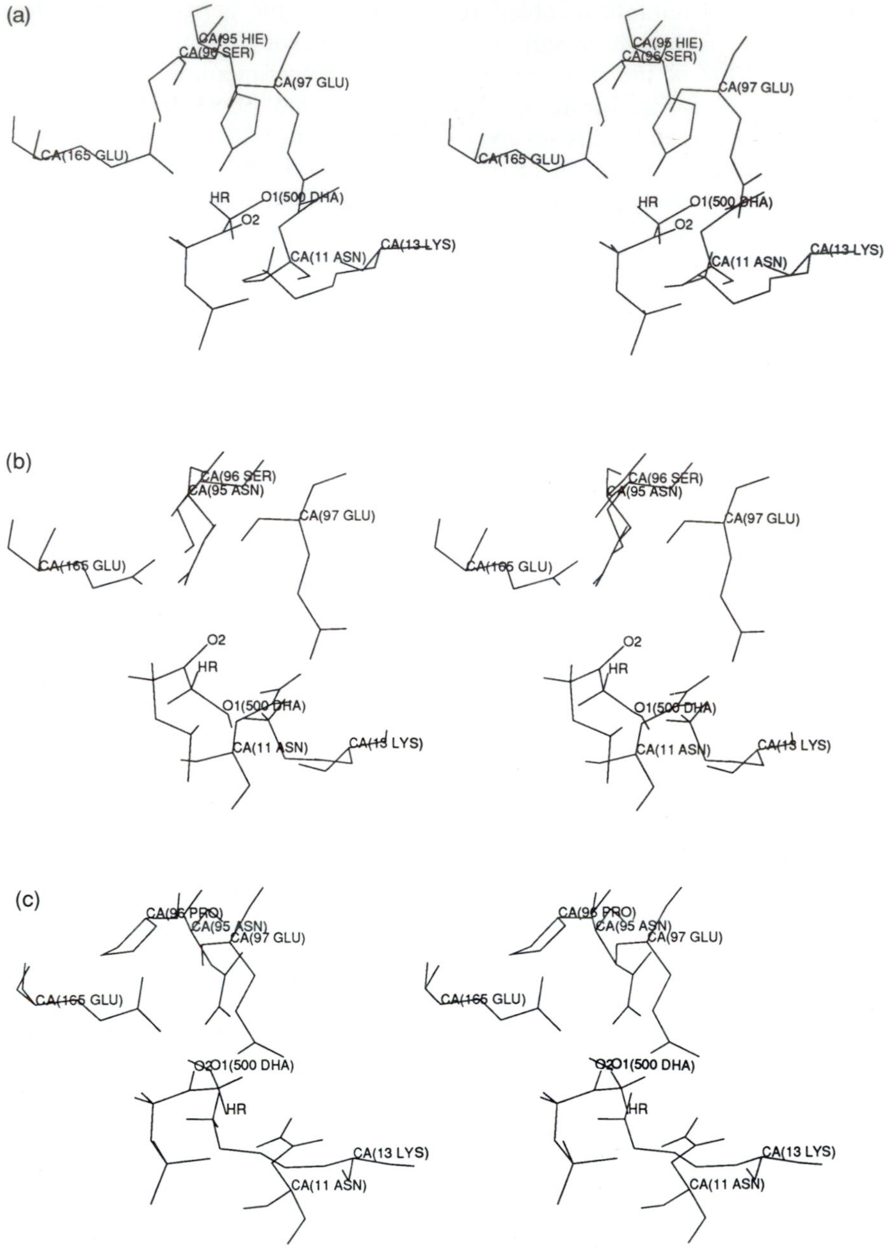

Figure 4. Stereoviews of the active-site region of non-covalent triosephosphate isomerase–substrate complexes after 20 psec of molecular dynamics: (a) wild-type triosephosphate isomerase; (b) the H95N mutant; (c) the double mutant H95N, S96P.

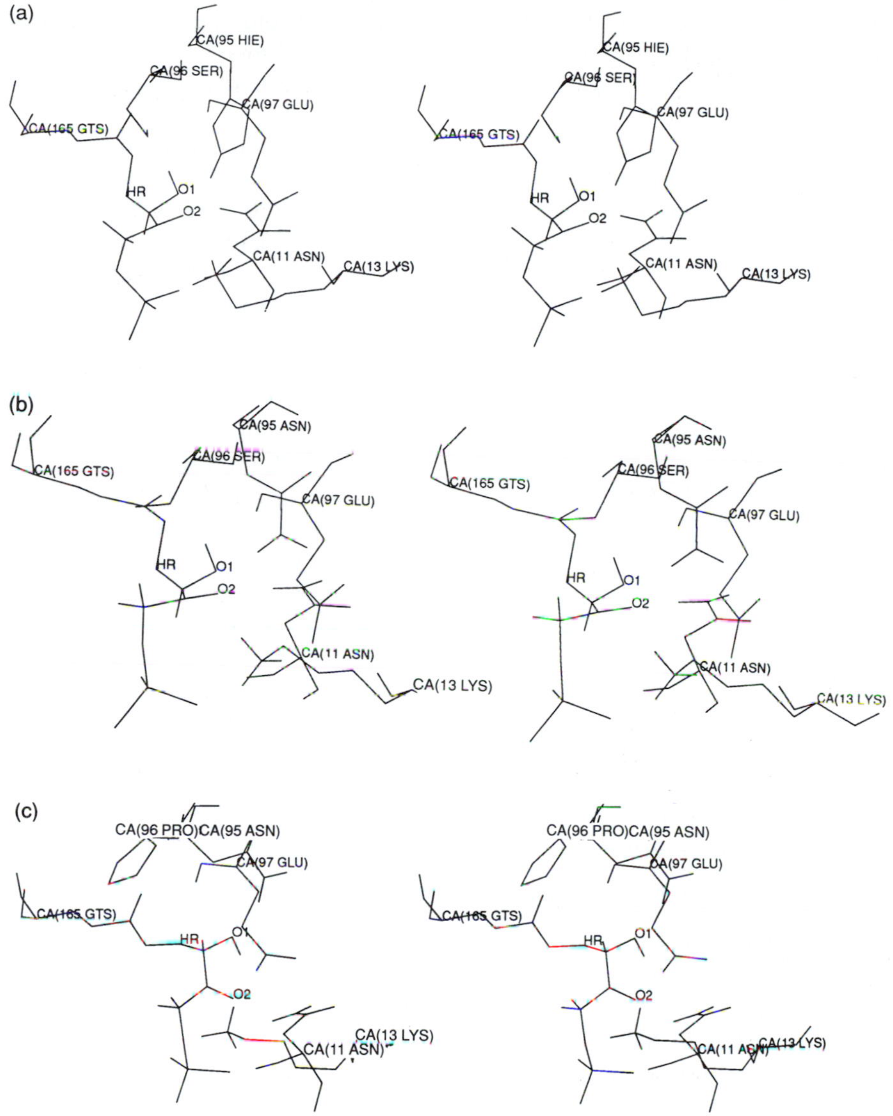

Figure 5. Stereoviews of the active-site region of triosephosphate isomerase complexes with covalently bound substrate after 20 psec of molecular dynamics: (a) wild-type TIM; (b) the H95N mutant; (c) the H95N, S96P pseudorevertant. These covalent complexes (with the oxygen of the catalytic base, Glu165, bonded to the pro-R hydrogen of the substrate dihydroxyacetone phosphate) are used as models for the transition state for enolization.

One must keep in mind that this method is really only reliable with small perturbations to the reference state. A schematic flow of the various steps involved in preparing a structure, and calculating free-energy changes is given in *Figure 3*.

In preparing to perform molecular dynamics, one begins by minimizing the crystal structure. Water molecules and counterions can be treated as described above. Following construction of the minimized model, the system is equilibrated at the appropriate temperature for 2–25 psec. Different lengths of equilibration are used to generate different starting conformations to check the dependence of the calculated free-energy changes on starting structure. After equilibration, perturbation from the wild-type protein ($\lambda = 0$) to the mutant form ($\lambda = 1$) is performed. Simulations with different numbers of steps (controlled by $\Delta\lambda$) for the conversion should be tested. Some simulations must be performed quickly because the structure may change drastically upon perturbation. Generally, one is confronted with the opposite problem, that of attaining sufficient sampling during the conversion. The structure resulting from a perturbation simulation ($\lambda = 1$) can then be minimized to relieve any strain, yielding a model for the mutant protein. Following perturbation to the mutant protein, the perturbation is run in the opposite direction to regenerate wild-type protein (*Figure 3*). The amount of hysteresis between the two runs and between different calculations beginning with different starting structures gives a measure of the error in the simulation. Hysteresis is a result, predominantly, of insufficient sampling or conformational changes such that the original structure is not reproduced (e.g. instead of running the reverse of the A → B transition, B → A, one may actually be simulating a transition from B to some new state, C) (23–24). At this time, there is no rigorous way to determine the uncertainties in the calculated free energies.

iv. Example: subtilisin and T4 lysozyme mutants

Rao and co-workers (25) have used the free-energy perturbation approach to calculate the differences in free energy of substrate binding and catalysis (*Scheme 1*) between native subtilisin and a mutant in which one of the hydrogen bonds in the oxyanion hole is obliterated (Asn 155 → Ala). The results of this study are consistent with the experimental results and, in fact, were performed prior to knowledge or publication of the experimental work. Independent free-energy calculations by Hwang and Warshel (26) also were successful in predicting the free-energy changes for this mutant and others.

Dang and co-workers (24) have estimated the relative stability of a T4 lysozyme mutant using the free-energy cycle in *Scheme 2*. They used a tripeptide to model the denatured state. The mutation involved Thr157 → Val. Their results are consistent with the experimental finding that this mutation destabilizes the protein by ~2 kcal/mole. Interestingly, the difference in stability between the two proteins is a van der Waals rather than a predominantly electrostatic effect. That is because the hydrogen bonding of the OH group of Thr157 is similarly stabilizing in both native and denatured proteins.

v. Free-energy component analysis

We have also described another, more qualitative application of the free-energy perturbation method for understanding the ramifications of mutations in proteins, referred to as a free-energy component analysis (27). Using this method, one changes particular interactions of interest to determine individual free energies of interaction contributing to substrate binding and transition-state stabilization. This type of approach is more qualitative in nature than when one residue is mutated directly into another, as a rigorous comparison to experimental free energies can only be made in the limit of performing the calculations for all interactions in the active site. The free-energy component approach was used to investigate the importance of particular electrostatic interactions in substrate binding and catalysis of wild-type triosephosphate isomerase and the E165D mutant mentioned above. On the basis of these results, we suggested a number of interactions that may play a role in determining the loss of catalytic efficiency upon mutation that had been heretofore unrecognized. Dang *et al.* (24) have also employed this method to address the role of particular hydrogen bonds in stabilizing T4 lysozyme.

vi. Conclusions

The free-energy perturbation method has now been used in many cases to predict successfully not only properties of proteins but also inhibitor binding to proteins and solvation properties of the amino acids and other compounds. However, this method is still in a developmental stage. Since free energy is determined by the conformational space accessible to the system, sampling is crucial in calculating accurate free energies. Also, this method works best when there are only local structural changes around a mutation site, but since all of the residues are allowed to move, most changes can probably be accounted for if the mutation is accomplished over a long period of time. However, this method is very computer intensive, even for straightforward mutations, making sampling the most serious concern with employing this approach. Also, one must keep in mind the possibility of calculating free-energy changes in good agreement with experiment for the wrong reasons, yielding an incorrect mutant structure. There have been some algorithmic developments in this area to improve sampling (28), and with even more powerful computers becoming available, we can be optimistic about future applications of this method. Ideally, theory and experiment can work together so that this method will play a predictive role in elucidating areas of interest for mutation that may not be obvious from experimental work alone, and in paring down the number of possible mutation sites to allow the experimentalist to concentrate on a manageable set.

3. Conformational search methods

Methods involving systematic searches of conformational space for some limited number of residues have been used successfully to model proteins from homologous proteins. There is every reason to believe that these methods will also be applicable to the problem of predicting a mutant protein structure. Generally, these methods are used to generate a number of possible structures, which are then evaluated using a detailed potential function as described above. (Molecular dynamics and the minimum perturbation procedure described above could also be considered conformational search methods.)

3.1 Systematic searches

Moult and James (29) have described a method for systematically searching conformational space for up to six residues by rotating the dihedral angles ϕ, ψ, and χ_n. The number of conformations to be considered further are those that obey a number of known protein properties (e.g. no bad van der Waals contacts, low electrostatic energy, ϕ and ψ in well-defined ranges determined from refined protein structures). The resulting low-energy structures can then be subjected to full energy minimization. In the ideal situation, the fully refined structures will converge to a single structure or closely related structures.

Snow and Amzel (30) have described a related method, the coupled perturbation procedure. Using this method, a dependency set is identified as any residue within van der Waals contact of either the wild-type or mutant residue. Conformational searches about the side-chain dihedrals (χ_n) of the residues in the dependency set are then performed in 15° increments. Low-energy conformations are identified by use of a potential energy function similar to the one described above. Only side-chains are considered in this method and, as such, its use is restricted to cases where the backbone is left relatively unchanged upon mutation.

3.2 Simulated annealing

Lee and Subbiah (31) have developed an automatic method, using simulated annealing, for predicting side-chain conformations and optimizing the resulting packing interactions. They have tested their method on nine proteins and achieved very good agreement with experiment. Better results are obtained for the core residues than those on the surface. This approach can easily accommodate fairly large conformational changes extending over many residues. But, this method assumes that the main-chain fold is unchanged upon mutation. For changes in surface residues, that are not well predicted with this method, one could add a term to account for solvation, as described below and by Schiffer and co-workers (20).

4. Knowledge-based methods

4.1 Side-chain preferences

Generally, knowledge-based methods refer to studies in which portions of the sequence of the protein of unknown structure are taken from a data-base of known structure. Applications of the studies of Ponder and Richards (32) and work from Blundell's group (33) provide straightforward examples of this approach. Ponder and Richards have constructed a rotamer library for most of the amino acids. This library describes the distributions of χ angles found in high-resolution crystal structures. Sutcliffe and co-workers (33) have compiled a similar data-base consisting of the most probable side-chain conformations for each amino acid in different types of secondary structure. These approaches are rapid for determining starting conformations, but the resulting structures will generally need further energy refinement by conventional force-field methods.

4.2 Empirical potential functions

Recently, Schiffer and co-workers (20) have reported an application of the data-base approach in which side-chain conformations are taken from the data-base of Ponder and Richards (32). The resulting conformers are then subjected to energy minimization. The conformer of lowest energy is chosen as the predicted structure. For their particular problem, they found that this procedure was sufficient for placement of internal side-chains, but the positions of external residues were most accurate when a term to account for the solvation energy (19), dependent upon the solvent accessibility of the residue, was included.

Sippl (34) has added another layer of complexity to the knowledge-based approach for predicting mutant structures by calculations of conformational ensembles from potentials of mean force. The potential functions are based on pairwise interatomic distances. These potentials are then used to evaluate all of the conformations in the data-base for a particular sequence of residues. The most probable conformations are then chosen from the most stable conformations. The net energy difference between the wild-type and mutant protein will be a free energy, since entropic effects are inherent in the crystal structures from the data-base, but a quantitative comparison to experiment has not yet been performed. Using this method, Sippl has shown the importance of the flanking residues in determining the secondary structure that a particular sequence adopts. Use of this method in its present form is probably best for situations where there is a change in backbone orientation upon mutation because it must be combined with other methods for building in the side-chains themselves. However, the method appears to be general enough that it can be adapted for prediction of side-chain conformations, as well.

4.3 Segment match modelling

Levitt has devised a novel method for predicting portions of protein structures (35). Using this approach, the main-chain is built first, given the positions of the α-carbons. One then moves through the main-chain, matching segments to the backbone from a data-base of high-resolution structures. The best segment is chosen based on root-mean-square deviation and energy, among other criteria. After completing the main-chain, the side-chains are added at random. The procedure is repeated 10 times and the mean of the coordinates is determined, with its associated standard deviations. This approach has not been tested for single-site mutations. But, it works well for building back entire proteins from just the C_α coordinates, giving root-mean-square deviations of the main-chains of between 0.4 and 0.7 Å from the crystal structure.

5. Conclusions

Any of the methods discussed above can be used to generate models of mutant proteins from the structure of the wild-type protein. For most of the methods, a test of whether the prediction is correct is only possible by comparing the model to the actual crystal structure or the structure determined using two-dimensional NMR techniques. Only the free-energy perturbation method can provide quantitative validation of the model when a mutant crystal structure is not available but when thermodynamic data are. The future of computer simulation methods in biochemistry is bright. As computers become faster, it should be possible to simulate mutations involving more drastic structural changes with the free-energy perturbation method. Methods relying on conformational searches will also benefit from larger computers, allowing one to include much more of the environment in the actual search procedures. Also, the projected future improvement in the ability to span conformational space will allow more extensive testing of the potential functions, which are used to some degree for almost all of the methods mentioned. Another promising area, that will be even more powerful when computer speed is less of a concern, involves methods combining quantum mechanics and molecular dynamics. In any case, sufficient computer power is now readily available to employ theoretical methods to elucidate the effects of mutations on proteins at the atomic level and to aid experimental practitioners of protein engineering in determining potential sites for mutation.

References

1. Kline, A. D., Braun, W., and Wuthrich, K. (1988). *J. Mol. Biol.,* **204,** 675.
2. Pflugrath, J., Wiegand, E., Huber, R., and Vertesy, L. (1986). *J. Mol. Biol.,* **189,** 383.

Valerie Daggett and Peter A. Kollman

3. Eigenbrot, C., Randal, M., and Kossikoff, A. A. (1990). *Prot. Engng,* **3,** 591.
4. Fermi, G. and Perutz, M. (1981). In *Atlas of molecular structures in biology,* (ed. D. C. Phillips and F. M. Richards). Clarendon, Oxford.
5. Matthews, B. W. (1987). *Biochemistry,* **26,** 6885.
6. Weiner, S. J., *et al.* (1984). *J. Am. Chem. Soc.,* **106,** 765.
7. Allinger, N. L. (1976). *Adv. Phys. Org. Chem.,* **13,** 1.
8. Metropolis, N., Rosenbluth, A., Rosenbluth, M., Teller, A., and Teller, E. (1953). *J. Chem. Phys.,* **21,** 1087.
9. Kirkpatrick, S., Gelatt, C. D., Jr, and Vecchi, M. P. (1983). *Science,* **220,** 671.
10. Warme, P. K., Momany, F. A., Rumball, S. V., Tuttle, R. W., and Scheraga, H. A. (1974). *Biochemistry,* **13,** 768.
11. Alagona, G., Desmeules, P., Ghio, C., and Kollman, P. A. (1984). *J. Am. Chem. Soc.,* **106,** 3623.
12. Nickbarg, E. B., Davenport, R. C., Petsko, G. A., and Knowles, J. R. (1988). *Biochemistry,* **27,** 5948.
13. Daggett, V. and Kollman, P. A. (1990). *Prot. Engng,* **3,** 677.
14. Levitt, M. (1983). *J. Mol. Biol.,* **170,** 723.
15. Shih, H. H.-L., Brady, J., and Karplus, M. (1985). *Proc. Natl. Acad. Sci.* **82,** 1697.
16. Alber, T. C., Davenport, R. C., Jr, Giammona, D. A., Lolis, E., Petsko, G. A., and Ringe, D. (1987). *Cold Spring Harbor Symp. Quant. Biol.,* **LII,** 603.
17. Warshel, A. and Levitt, M. (1976). *J. Mol. Biol.,* **103,** 227.
18. McCammon, J. A., Gelin, B. R., Karplus, M., and Wolynes, P. G. (1976). *Nature,* **262,** 325.
19. Eisenberg, D. and McLachlan, A. D. (1986). *Nature,* **319,** 199.
20. Schiffer, C. A., Caldwell, J. W., Kollman, P. A., and Stroud, R. M. (1990). *Proteins Struct. Func. Genet.,* **8,** 30.
21. Daggett, V., Kollman, P. A., and Kuntz, I. D. (1991). *Biopolymers,* **31,** 285.
22. Blacklow, S. C. and Knowles, J. R. (1990). *Biochemistry,* **29,** 4099.
23. Daggett, V., Kollman, P. A., and Kuntz, I. D. (1989). *Chemica Scripta,* **29A,** 205.
24. Dang, L. X., Merz, K. M., Jr, and Kollman, P. A. (1989). *J. Am. Chem. Soc.,* **111,** 8505.
25. Rao, S. N., Singh, U. C., Bash, P. A., and Kollman, P. A. (1987). *Nature,* **328,** 551.
26. Hwang, J.-K. and Warshel, A. (1987). *Biochemistry,* **26,** 2669.
27. Daggett, V., Brown, F., and Kollman, P. A. (1989). *J. Am. Chem. Soc.,* **111,** 8247.
28. Pearlman, D. A. and Kollman, P. A. (1989). *J. Chem. Phys.,* **90,** 2460.
29. Moult, J. and James, M. N. G. (1986). *Proteins Struc. Func. Genet.,* **1,** 146.
30. Snow, M. E. and Amzel, L. M. (1986). *Proteins Struct. Func. Genet.,* **1,** 267.
31. Lee, C. and Subbiah, S. (1991). *J. Mol. Biol.,* **217,** 373.
32. Ponder, J. W. and Richards, F. M. (1987). *J. Mol. Biol.,* **193,** 775.
33. Sutcliffe, M. J., Hayes, F. R. F., and Blundell, T. L. (1987). *Prot. Engng,* **1,** 385.
34. Sippl, M. J. (1990). *J. Mol. Biol.,* **213,** 859.
35. Levitt, M. (1992). *J. Mol. Biol.,* in press.

PART II

Protein stability and physico-chemical analysis of mutants

7

Principles of protein stability. Part 1—reversible unfolding of proteins: kinetic and thermodynamic analysis

BRYAN E. FINN, XIAOWU CHEN, PATRICIA A. JENNINGS, SUSANNE M. SAALAU-BETHELL, and C. ROBERT MATTHEWS

1. Introduction

Over 30 years ago Anfinsen and his colleagues demonstrated that the amino acid sequence is the primary determinant of the three-dimensional structure of a folded protein (1). This seminal observation stimulated numerous efforts to define the rules that govern the folding reaction (2–6). Unfortunately, little progress on solving the folding code has been made until recently. The main obstacle has been the high co-operativity of the unfolding transition. Only the native and unfolded forms are highly populated under equilibrium conditions; stable, partially folded forms are generally not detected. Transient intermediates, when they do appear, typically have lifetimes in the millisecond range, making it difficult to characterize their structures.

The importance of deciphering the folding code lies in the potential applications:

(a) Prediction of the three-dimensional structure of a protein from its amino acid sequence. The accumulation of DNA and protein sequence information far outpaces the determination of protein structures by X-ray crystallography and NMR spectroscopy. With the advent of the human genome sequencing project, this discrepancy will rapidly increase.

(b) Alteration of the amino acid sequence of naturally occurring proteins to increase stability, enzymic activity, or to alter molecular recognition properties.

(c) Design of new proteins which catalyse reactions not detected in nature.

(d) Optimization of the recovery of recombinant proteins from inclusion bodies by denaturation and renaturation under conditions favouring productive folding pathways.

A major part of the effort to solve the folding problem has involved *in vitro* studies on purified proteins. The *in vitro* approach attempts to reduce this complex problem to its simplest elements and has the potential to provide general information which is useful in understanding the folding of any polypeptide. Although the *in vivo* folding reaction may differ in some respects, we presume that the same basic biophysical principles pertain in both cases. We base this presumption on the high recovery of the native, active conformation from the unfolded form for *in vitro* studies on many proteins and on the time-scale observed, 10^{-1}–10^2 sec, which is biologically feasible.

In this chapter we describe the use of *in vitro* kinetic and equilibrium folding experiments to probe folding mechanisms in systems where the reaction is reversible. Using the data obtained from these experiments, measurements of the stability of the protein and the various rate constants in the active process of folding can be made. This information can be used to assemble a working model for the folding reaction and to determine the relative energies of various species along the pathway. We then discuss how mutagenesis can be used to probe the role of individual side-chains in the process. The case of irreversible unfolding of proteins is discussed in Chapter 8.

2. Design of the folding experiment

There are a number of aspects to consider before one begins a study of a protein folding mechanism:

(a) the suitability of the target protein for folding studies

(b) the method of denaturation and the solvent conditions

(c) the techniques employed to monitor folding

2.1 Selection of the protein

If one has some latitude in selecting the system to be studied, there are several desirable features:

(a) Availability of pure protein. 10–50 mg is usually required for a minimal study, and gram quantities for a complete analysis.

(b) Reversible folding reaction. This property permits one to use thermodynamics as a means of measuring the free-energy differences between stable species in the folding reaction.

(c) Absence of post-translational modification. Covalent modifications may alter the folding information present in the original polypeptide.

(d) Availability of high-resolution structural information. This is especially important for the design and interpretation of mutagenesis experiments.

2.2 Selection of the denaturation method

The non-covalent interactions which stabilize the folded forms of proteins, e.g. van der Waals interactions, hydrogen bonds, and salt bridges, can be disrupted by a variety of methods.

(a) Chemical denaturation. Chaotropic reagents, such as urea or guanidine hydrochloride, are thought to act by disrupting the intra-protein hydrogen bonds and by increasing the solubility of hydrophobic side-chains, which are buried in the native conformation.

(b) Acid or base denaturation. Titrating the acidic or basic side-chains can disrupt salt bridges and significantly increase the net charge of the protein. Either change can destabilize the native conformation and shift the equilibrium to favour the unfolded form.

(c) Thermal denaturation. The endothermic nature of the high-temperature unfolding reaction permits one to unfold the protein by increasing the temperature. In some cases low-temperature unfolding cold denaturation has also been observed (7, 8).

The choice of the denaturation method is often coupled with the properties of the protein and the method of observation (Section 2.3). Chemical denaturation has the advantages of high reversibility, low propensity for chemical modification, and effectiveness at the disruption of secondary and tertiary structure. Thermal unfolding can provide direct measurements of the enthalpy and heat capacity changes, but proteins often suffer from chemical damage during thermal unfolding at and above neutral pH (9). Also, residual structure has been observed in the high-temperature baseline region (10–12). Alkaline pH enhances deamidation reactions and acid pH is not sufficient to unfold completely a significant number of proteins. Solubility problems near the isoelectric point of a protein also complicate the use of pH denaturation. The experiments described in this contribution focus exclusively on chemical denaturation.

2.3 Selection of the observation method

The protein folding reaction can be monitored by techniques that are sensitive to spectroscopic, functional, hydrodynamic, or thermodynamic properties of a protein.

Spectroscopic techniques, such as circular dichroism, absorbance, or fluorescence, are used to monitor changes in secondary and tertiary structure during folding. These methods are described in detail elsewhere in this series (13). Functional properties, such as the recovery of enzymic activity or the binding of inhibitors, can also be used to follow folding, with the focus being on the formation of the active site. Immunoreactivity has also been used to monitor formation of native-like epitopes during folding (14).

169

Hydrodynamic properties of the protein can be exploited with size exclusion chromatography (15), urea gradient polyacrylamide gel electrophoresis (16), and viscosity measurements (2). Calorimetry provides a direct measurement of the enthalpy and heat capacity changes for the thermal unfolding process. The entropy, transition temperature, and free-energy changes can be calculated from these parameters and together provide a detailed thermodynamic analysis of the reaction (8, 17).

Hydrogen-exchange methods have been used in combination with partial proteolytic degradation and rapid separation to identify folding intermediates (18, 19). Coupled with NMR spectroscopy, pulse-labelling hydrogen-exchange techniques can now be used to follow the acquisition of specific hydrogen bonds in the millisecond time-range (20–22). This very important advance allows one to monitor the development of secondary structures during folding.

In this report, discussion will be confined to optical spectroscopic methods because the equipment is commercially available, the sensitivity is sufficient to generate real-time data on kinetic processes, and the methods are applicable to nearly all proteins.

3. Collection and analysis of data

3.1 Equilibrium studies

Equilibrium experiments measure the free-energy difference between the stable species in a folding reaction. For most proteins, this means between the native and unfolded forms. In some instances, such studies have provided evidence for the presence of stable folding intermediates (23, 24, B. E. Finn and C. R. Matthews, in preparation).

3.1.1 Design of the equilibrium experiment

Pace and colleagues have discussed how to collect and analyse unfolding data from fluorescence and circular dichroism spectroscopies (25). A detailed description of the difference absorbance spectroscopy technique and its application to unfolding reactions at equilibrium has been presented by Schmid (13). Two points are worth emphasizing:

(a) The unimolecular nature of the folding reaction expected for monomeric proteins should be tested by comparing the transition curves at two different protein concentrations. At the protein concentrations required for spectroscopic measurements (10^{-6}–10^{-4} M), some proteins have been observed to associate or aggregate in the transition region (26). This problem is obviously exacerbated at the concentrations required for NMR spectroscopy (10^{-3} M). Coincidence of the curves measured at different protein concentrations ensures that the reaction is indeed unimolecular under the conditions employed.

(b) The unfolding transition should be monitored by more than a single technique and the results compared on a normalized plot of F_{app}, the apparent fraction of unfolded protein, versus denaturant concentration. F_{app} is calculated from the data by:

$$F_{app} = \frac{Y_O - Y_N}{Y_U - Y_N} \tag{1}$$

where Y_O is the observed value of the spectroscopic parameter at a given denaturant concentration and Y_N and Y_U are the values of the native and unfolded baselines at the same denaturant concentration. The values of Y_N and Y_U in the transition region are obtained by linear extrapolation of the native and unfolded baseline regions, respectively. An example is shown in *Figure 1*. If the reaction follows a two-state model, the curves determined by different techniques will be coincident within experimental error. If the curves are non-coincident, additional species must be present at equilibrium. This comparison test appears to be more sensitive than simply evaluating the quality of the fit of an individual data-set to a two-state model (27, 28).

3.1.2 Fitting equilibrium data

If the unfolding transition follows the behaviour expected for a two-state model,

$$N \underset{k_{-1}}{\overset{k_1}{\rightleftharpoons}} U \qquad \text{(Model I)}$$

i.e. a smooth, sigmoidal change in the observed property, the free-energy difference between the native and unfolded forms can be calculated. The apparent fraction of unfolded protein, F_{app}, is related to the equilibrium constant, $K_{NU} = [U]/[N]$, by:

$$F_{app} = \frac{K_{NU}}{1 + K_{NU}} \tag{2}$$

Since $K_{NU} = \exp(-\Delta G_{NU}/RT)$,

$$F_{app} = \frac{\exp(-\Delta G_{NU}/RT)}{1 + \exp(-\Delta G_{NU}/RT)} \tag{3}$$

Assuming that the stability depends linearly on the denaturant concentration, [D], i.e., $\Delta G_{NU} = \Delta G_{NU}^{H_2O} + A[D]$ (29),

$$F_{app} = \frac{\exp\{-(\Delta G_{NU}^{H_2O} + A[D])/RT\}}{1 + \exp\{-(\Delta G_{NU}^{H_2O} + A[D])/RT\}} \tag{4}$$

Non-linear least squares fitting programs can be used to obtain the free-energy difference in the absence of denaturant, $\Delta G_{NU}^{H_2O}$, and the parameter, A, which reflects the co-operativity of the reaction.

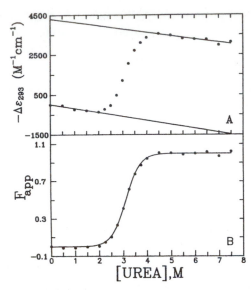

Figure 1. (A) The urea dependence of the molar extinction coefficient at 293 nm of wild-type dihydrofolate reductase at pH 7.8, 15°C. The solid lines represent linear least squares fits of the native and unfolded baselines to Equations 6 and 7. (B) The urea dependence of F_{app} calculated from the data in panel (A). The solid line is the predicted curve for a fit of the data to a two-state model.

If the equilibrium unfolding transition shows an inflection, the data may fit better to a three-state model involving a stable intermediate, I:

$$N \underset{k_{-1}}{\overset{k_1}{\rightleftharpoons}} I \underset{k_{-2}}{\overset{k_2}{\rightleftharpoons}} U \qquad \text{(Model II)}$$

where $K_{NI} = [I]/[N]$ and $K_{IU} = [U]/[I]$. In this case:

$$F_{app} = K_{NI}(Z_I + K_{IU})/\{1 + K_{NI}(1 + K_{IU})\} \qquad (5)$$

where $Z_I = (Y_I - Y_N)/(Y_U - Y_N)$. This latter parameter normalizes optical properties of the intermediate, Y_I, to those of the native, Y_N, and unfolded, Y_U, forms. By assuming a linear dependence of ΔG_{NI} and ΔG_{IU} on the denaturant concentration, the F_{app} curve can be fitted with a non-linear least squares program. Values for the free-energy difference in the absence of denaturant between N and I, $\Delta G_{NI}^{H_2O}$, and between I and U, $\Delta G_{IU}^{H_2O}$, can be obtained.

Protocol 1. Fitting equilibrium unfolding data

1. Plot the dependence of the observed property, Y_O, on the denaturant concentration (*Figure 1A*).

2. Fit the native and unfolded baseline regions to linear equations:

$$Y_N = Y_N^{H_2O} + m_N[D] \tag{6}$$

$$Y_U = Y_N^{H_2O} + m_U[D] \tag{7}$$

where Y^{H_2O} is the value in the absence of denaturant, and m reflects the denaturant dependence of the optical property for the native and unfolded forms.

3. Convert the plot of the raw data to a plot of F_{app} versus the denaturant concentration (*Figure 1B*) using Equation 1.

4. Fit the F_{app} curve to the equation appropriate for the model being tested and extract the free-energy and co-operativity parameters, ΔG^{H_2O} and A.

5. Generate a predicted F_{app} curve from these parameters and compare it to the actual data. The simplest model that fits the data within experimental error is selected unless other evidence dictates a more complex model.

Alternative methods of fitting the equilibrium data to extract the stability have been described by Pace *et al.* (25) and Santoro and Bolen (30). Both of these groups, and most others, assume a linear dependence of the free energy of folding on the denaturant concentration. Pace prefers to fit ΔG as a function of the denaturant concentration in the transition region, rather than F_{app} over the entire range as we do. Santoro and Bolen have suggested fitting the raw data to a single equation which incorporates the denaturant dependence of the native and unfolded baseline regions in addition to the transition region. The values of ΔG from the three methods generally agree within experimental error; however, the estimates of the error by Santoro and Bolen are significantly larger.

3.2 Kinetic studies

Kinetic experiments can provide detailed information about the folding pathway of a protein, including the presence of transient folding intermediates. When combined with equilibrium results, they can be used to construct a working model for the folding reaction.

3.2.1 Design of the kinetic experiment

For absorbance spectroscopy, the choice of the detection wavelength is determined by extrema in the difference spectrum. The optimum wavelength is usually at or near 287 nm for proteins which contain tyrosine but not tryptophan. When tryptophan is present, a useful extremum occurs at approximately 292 nm. Reactions with relaxation times greater than 10 sec can be studied with manual mixing techniques and a standard spectrometer. The accurate measurement of folding reactions that exceed 500 sec requires a very stable instrument. From our experience, a double-beam absorbance spectrometer with an instrument drift of less than 0.001 a.u./h is useful in this regard. Relaxation times of less than 10 sec require stopped-flow techniques.

Protocol 2. Procedures for manual mixing kinetic measurements

1. Using a 10× concentrated buffer stock, prepare stock solutions of 1× buffer and 9 M urea in 1× buffer. Filter the solutions through a 0.22 μm membrane to remove particulate matter and degas the solutions by aspiration to remove dissolved oxygen. If urea is used as the denaturant, solutions should be prepared fresh daily to minimize the decomposition into ammonium and cyanate. Protocols for the preparation of such solutions are given by Pace *et al.* (25).

2. Prepare a protein stock solution by extensive dialysis against 1× buffer. If a protein concentration of greater than 10 mg/ml is necessary, a commercially available concentrator can be used. Following a concentration step, it is often necessary to filter the sample to remove any undissolved materials. The final protein concentration required for an adequate signal depends upon the number of tyrosine and tryptophan residues that are exposed to solvent upon refolding. For absorbance spectroscopy, a typical concentration range is 10^{-5}–10^{-4} M. Fluorescence spectroscopy is intrinsically more sensitive, allowing one to work at 10^{-6}–10^{-5} M.

3. Incubate all solutions at the desired temperature until equilibrium is reached. For refolding experiments, several hours' incubation at room temperature is usually sufficient to unfold the protein completely; however, one should actually check to be certain that this is adequate.

4. Prepare a reference solution according to *Table 1* and seal with a stopper to minimize evaporation. Determine the protein concentration in the experiment by taking a wavelength scan of this reference solution against buffer.

5. Add the appropriate amounts of urea and buffer to the test solution cuvette and mix. Typical volumes are given in *Table 1*. Place the test and reference cuvettes in the spectrometer and allow them to re-equilibrate to the desired temperature.

6. Add the predetermined amount of protein stock to the test cell with a microsyringe and simultaneously initiate data collection. Mix by importing and exporting the solution several times with a clean glass pipette. Be careful to avoid the introduction of air bubbles during mixing. Alternatively, mix with a small Teflon stirring paddle or a magnetic stirring bar if the spectrometer is so equipped. Seal the cuvette with a stopper. The dead time for this procedure can be held to 10 sec.

7. Collect data until no further change in signal is observed (see *Figure 2a*). It has been shown that the absence of even the last few per cent of the trace can lead to ambiguities in fitting (31).

8. Analyse the data as discussed in Section 3.2.2.

Table 1. Kinetic sample recipes[a]

A. For unfolding starting from native protein

	Sample cell			Reference cell	
[Urea][b]	Protein	Buffer	Urea[c]	Buffer	Protein
4.00	100	400	400	800	100
5.00	100	300	500		
6.00	100	200	600		

B. For refolding starting from 6 M urea

	Sample cell			Reference cell	
[Urea][b]	Protein in urea	Buffer	Urea	Buffer	Protein[d]
1.00	100	766.7	33.3	866	33.3
1.50	100	716.7	83.3		
2.00	100	666.7	133.3		

[a] Total volume is 900 µl for all samples, all volume units are in µl.
[b] Final urea concentration in M.
[c] 9 M urea stock.
[d] 3× concentrated stock used to make unfolded protein stock.

Fluorescence and circular dichroism measurements are made in a similar fashion, with the exception that there is no reference cuvette. Fluorescence studies must be done in the protein concentration range where the signal is directly proportional to concentration. Inner-filter effects can lead to an undesirable reduction in signal intensity (13). For circular dichroism measurements, high absorbance by the sample can reduce the signal-to-noise ratio. This problem can be avoided by using short pathlength (1 or 2 mm) sample cells. The total absorbance of protein, buffer, and denaturant should be less than 1 a.u. at 220 nm.

For reactions faster than 10 sec, stopped-flow techniques must be used to monitor folding reactions. The following protocol was developed using a Durrum 110 absorbance/fluorescence stopped-flow spectrometer interfaced to a microcomputer for data collection.

Protocol 3. Procedure for stopped-flow kinetic experiments

1. Prepare the same solutions as for manual mixing experiments.
2. The Durrum–Gibson stopped-flow instrument has a pair of over and under horizontal drive syringes. To minimize convection artefacts following mixing, place the solution with the greater density (higher denaturant concentration) in the lower syringe. Several of the available instruments have vertical drive syringes which also have convection problems when mixing solutions of different densities. Check with the manufacturer for hardware modifications required to eliminate this artefact.
3. Flush the syringes and the flow cell thoroughly with degassed ultrapure water to clean them and remove any air trapped in the system.
4. Separate water baths should be used to control the temperature of the drive syringes and the temperature of the observation cell. The observation cell should be set at the experimental temperature while that for the drive syringes should be set slightly higher ($\sim$1°C). The differential temperature settings are required to offset the decrease in temperature which accompanies the dilution of concentration solutions of urea and, especially, guanidine hydrochloride. To minimize the Schlieren effects, which can arise in this situation, the temperature of the drive syringes should be increased until the artefact disappears. Both the density and temperature problems can be examined and resolved with the aqueous solutions which do not contain protein.
5. Load the test solutions into the syringes and allow them to equilibrate to the required temperature. The choice of the relative mixing volumes, i.e. the diameters of the two drive syringes, is determined by the desired changes in denaturant concentration. The absolute volumes must be sufficient to allow complete mixing.
6. Open the entry valves to the flow cell and start the run by pressing the actuation button. Close the entry valves to the cell as soon as possible to minimize convection artefacts. Data collection is initiated by triggering the microcomputer from a switch attached to the stop syringe.
7. Open the drain valve and flush the observation cell after the data collection is complete.
8. Repeat the above procedure several times at each final denaturant concentration to minimize errors. The first few runs at each denaturant concentration are usually discarded because they reflect wash-out artefacts. Only after consistent traces are observed are the data retained for analysis.
9. The dead-time of the instrument can be determined by published procedures (32). This value should be obtained under conditions similar to those employed in the experiments with respect to temperature and solvent viscosity.

Aggregation is a potential problem for all kinetic experiments, especially for refolding jumps ending in or below the transition region. When the unfolded protein is rapidly introduced into solutions containing a low concentration of denaturant, marginally soluble folding intermediates can become highly populated and precipitate. This problem may be alleviated by lowering the protein concentration, altering solvent conditions such as pH or ionic strength, or maintaining sufficient denaturant in the final solution.

3.2.2 Fitting kinetic data

i. *Extraction of relaxation times and amplitudes from the data*
For unimolecular folding reactions the concentration of each species at any time $c_i(t)$, can be expressed as:

$$c_i(t) = \sum_j C_{ij} \exp(-t/\tau_j) \tag{8}$$

where the coefficients, C_{ij}, are determined by the microscopic rate constants at the final conditions and by the initial concentrations. The relaxation times, τ_j, are determined only by the rate constants under the final conditions (2, 32). Any property that is linear in protein concentration, e.g. absorbance, fluorescence, or circular dichroism, can be expressed as:

$$Y(t) = \sum_i Y^\circ{}_i c_i(t) = Y(\infty) + \sum_j Y_j \exp(-t/\tau_j) \tag{9}$$

$Y^\circ{}_i$ is molar optical property of species C_i, $Y(\infty)$ is the value of the optical property at infinite time, and Y_j is the amplitude associated with kinetic phase τ_j. Once again, the amplitudes are determined by both the final rate constants and initial concentrations, while the relaxation times are determined by the final rate constants. Detailed discussions on the analysis of protein folding reaction kinetics are available (2, 3, 33–36).

Relaxation times and their associated amplitudes for a folding reaction can be extracted from experimental data by fitting the data to the minimum number of exponentials required using Equation 9. Three criteria are useful in this regard:

(a) The distribution of the differences between the observed and predicted kinetic traces, the residual error, should be random (*Figure 2*).

(b) The error in the fitting parameters decreases as one approaches the optimum set of exponential terms. The error begins to increase as one adds additional, unnecessary parameters.

(c) The rate constants should be independent of the detection wavelength.

If the relaxation times vary with wavelength, the number of exponentials should be increased.

ii. *Evaluating the relaxation times*
A systematic study of the dependences of the relaxation times and amplitudes of the various phases detected in folding upon the denaturant concentration

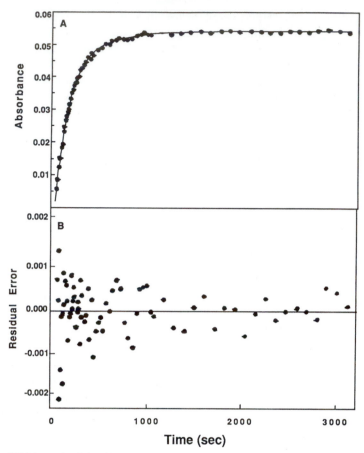

Figure 2. (A) Manual mixing kinetic trace of the change in absorbance accompanying the unfolding of wild-type dihydrofolate reductase from 0 to 4 M urea at pH 7.8, 15°C. The solid symbols are a representative subset of the entire data-set and the solid line represents the fit to two exponentials with relaxation times of 127 and 399 sec. (B) The residual error of the fit of the data to two exponentials calculated from $R = Y_{obs} - Y_{pred}$, where the latter two terms correspond to the observed and predicted absorbance at any time.

can provide valuable information for the construction of a folding model. Several detailed descriptions of the development of kinetic folding models are available (32, 34, 35).

In this discussion, we choose to focus on a common feature of a number of folding mechanisms, namely, that the unfolding reaction is usually controlled by a single exponential whose relaxation time decreases logarithmically with increasing denaturant concentration. Refolding is almost always more complex, consisting of a series of exponential phases. The relaxation times of some of these phases decrease logarithmically with decreasing denaturant

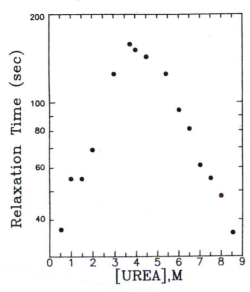

Figure 3. A semi-log plot of the urea dependence of the relaxation time for a typical reversible folding reaction. The source of the data is the major N ↔ I reaction for wild-type dihydrofolate reductase at pH 7.8, 15°C.

concentration, while others may be independent of denaturant. The former have behaviour consistent with reactions limited by protein folding while the latter may be limited by proline isomerization which occurs in the unfolded protein (37). Inspection of a semi-log plot of the relaxation times versus the denaturant concentration for a number of proteins shows that the single phase in unfolding connects smoothly with one of the slower phases in refolding. The data has the appearance of an inverted 'V' or chevron (*Figure 3*). A description of this behaviour and its interpretation has been provided elsewhere (38).

The correlation of the observed relaxation times with the microscopic rate constants depends upon the particular folding model. For a two-state reaction, as shown above in Model I, $\tau^{-1} = k_1 + k_{-1}$. At high denaturant concentrations, $k_1 >> k_{-1}$ and $\tau^{-1} \simeq k_1$; at low denaturant concentrations, $k_1 >> k_{-1}$ and $\tau^{-1} \simeq k_{-1}$. In the transition zone, τ depends on both rate constants. The urea dependence of the relaxation time reflects the urea dependence of the unfolding and refolding rate constants. These same rate constants dictate the urea dependence of the unfolding equilibrium constant, K_{NU}, because $K_{NU} = k_1/k_{-1}$.

To illustrate further, consider a three-state model, similar to Model II, but which involves a transient rather than a stable intermediate. Let us further stipulate that the folding of this transient intermediate to the native conformation is rate limiting, and therefore $k_2, k_{-2} >> k_1, k_{-1}$. This mechanism has two relaxation times, which depend upon the microscopic rate constants as follows:

$$\tau^{-1}{}_{\text{fast}} = k_2 + k_{-2} \tag{10}$$

$$\tau^{-1}{}_{\text{slow}} = k_1 + k_{-1}\{k_{-2}/(k_2 + k_{-2})\} \tag{11}$$

or

$$\tau^{-1}{}_{\text{slow}} = k_1 + k_{-1}/(K_{\text{IU}} + 1) \tag{12}$$

where $K_{\text{IU}} = [\text{U}]/[\text{I}] = k_2/k_{-2}$. The microscopic rate constants for the slow step can, once again, be obtained by measurements of the rate-limiting relaxation time in the native and unfolded baseline regions. Note that under strongly folding conditions, where $k_{-2} \gg k_2$ and $k_{-1} \gg k_1$, the term $k_{-1}/(K_{\text{IU}} + 1)$ reduces to k_{-1}, so that $\tau^{-1}{}_{\text{slow}} \simeq k_{-1}$. Under strongly unfolding conditions, where $k_2 \gg k_{-2}$ and $k_1 \gg k_{-1}$, the term $k_{-1}/(K_{\text{IU}} + 1)$ is much less than k_1, so that $\tau^{-1}{}_{\text{slow}} \simeq k_1$. The k_{-2} rate constant may be determined from refolding experiments; under strongly folding conditions, $k_{-2} \gg k_2$ so that $\tau^{-1}{}_{\text{fast}} \simeq k_{-2}$.

iii. Reaction coordinate diagrams

The data from both equilibrium and kinetic experiments can be displayed in terms of a reaction coordinate diagram. The thermodynamic data permits one to plot the relative free energies of stable species, e.g. the native and unfolded forms in a two-state model. The kinetic data determines the relative free energies of transient intermediates and transition states. The equations required to convert the equilibrium and rate constants to free energies are:

$$\Delta G_{\text{NU}} = -RT \ln K_{\text{NU}} \tag{13}$$

$$\Delta G_1^{\ddagger} = -RT \ln (k_1 h/k_{\text{B}}T) \tag{14}$$

$$\Delta G_{-1}^{\ddagger} = -RT \ln (k_{-1} h/k_{\text{B}}T) \tag{15}$$

$$\Delta G_2^{\ddagger} = -RT \ln '(k_2 h/k_{\text{B}}T) \tag{16}$$

$$\Delta G_{-2}^{\ddagger} = -RT \ln (k_{-2} h/k_{\text{B}}T) \tag{17}$$

where h, k_{B}, and R are the Planck, Boltzmann, and gas constants and T is the absolute temperature.

The linear dependence of $\log \tau$ on denaturant concentration in the native and unfolded baseline regions (*Figure 3*) can be explained in terms of a linear dependence of the activation free energies $\Delta G_1^{\ddagger}$, and $\Delta G_{-1}^{\ddagger}$ on the denaturant concentration (38). Schellman has provided a justification for this behaviour for ΔG_{NU} (29) and it appears that a similar explanation can be applied to the activation free energies (39). Thus, the values of these parameters in the absence of denaturant can be obtained from the following equations:

$$\Delta G_{\text{NU}} = \Delta G_{\text{NU}}^{\text{H}_2\text{O}} + A_{\text{NU}}[\text{D}] \tag{18}$$

$$\Delta G_1^{\ddagger} = \Delta G_1^{\ddagger H_2O} + A_1[D] \tag{19}$$

$$\Delta G_{-1}^{\ddagger} = \Delta G_{-1}^{\ddagger}{}^{H_2O} + A_{-1}[D] \tag{20}$$

$$\Delta G_2^{\ddagger} = \Delta G_2^{\ddagger H_2O} + A_2[D] \tag{21}$$

$$\Delta G_{-2}^{\ddagger} = \Delta G_{-2}^{\ddagger}{}^{H_2O} + A_{-1}[D] \tag{22}$$

The relative magnitudes of these free energies are most easily displayed in a reaction coordinate diagram (*Figure 4*). If the energy of the unfolded form is arbitrarily set to zero, the relative energies of the other species in the folding reaction can be calculated from the thermodynamic and kinetic data as follows:

$$\Delta G_N = -\Delta G_{NU} \tag{23}$$

$$\Delta G_{TSI} = \Delta G_N + \Delta G_1^{\ddagger} \tag{24}$$

$$\Delta G_I = \Delta G_{TSI} - \Delta G_{-1}^{\ddagger} \tag{25}$$

$$\Delta G_{TS2} = \Delta G_{-2}^{\ddagger} \tag{26}$$

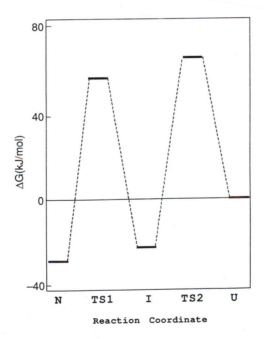

Figure 4. A reaction coordinate diagram for the major folding channel of dihydrofolate reductase in the absence of denaturant at pH 7.8, 15°C. The folding follows the three-state folding model involving the native (N), intermediate (I), and unfolded forms (U), linked by the respective transition states (TS1 and TS2). Although TS2 has a higher energy than TS1, the rate-limiting step in folding is the conversion of the intermediate to the native conformation via TS1. In mathematical terms, $\Delta G_{-1}^{\ddagger} > \Delta G_{-2}^{\ddagger}$.

The effort required to construct such a diagram for a protein-folding reaction is rewarded by the information available on the role of a particular amino acid side-chain in the process. By comparing the diagrams for wild-type and mutant proteins, one can determine the species whose energies are altered by the mutation. Because the perturbation of the energy implies an involvement with structure, this approach provides a means of pin-pointing the location at which the side-chain becomes involved in the folding reaction.

4. Probing the folding mechanism by mutagenesis

A kinetic folding model highlights the species that play important roles when a protein is displaced from equilibrium. In the past this level of characterization has been difficult to extend to a detailed structural description of the proposed species, information which is critical to the development of generalized rules for folding reactions. New developments in NMR spectroscopy now permit one to monitor the formation of stable hydrogen bonds, i.e. secondary structure, in the millisecond time range (20–22). Site-directed mutagenesis can provide insight into the role of specific side-chains in folding, i.e. the development of tertiary structure. The latter technique also has the potential to elucidate the structures of transition states in folding reactions (40).

4.1 Methods and strategies of mutagenesis

A number of methods for introducing mutations into proteins have been developed (41, 42). There are two basic strategies: random and site-directed mutagenesis.

4.1.1 Random mutagenesis

Given the availability of an appropriate selection or screen, random mutagenesis provides a way to determine the key factors in folding which is unfettered by any preconceived notions. King and co-workers (43) have developed a selection to detect mutants in the phage P22 tailspike protein which induce misfolding *in vivo* at elevated temperatures. At lower temperatures, the mutant proteins fold properly and have a stability similar to that of the wild-type protein. These temperature-sensitive folding mutations are interpreted as identifying residues that play essential roles in directing early events in folding. Sauer and colleagues have used a semi-random multiple replacement strategy to identify amino acids that play important roles in stabilizing λ repressor and other DNA-binding proteins (44–46). Multiple replacements in the hydrophobic core of λ repressor test the adaptability of the interior of these proteins to changes in volume (44). Extracellular assays have been developed to screen for temperature-stable mutants in T4 lysozyme (47) and temperature-sensitive mutants in bovine pancreatic trypsin inhibitor (48). In these latter two cases, the protein was released from its intracellular environment and its function probed by a plate assay. This approach allows one to monitor functional properties of proteins which are not essential to the host bacteria. Presumably, a subset of the mutants with altered stabilities will also perturb the folding reaction in interesting ways.

4.1.2 Site-directed mutagenesis

The maximum utility of site-directed mutagenesis can be realized if a high-resolution structure is available from X-ray crystallography or NMR spectroscopy. This information is important for the rational design of experiments and for the interpretation of the results. Our approach in constructing mutations has centred on a systematic analysis of the effects of replacements in elements of secondary structure. This choice was influenced by the framework model of folding (4) which proposes that elements of secondary structure arise early in folding and guide the progressive formation of the native form. Given the significant role for the hydrophobic interaction in stabilizing proteins (49, 50), one could equally well target buried hydrophobic side-chains for mutagenesis (44, 51).

4.2 Interpretation of the effects of mutations on folding and stability

Comparison of the reaction coordinate diagrams for wild-type and mutant proteins raises the issue of a proper reference state. Equilibrium and kinetic studies of the folding reaction can only measure the differences in free energy between a pair of states, not the absolute free energy of either. Thus, one cannot determine with complete certainty which of the two states (or perhaps both) is altered by the mutation. A reasonable convention is to use the unfolded form as the reference state for that particular protein; the energies of other species in the folding reaction are then related to that value. This does not assume that the energy of the unfolded form is unaffected outside of the obvious contribution from the change in primary structure. Rather, this convention simply examines whether the mutation affects other species in the reaction differently than the unfolded form. Assuming that the unfolded form is devoid of secondary and tertiary structure, the perturbations in the energies of intermediates and transition states (relative to the unfolded form) reflect the involvement of the side-chain in the development of higher-order structure.

Arguments have been advanced which focus attention on the effects of mutations on the structure and energy of the unfolded form (52). We note that it seems more reasonable to assume that the greater perturbation in energy will be on the native form. Its secondary and tertiary structure provide an environment which is expected to be sensitive to the features of the side-chain at a given position. This view is supported by experiments on T4 lysozyme which show that all of 25 randomly generated temperature-sensitive mutants at 20 sites occur at positions of low solvent accessibility and low mobility (53).

4.3 Application of the mutagenic analysis to the folding and stability of dihydrofolate reductase (DHFR)

Wild-type DHFR from *Escherichia coli* has been proposed to unfold via two parallel channels designated by their relaxation times as the τ_1 and τ_2 channels (54). These channels reflect the independent unfolding of two stable native conformers. Refolding proceeds from four unfolded forms through four corresponding intermediates to the two native forms or two native-like intermediates. Folding through the channels is proposed to be faster than interconversions between the channels. The latter may reflect proline isomerization reactions (37). Each channel in the model obeys the three-species kinetic model described in Section 3.2.2, with a rapid collapse of the unfolded form to the intermediate (τ_5 phase) followed by the rate-limiting conversion to the native or native-like form (τ_1–τ_4 phases). Only the τ_1 and τ_2 reactions are detected in both unfolding and refolding. The relaxation times for both of these phases follow the urea dependence expected for protein-folding reactions (*Figure 3*).

The replacement of Leu28 with arginine stabilizes the protein by 7.1 ± 1.7 kJ/mol; $\Delta G_{NU}^{H_2O}$ (WT) = 24.7 kJ/mol and $\Delta G_{NU}^{H_2O}$ (L28R) = 31.8 kJ/mol (*Figure 5*) (55). Manual mixing kinetic experiments revealed that the unfolding relaxation times of both the slower, minor phase (τ_1), and the faster, major phase (τ_2) were increased by a factor of three at 6 M urea (*Figure 6*). The τ_1 refolding relaxation times were essentially unchanged while the urea dependence of the τ_2 phase was increased by a factor of two. Even with the increase in slope, the magnitudes of the τ_2 relaxation times in the native baseline region (0–2 M urea) are nearly identical for the wild-type and L28R mutant proteins.

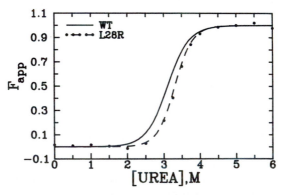

Figure 5. The urea dependence of F_{app} for the wild-type (WT) and L28R mutant of dihydrofolate reductase at pH 7.8, 15°C.

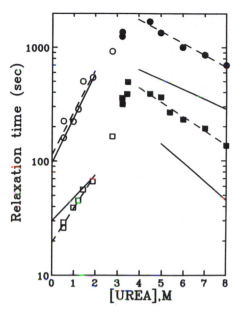

Figure 6. The urea dependence of the $\tau 1$ ($\bigcirc$, $\bullet$) and $\tau 2$ ($\square$, $\blacksquare$) relaxation times for unfolding (closed symbols) and refolding (open symbols) for the L28R mutant DHFR at pH 7.8, 15 °C. The dashed lines indicate the fits of the data in the baseline regions to Equations 19 and 20. The solid lines represent the data for the wild-type protein under the same conditions.

Using the methodology discussed in Section 3, we can calculate the free energies of the various species and construct reaction coordinate diagrams for both proteins. To simplify this discussion, we will only examine the effects on the major, τ_2 folding channel obtained from manual mixing kinetics. The results for the τ_1 channel are similar. Stopped-flow kinetic studies have not been performed on the L28R mutant, precluding an analysis of the effects of this mutation on the transition state linking I and U, TS2.

As can be seen in *Figure 7*, the energies of the native, transition state, and intermediate forms in the τ_2 channel all appear to be stabilized relative to the unfolded form by the L28R replacement. The magnitudes of these effects on each of these species can be calculated from:

$$\Delta\Delta G_N = \Delta G_N \text{ (wild type)} - \Delta G_N \text{ (mutant)} \tag{27}$$

$$\Delta\Delta G_{TS1} = \Delta G_{TS1} \text{ (wild type)} - \Delta G_{TS1} \text{ (mutant)} \tag{28}$$

$$\Delta\Delta G_I = \Delta G_I \text{ (wild type)} - \Delta G_I \text{ (mutant)} \tag{29}$$

The increases in stability caused by the mutation are $\Delta\Delta G_N = -4.9 \pm 1.7$ kJ/mol, $\Delta\Delta G_{TS1} = -5.4 \pm 1.7$ kJ/mol, $\Delta\Delta G_I = -4.2 \pm 1.7$ kJ/mol. The uncertainty is the standard error obtained from a propagation analysis. These results show that, during the folding reaction, the side-chain at position 28 is

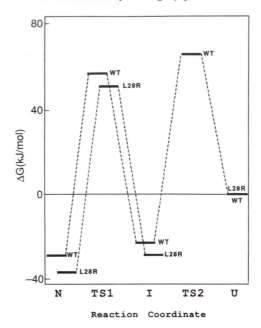

Figure 7. Reaction coordinate diagrams in the absence of urea for the wild-type (WT) and the L28R mutant of DHFR.

involved in structure in the intermediate, rate-limiting transition state, and native conformations. The errors are sufficiently large that differential effects cannot be discerned.

An answer to the question of differential effects can be obtained by focusing solely on the kinetic data. Because extrapolations to zero molar denaturant concentration can once again introduce error comparable to the perturbations on the energies, an alternative procedure is to compare the relaxation times for the wild-type and mutant protein in the native and unfolded baseline regions. Under such conditions, the relaxation times are simply related to either the unfolding or refolding rate constants and can be calculated using Equations 17 and 18.

If this approach is applied to the wild-type and L28R mutant DHFRs, $\Delta\Delta G^{\ddagger}_1 = -2.1 \pm 1.3$ kJ/mol at 6 M urea and $\Delta\Delta G^{\ddagger}_{-1} = 0.8 \pm 0.8$ kJ/mol at 1 M urea. The replacement of Leu28 by arginine lowers the energy of the native conformation by 2.1 ± 1.3 kJ/mol relative to that of the transition state. The relative energies of the intermediate and the transition state are not significantly perturbed. From this analysis, we can conclude that the side-chain at position 28 changes its environment after the transition state for the rate-limiting step in folding. Similar results for other mutations in the sequence from residues 21 to 31 in DHFR suggest that this region acts in a concerted fashion (55; P. A. Jennings, S. M. Saalau-Bethell, J. J. Onuffer,

K. M. Perry, E. E. Howell, and C. R. Matthews, in preparation). By analysing replacements in an element of secondary or tertiary structure, we can test
the possibility that a group of amino acids acts as a unit during folding.

The above procedure is reasonable in cases where the slopes of the log τ
versus denaturant concentration are not changed significantly by the mutation. If such changes do occur, comparisons of the free energy changes should
be made by extrapolation to zero molar denaturant. Given that the slopes
reflect structural properties of the stable states and the intervening transition
states (39), perturbations of the slopes also indicate a significant role for the
side-chain in the folding reaction. However, this type of effect does not lend
itself to interpretation in terms of reaction coordinate diagrams. Application
of the extrapolation method to the τ_2 refolding reaction for L28R, where the
slope is substantially increased relative to that for the wild type (*Figure 6*),
does not change the results. In the absence of denaturant, $\Delta\Delta G^{\ddagger}_{-1}{}^{H_2O} = 0.8$
± 0.4 kJ/mol.

If only a single replacement is examined at a given site, the possibility exists
that the observed effects reflect the property of the particular side-chains and
not the behaviour of the position in the folding reaction. This can be tested by
making multiple replacements and looking for common effects on folding and
stability.

This approach was applied to a study of the role of Val75 in the folding of
DHFR (24). Although the V75A and V75C mutants had thermodynamic and
kinetic properties identical to the wild-type protein, the V75S, V75H, V75Y,
V75I, and V75R mutants all had decreased stabilities. Kinetic studies on the
V75H, V75Y, and V75I mutants all showed increased rates of unfolding and
decreased rates of refolding. It was concluded that position 75 can play a
critical role in the rate-limiting step in folding. If only the alanine or cysteine
mutants had been examined, one would have had quite a different view of the
role of this position in folding.

5. Conclusion

Using a combination of equilibrium and kinetic folding experiments one can
characterize quantitatively the energetics of the significant species in a folding
reaction. By determining the species whose energies are perturbed by amino
acid replacement, one can pin-point the step at which that side-chain becomes
involved in tertiary structure. When this information is combined with that
available from NMR experiments on the formation of secondary structure,
one now has the tools required to obtain detailed structural information on
the intermediates that guide, and the transition states that limit, folding
reactions. The possibility of deciphering the folding code seems more real
than at any time in the past three decades.

References

1. Anfinsen, C. B., Haber, E., Sela, M., and White, F. H. (1961). *Proc. Natl Acad. Sci. USA*, **47**, 1309.
2. Tanford, C. (1968). *Adv. Prot. Chem.*, **23**, 121.
3. Tanford, C. (1970). *Adv. Prot. Chem.*, **25**, 1.
4. Kim, P. S. and Baldwin, R. L. (1982). *Ann. Rev. Biochem.*, **51**, 459.
5. Jaenicke, R. (1987). *Prog. Biophys. Mol. Biol.*, **49**, 117.
6. Kim, P. S. and Baldwin, R. L. (1990). *Ann. Rev. Biochem.*, **59**, 631.
7. Chen, B. and Schellman, J. A. (1989). *Biochemistry*, **28**, 685.
8. Privalov, P. L. (1989). *Ann. Rev. Biophys. Biophys. Chem.*, **18**, 47.
9. Thannhauser, T. W. and Scheraga, H. A. (1985). *Biochemistry*, **24**, 7681.
10. Labhardt, A. M. (1979). In *Protein folding*, (ed. R. Jaenicke), pp. 401–25. Elsevier/North-Holland, Amsterdam.
11. Matthews, C. R. and Westmoreland, D. G. (1975). *Biochemistry*, **14**, 4532.
12. Matthews, C. R. and Froebe, C. L. (1981). *Macromolecules*, **20**, 2700.
13. Schmid, F. X. (1989). In *Protein structure: a practical approach*, (ed. T. E. Creighton), pp. 251–85. IRL Press, Oxford.
14. Friguet, B., Djavadi-Ohaniance, L., and Goldberg, M. E. (1989). In *Protein structure: a practical approach*, (ed. T. E. Creighton), pp. 287–310. IRL Press, Oxford.
15. Shalongo, W., Ledger, R., Jagannadham, M. V., and Stellwagen, E. (1987). *Biochemistry*, **26**, 3135.
16. Goldenberg, D. P. (1989). In *Protein structure: a practical approach*, (ed. T. E. Creighton), pp. 225–50. IRL Press, Oxford.
17. Privalov, P. L. and Gill, S. J. (1988). *Adv. Prot. Chem.*, **39**, 191.
18. Rosa, J. J. and Richards, F. M. (1979). *J. Mol. Biol.*, **133**, 399.
19. Beasty, A. M. and Matthews, C. R. (1985). *Biochemistry*, **24**, 3547.
20. Roder, H., Elöve, G. A., and Englander, S. W. (1988). *Nature*, **335**, 700.
21. Udgaonkar, J. B. and Baldwin, R. L. (1988). *Nature*, **335**, 694.
22. Bycroft, M., Matouschek, A., Kellis, J. T., Serrano, L., and Fersht, A. R. (1990). *Nature*, **346**, 488.
23. Matthews, C. R. and Crisanti, M. M. (1981). *Biochemistry*, **20**, 784.
24. Garvey, E. P. and Matthews, C. R. (1989). *Biochemistry*, **28**, 2083.
25. Pace, C. N., Shirley, B. A., and Thomson, J. A. (1989). In *Protein structure: a practical approach*, (ed. T. E. Creighton), pp. 311–30. IRL Press, Oxford.
26. Havel, H. A., Kauffman, E. W., Plaisted, S. M., and Brems, D. N. (1986). *Biochemistry*, **25**, 6533.
27. Perry, K. M., Onuffer, J. J., Gittelman, M. S., Barmat, L., and Matthews, C. R. (1989). *Biochemistry*, **28**, 7961.
28. Borden, K. L. B. and Richards, F. M. (1990). *Biochemistry*, **29**, 3071.
29. Schellman, J. A. (1978). *Biopolymers*, **17**, 1305.
30. Santoro, M. M. and Bolen, D. W. (1988). *Biochemistry*, **27**, 8063.
31. Johnson, M. L. and Frasier, S. G. (1985). In *Methods in enzymology*, (ed. C. H. W. Hirs and S. N. Timasheff), Vol. 117, pp. 301–42. Academic Press, London.
32. Utiyama, H. and Baldwin, R. L. (1986). In *Methods in enzymology*, (ed. C. H. W. Hirs and S. N. Timasheff), Vol. 131, pp. 51–70. Academic Press, London.

33. Hagerman, P. J. and Baldwin, R. L. (1975). *Biochemistry*, **15**, 1462.
34. Ikai, A. and Tanford, C. (1973). *J. Mol. Biol.*, **73**, 145.
35. Hagerman, P. J. (1977). *Biopolymers*, **16**, 731.
36. Benson, S. W. (1960). *The foundations of chemical kinetics*, pp. 39–42. McGraw-Hill, New York.
37. Brandts, J. F., Halvorson, H. R., and Brennan, M. (1975). *Biochemistry*, **14**, 4953.
38. Matthews, C. R. (1987). In *Methods in enzymology*, (ed. R. Wu and L. Grossman), Vol. 154, pp. 498–511. Academic Press, London.
39. Chen, B., Baase, W. A., and Schellman, J. A. (1989). *Biochemistry*, **28**, 691.
40. Matouschek, A., Kellis, J. T., Serrano, L., Bycroft, M., and Fersht, A. R. (1990). *Nature*, **346**, 440.
41. Refer to companion volume: *Directed mutagenesis: a practical approach* (1991) (ed. M. J. McPherson). IRL Press, Oxford.
42. Smith, M. (1985). *Ann. Rev. Genet.*, **19**, 423.
43. King, J. and Yu, M.-H. (1986). In *Methods in enzymology*, (ed. C. H. W. Hirs and S. N. Timasheff), Vol. 131, pp. 250–66. Academic Press, London.
44. Lim, W. A. and Sauer, R. T. (1989). *Nature*, **339**, 31.
45. Bowie, J. U. and Sauer, R. T. (1989). *Proc. Natl Acad. Sci. USA*, **86**, 2152.
46. Pakula, A. A. and Sauer, R. T. (1989). *Ann. Rev. Genet.*, **23**, 289.
47. Alber, T. A. and Wozniak, J. A. (1985). *Proc. Natl Acad. Sci. USA*, **82**, 747.
48. Coplen, L. J., Frieden, R. W., and Goldenberg, D. P. (1990). *Proteins: Struc. Func. Genet.*, **7**, 16.
49. Kauzmann, W. (1959). *Adv. Prot. Chem.*, **14**, 1.
50. Dill, K. A. (1990). *Biochemistry*, **29**, 7133.
51. Sanberg, W. S. and Terwilliger, T. C. (1989). *Science*, **245**, 54.
52. Shortle, D. and Meeker, A. K. (1986). *Proteins: Struc. Func. Genet.*, **1**, 81.
53. Alber, T., Dao-pin, S., Nye, J. A., Muchmore, D. C., and Matthews, B. W. (1987). *Biochemistry*, **26**, 3754.
54. Touchette, N. A., Perry, K. M., and Matthews, C. R. (1986). *Biochemistry*, **25**, 5445.
55. Perry, K. M., *et al.* (1987). *Biochemistry*, **26**, 2674.

8

Principles of protein stability. Part 2—enhanced folding and stabilization of proteins by suppression of aggregation *in vitro* and *in vivo*

RONALD WETZEL

1. Introduction

Protein folding is most conveniently studied under conditions where unfolding is reversible, and much of what we know about the forces that control the formation of native protein structure is derived from studies on reversible systems. But outside of the biophysical laboratory—and, for that matter, often within it—proteins are more mortal. Irreversible side-reactions limit the stabilities and folding efficiencies of many proteins under native conditions *in vitro*. Similar side-reactions also play roles in protein folding *in vivo*.

A major class of irreversible side-reactions, which can occur both *in vivo* and *in vitro*, is chemical modification. Proteolysis is of major importance *in vivo*, and can also occur during *in vitro* folding of proteases. Other chemical reactions occur as well, especially in the *in vitro* inactivation of proteins (1). These pathways will not be discussed in this chapter. The topic under consideration here is aggregation, which in many ways is more subtle than other non-productive pathways. Aggregation can take place under more mild conditions, and generally leaves the protein intact chemically. *Aggregation and precipitation are often thought of as the consequences of failed folding experiments, whereas they are often the cause of failure.* An appreciation of the roles of aggregation and precipitation is a critical prerequisite for effectively designing and monitoring the progress of formal folding and stability experiments in the biophysics lab, as well as for the everyday production and handling of proteins in the laboratory and the factory. This problem thus touches protein engineering both in the mundane aspect of obtaining well-behaved protein variants from heterologous expression systems, and in the use of such variants in folding and stability studies.

In this chapter I discuss the measurement of aggregation, and possibilities for suppressing it, in:

- inactivation of proteins stressed by heat, pH, and/or solute denaturation
- folding of proteins from denaturant, as in attempts to recover active material from inclusion bodies
- folding *in vivo* in bacteria, with respect to the side-reaction of inclusion body formation

A more thorough discussion of some fundamental aspects of *in vivo* protein aggregation, as manifested in bacterial inclusion bodies and mammalian amyloid deposits, can be found elsewhere (2). Other discussions emphasizing more practical aspects of protein aggregation during folding have appeared recently (3, 4).

Proteins are individualistic in their folding behaviour. This may be especially true in the kinetic competition between productive folding and aggregate formation, since similar forces, to a large extent, drive both processes. This means that conditions known to enhance or weaken hydrophobic interactions, for example, might have different effects on the refolding of different proteins, depending on the relative importance of hydrophobic interactions in the transition states for processes competing for a common folding intermediate. This ambiguity or individuality in protein behaviour not only limits the utility of general protocols, it also should limit our faith in those seemingly general rules that have been formulated. Perhaps the most useful contribution of a practical guide to protein folding may be to draw attention to the important *parameters* which should be freshly explored in studying and optimizing conditions for a previously uncharacterized protein.

1.1 General mechanism

It is useful to think in terms of the following very general model for protein folding, where N is the native, folded state; U is the unfolded state; I_1, etc. are a series of folding intermediates on the productive folding pathway; X_1, etc. are a series of intermediates or associated states off the productive folding pathway; and Ag (aggregated) and C (covalently modified) are irreversibly formed products. Particular proteins may have fewer or more intermediates, or may have more than one productive pathway connecting the native and unfolded states. U may be a statistical random coil or an unfolded state containing some residual structure. *In vivo*, U might also be nascent polypeptide chain exiting the ribosome or passing through a membrane. All these states should be considered collections of microstates of similar structures and properties rather than discrete structures. Folding intermediates have now been observed in many systems by a number of methods (5). Mechanisms similar to Scheme 1 have been proposed to account for *in vitro* folding properties of some proteins (6), but it should be cautioned that for most proteins the extents and causes of aggregation in incomplete folding and inactivation have not been determined. None the less, Scheme 1 provides a useful framework for the design of methods to enhance folding and stability, as will be discussed in the following sections. The important points of the general model are:

(a) This model should be useful in describing the stability of native proteins *in vivo* and *in vitro* (the N → U direction), as well as the folding of denatured proteins *in vitro* and nascent polypeptide chains *in vivo* (the U → N direction).

(b) Aggregation/precipitation and proteolysis represent irreversible steps from which protein cannot recover under reaction conditions; *soluble aggregate formation not followed by precipitation might well be reversible.*

(c) If the dominant cause of irreversibility in a system is the aggregation and/or precipitation of N, X_n, I_n, or U, then the important factors are the solubilities and kinetics of formation and decay of these states, and these will be influenced by various structural and environmental factors.

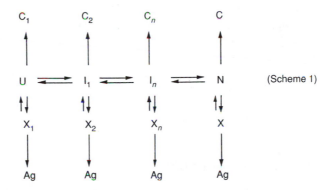

(Scheme 1)

1.2 Aggregation

The word 'aggregation' is used in protein chemistry to describe both native interactions between subunits of oligomeric proteins and non-native interactions between proteins. While the former is highly specific, the latter is generally less well-defined. Both, however, are driven by the same forces which determine the intramolecular folding and stability of proteins.

The strengths of subunit interactions of oligomeric proteins correlate well with the extent of the subunit's surface area that is buried in the interaction, suggesting a major role for hydrophobic forces in native aggregation phenomena (7). At the same time, other interactions also play roles. For example, H-bonded β-barrel formation is an important part of the subunit interfaces of many oligomeric proteins (7). Other H-bonds and salt bridges are also found at protein interfaces (7). In the association of stably folded protein subunits, complementarity of fit at the interface is very important (7), since trapped water molecules significantly reduce the energy gained from the hydrophobic effect, and other interactions such as H-bonds have strict geometric constraints.

Non-native aggregation of *folded* proteins is a common observation in protein biochemistry, being a classical method for protein isolation and purification. The phenomenon of isoelectric precipitation, as well as the precipitation of protein at very low or very high ionic strengths, can be interpreted on the basis of the surface charge and hydrophobicity of folded proteins (8). Salting out of proteins by ammonium sulphate and other neutral salts is driven largely by the hydrophobic effect (8, 9), but other forces can also contribute to aggregation of folded proteins. This class of protein aggregation is not very specific, as evidenced by the poor resolution and dependence on protein background of ammonium sulphate precipitations: proteins that precipitate in the pure state at one ammonium sulphate concentration may precipitate at a different salt concentration when contaminating proteins are present (8). Another distinguishing feature of aggregation of native, folded proteins is the ease with which precipitated aggregates can be redissolved in appropriate native buffer. This behaviour sets such aggregates apart from those observed in protein stability and refolding experiments.

Protein aggregates arising from thermal or denaturant-induced inactivation, or from protein refolding experiments, or as represented in bacterial inclusion bodies, are generally resistant to dissolution in native buffer. In general, strongly denaturing conditions, such as high concentrations of urea or guanidine hydrochloride, or extreme pH conditions, are required to break and dissolve such aggregates. This implies a fundamental difference between these aggregates and those of native, folded proteins. Although this poor solubility may be related to their mode of formation (i.e. from non-native states), it does not necessarily mean that different forces are at work in stabilizing the structures of the aggregates. In fact, hydrophobic forces probably also play a critical role in this type of aggregation. Although the aggregating species in (mis)folding-related aggregation may be less highly structured than native proteins, poor solubility also does not mean that the interactions involved are non-specific. Such aggregation processes display a surprisingly high selectivity, in that aggregate is enriched in one particular protein (10) or in that its formation is not accelerated by the presence of high concentrations of other proteins (11). This also argues for some significant level of structure in the aggregating species.

2. *In vitro* studies

2.1 Quantifying aggregation

2.1.1 Rationale

Protein folding in well-defined, reversible systems can be studied by observing the formation of the native state spectroscopically. These methods can sometimes also be used in folding and stability studies of proteins which are less well-defined and/or well-behaved, but in such instances they can suffer from a number of limitations:

(a) some systems may be chemically less well-defined, proteins are impure, etc.

(b) subject proteins may not be well-characterized spectroscopically

(c) light scattering of aggregated protein compromises the measurements

When folding or stability cannot be monitored spectroscopically, the amount of native protein product is often assessed by activity measurements, with unrecovered activity simply considered as 'not properly folded' or 'inactivated' without further characterization. In fact, that part of the expected activity which is not recovered in a stability or folding experiment is often to be found in an aggregated—or absorbed (12)—state, although losses due to chemical modification, proteolysis, etc. are clearly also possible.

The kinetic model for protein folding suggests that native states of some proteins may exist in local free-energy minima, and that there may be soluble, well-behaved, non-native states of lower free energy accessible by other pathways (7). Few stable states have been characterized that differ in major ways from the native state in their basic folding pattern. One might consider the microheterogeneity of disulphide-bonded states observed in soluble preparations of bovine mercaptalbumin and other serum proteins (13) to be examples of soluble, stable misfolded states. The paucity of such reported examples suggests that most non-native folding states are poorly behaved in solution and will probably be lost to solution via aggregation. Stated another way: in many—but not all—systems, any protein left *clearly* in solution at the end of a stability or refolding experiment will probably be in its native state. To the extent that this is true, direct or indirect measurement of aggregates should be a generally useful initial screen for correctly folded proteins.

Light scattering can be used directly to measure aggregation qualitatively or quantitatively. Alternatively, simple methods can be used to assess the amount of protein remaining in solution, which in many cases will be no worse a measure of refolding than a specific activity measurement which senses the native state of the protein of interest. Clearly, the final measure of refolding of a protein is its ability to produce its specific biological activity (keeping in mind that there may be crippled active-site mutants which exhibit low activity even though they fold in the same way as wild type), and such activity measurements will normally be required at some point in a refolding study. On the other hand, there is no point in carrying out an expensive, time-consuming assay on a sample that contains only aggregated or precipitated protein.

2.1.2 Quantifying protein lost from solution—light scattering

A key requirement in any method for measuring light scattering is to minimize the level of dust and other particles, including pre-existing protein aggregates, in the starting mixture. These can not only give a high background scattering at zero time, but also can influence negatively the course of the folding/unfolding experiment. For example, a small percentage of protein present as aggregate in a thermal stability experiment can greatly enhance the rate of thermally induced aggregation in the rest of the sample (Mulkerrin and Wetzel, unpublished). Aggregate and other particulates can be minimized by microfuge centrifugation, or, preferably, filtration through a 0.2 μm membrane.

One convenient method for following light scattering is to monitor the total intensity loss, or turbidity (14), of the sample at a wavelength where the properly folded sample does not absorb; for most proteins, 300–400 nm is a convenient range. The disadvantage of turbidity measurements is that the results cannot be used to generate data on the size, shape, and number of protein aggregates (14). The advantage of this method, however, is that it requires a conventional ultraviolet/visible (UV/VIS) spectrophotometer. Turbidity measurements are often adequate in cases where a crude measure of aggregation is sufficient to characterize the effects of variables on folding yields or stability. In one set of experiments (15), the thermal stability of human interferon-γ was followed by determining the kinetics of turbidity increase with respect to temperature; the data agree very well with the temperature of the midpoint of the unfolding transition (T_m) (see *Figure 1*). In other experiments, turbidity has been used to assess the extent of aggregation and precipitation of folding intermediates in the refolding of bovine growth hormone (16) and carbonic anhydrase (17).

Light scattering can also be measured by the more sophisticated technique of dynamic, or quasi-elastic, light scattering (14). In this method the intensity change due to scattering is considered with respect to the angle of scatter. The data from such experiments can be related not only to the number of scattering particles in solution/suspension, but also to the shapes and sizes of the particles. This introduces the possibility of studying some of the details of the aggregation process. For example, quasi-elastic light scattering has been used to characterize the aggregation state of carbonic anhydrase under various refolding conditions (18). This analysis allowed formulation of a detailed folding mechanism including off-pathway, reversible formation of aggregates which, under appropriate conditions, can also irreversibly precipitate.

2.1.3 Quantifying protein retained in solution

Although formation of soluble aggregates can be an important side-reaction in folding and stability experiments, a more typical result is the precipitation of aggregated protein. Thus, an alternative measure of the success of a folding/stability experiment is to determine the amount of the subject protein that remains in solution as the experiment proceeds. This is the basis of the use of activity measurements to follow thermal stability, for example. Activity is often lost because protein has disappeared from solution; in the thermal inactivation of T4 lysozyme and interferon-γ (*Figure 1*), loss of soluble

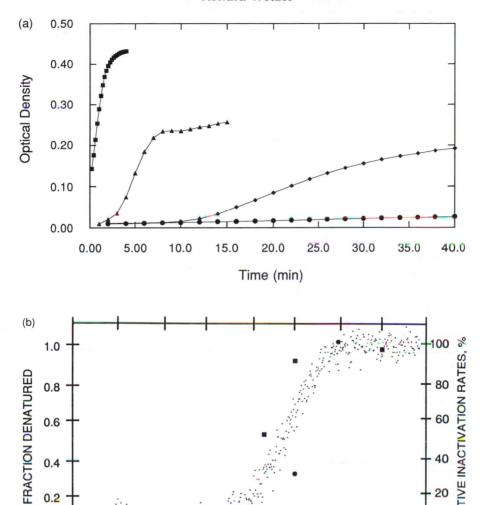

Figure 1. Aggregation and thermal stability in human interferon-γ. (a) Time course of generation of light scattering on heating interferon at pH 6 at various temperatures: 45°C (●), 50°C (◆), 55°C (▲), 60°C (■). (b) Temperature dependence of aggregation-mediated inactivation for interferon, as monitored by rates of generation of light-scattering particles (a) (●) and residual immunochemical activity (■), compared to its reversible unfolding as monitored by circular dichroism (•). (See ref. 15.)

protein—as measured by enzyme activity or immunoassay—is proportional to generation of light scattering particles (12). Immunochemical assays can be used to assess the amount of the protein of interest that remains in solution; if the protein is still present, but in an aggregated state, the response in an enzyme-linked immunosorbent assay (ELISA) is normally dramatically diminished. A similar method has also been used to screen for inclusion body formation in cells (see Section 3).

If folding/stability experiments are conducted with purified protein, a simple micro-assay for protein concentration, such as that described by Bradford (19), can be used to determine the amount of protein remaining in solution at various reaction times. In this case protein aggregates must be cleared from the suspension by physical means before it is assayed. We have used this method in screening for suitable refolding conditions for proteins made in inclusion bodies in *Escherichia coli*. The procedure is as follows:

Protocol 1. Micro-assay for protein concentration

1. Initiate the folding/stability reaction in a series of Eppendorf tubes, one tube per time-point, 10 µg protein/tube.
2. At assay time, centrifuge for 10 min and carefully decant part of the supernatant (alternatively, filter through a 0.2 µm filter).
3. Assay the supernatant for protein content.
4. Compare the result to that obtained from a freshly prepared, non-centrifuged sample, containing the original starting amount of protein.

2.2 Aggregation and protein stability

Traditionally, protein stability—against most modes of inactivation, not just aggregation—has been addressed in terms of a much simpler mechanism than that shown in Scheme 1, a mechanism considering only a two-state folding equilibrium between N and U. In this mechanism (Scheme 2), N is relatively stable, while U is a statistical coil state which is much more susceptible to the irreversible reactions that contribute to overall stability. Although the folding and inactivation of many proteins is not fully described by such a mechanism—requiring more complex ones such as versions of Scheme 1—it is possible, even in many of these complex cases, to rationalize or model the inactivation results, as well as to engineer improved stability, by assuming the simplistic model shown in Scheme 2. According to this mechanism, a necessary and sufficient means of stabilizing the protein is to shift the unfolding equilibrium toward N: a lower concentration of U means a lower rate of inactivation. Two examples of this straightforward approach to increasing stability are given in the next section. It has also recently become clear that there are other ways in which proteins can be stabilized against irreversible inactivation by aggregation which do not rely on direct effects on the unfolding equilibrium. Several examples of such approaches are given in Section 2.2.2.

This section deals only with structural changes to the protein itself as a means for improving stability. However, the protein's concentration and environment clearly also play major roles in defining stability against unfolding-induced aggregation (12). In general, similar parameters are involved in both unfolding-induced inactivation and in the *in vitro* refolding of proteins. To limit duplication, such environmental factors are discussed only later, in the context of *in vitro* refolding (Section 2.3), even though the same factors and conclusions can also apply to *in vitro* stability.

$$N \rightleftarrows U \longrightarrow Ag \qquad \text{(Scheme 2)}$$

2.2.1 Stabilization by influencing the folding equilibrium

i. *T4 lysozyme and structure-based directed mutagenesis*

T4 lysozyme undergoes irreversible thermal inactivation in the neutral pH range when heated at about 65°C, and the various mechanisms by which activity is lost under these conditions have been described (12):

- unfolding-dependent adsorption to the vessel
- unfolding-dependent disulphide bond formation
- unfolding-dependent non-covalent aggregation and precipitation
- chemical decomposition

At low phosphate concentrations and at a protein concentration of 30 μg/ml, the two major mechanisms of inactivation are non-covalent aggregation/precipitation and disulphide bond-mediated aggregation. Since the rates of both these processes become significant only under conditions at which the protein is unfolded, changes in the molecule that increase its T_m should decrease the inactivation rate by reducing the concentration of U at the reaction temperature. Such a relationship between inactivation rate and T_m was observed for a series of T4 lysozyme mutants under the above conditions (12, 20).

The X-ray crystal structure of T4 lysozyme has been known for some time, and Matthews' group has been very successful at building thermal stability into the molecule by modifying the structure, via mutagenesis, according to the theoretical considerations of the basis of folding stability:

- decreasing configurational entropy of U (21)
- increasing hydrophobic stabilization (decrease in solvated surface of N) (22)

- electrostatic stabilization of helix dipoles (23)
- enhancement of helix-forming potential (24)

In one study, mutations were constructed which were expected to decrease the configurational entropy of the unfolded state, and thus shift the unfolding equilibrium toward the folded state (21). An Ala82 → Pro mutation, expected to be tolerated by the native structure but to constrain U of the mutant more than the wild type, increased T_m by 2.1 °C at pH 6.5. Similarly, Gly77 → Ala provided an increase of 0.9 °C in the T_m. In this series of mutants, the rate of irreversible inactivation at 65 °C is decreased in proportion to the increase in the mutant's T_m. In another series of (destabilized) T4 lysozyme mutants (12), the inactivation kinetics with respect to temperature clearly sweep out sigmoidal curves resembling co-operative melting transitions, with midpoints corresponding to approximate T_ms of the mutants (*Figure 2a*).

Such experiments support the notion, based on the model in Scheme 2, that any change which increases the T_m of the protein should decrease its rate of inactivation, *when experiments are conducted in the unfolding transition temperature range* (of course, increasing T_m from 65° to 68 °C will not substantially change the inactivation rate by the Scheme 2 mechanism when inactivation is at a temperature, such as 80 °C, where all variants are totally unfolded). One type of stabilizing mutation, the use of disulphide cross-links, does not exhibit a correlation between T_m effect and inactivation kinetics; this is discussed in Section 2.2.2.

ii. *Kanamycin nucleotidyltransferase (KNTase): random mutagenesis with selection*

The KNTase encoded on bacterial plasmid pUB110 undergoes irreversible thermal inactivation at about 55 °C. Transformed into the thermophilic *Bacillus stearothermophilis*, the plasmid confers resistance to kanamycin when cells are grown at 55 °C, but not at higher temperatures, where the encoded protein is inactivated. This provides an efficient selection system for mutant enzymes of enhanced stability. Using such a system, two groups have isolated stabilizing mutations in the protein (25, 26), whose three-dimensional structure is not known. These experiments demonstrate the potential for moving mesophilic enzymes into thermophilic organisms to provide a selection for temperature-stable mutants. Screening has also been used to identify thermostable mutants in enzymes such as T4 lysozyme (27).

The mechanism by which KNTase inactivates *in vivo* or *in vitro* is not clear. Presumably, inactivation is dependent on folding stability, since the temperature dependence of inactivation *in vitro* (28) exhibits the large temperature coefficient characteristic of co-operative folding of proteins. It may be that the *in vitro* and *in vivo* mechanisms differ, only sharing the feature that inactivation rate depends on the extent of protein unfolding. For example, *in vivo* inactivation might result from proteolysis, while the *in vitro* source of

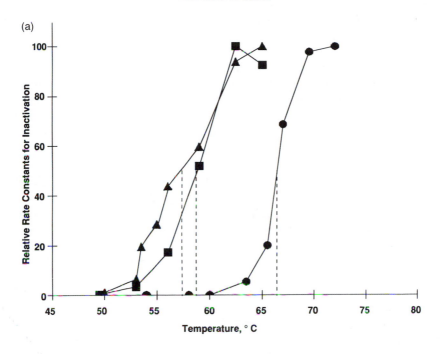

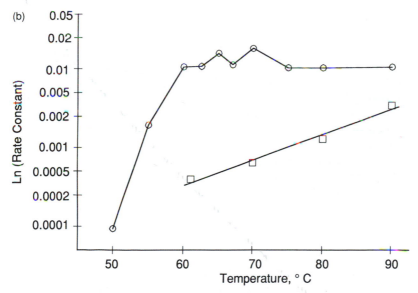

Figure 2. Dependence of aggregation-mediated inactivation rates of T4 lysozyme variants on temperature. (a) Non-crosslinked variants, wild type (●), I3C/C54T/A146T (■), I3C/C54T/R96H (▲). Temperatures of half-maximal rate for each mutant agree well with expected T_m values (see ref. 12). (b) Non-crosslinked variant C54V/C97S (○) and disulphide cross-linked variant I3C-97C/C54V (□) (see ref. 12).

irreversibility in the purified protein might be aggregation. In any event, the kinetics of inactivation correlate well with ΔG_{stab} in a series of mutants, indicative of the fundamental control exerted by the folding equilibrium (29).

Whether temperature stabilizing mutations are identified by genetics or by a structure-based approach, it is possible to combine a number of them in the same molecule with some expectation that their effects will be roughly additive. Additivity was demonstrated in the effects of mutations on inactivation rate in KNTase (28), as well as in some cases of reversible unfolding of proteins (30), and in other protein engineering experiments (31).

2.2.2 Stabilization by affecting other properties

i. Disulphide-bonded T4 lysozyme

Disulphides and other cross-links fall into the general class of mutations which are expected to stabilize proteins against reversible unfolding by decreasing the configurational entropy of the unfolded molecule (see Section 2.2.1.*i*). Although the details of stabilization by disulphides are probably more complex than this simple theory predicts (32, 33), a number of examples of added stability to reversible unfolding by disulphide engineering have been reported.

In one case, the effect of the disulphide bond on *irreversible inactivation* was also studied. The results suggest that the disulphides can stabilize against irreversible thermal inactivation by a mechanism unrelated to its effect on ΔG_{stab} or T_m. In these experiments (20), a disulphide added to T4 lysozyme increased the T_m by only 3°C, but provided appreciably more stability against irreversible thermal inactivation, by suppressing aggregation/precipitation under reaction conditions. The most likely explanation for this effect is that the cross-link either improves the solution properties of the thermally induced non-native state (possibly a folding intermediate) responsible for irreversible aggregation, or alters the unfolding kinetics (34). One important feature of this stabilization is the shape of the curve relating inactivation rate to temperature (*Figure 2b*). While all non-crosslinked T4 lysozymes, like many other proteins (12), exhibit sigmoidal inactivation curves consistent with a dependence on a co-operative unfolding equilibrium, the cross-linked lysozyme exhibits a linear temperature dependence of inactivation rate in a semi-log plot characteristic of chemical mechanisms of inactivation and chemical reactions in general. One important consequence of this kind of stabilization, which removes the strict dependence of inactivation rate on the folding state, is that *stabilization via such mutations can be extended to well outside the thermal unfolding transition region* (*Figure 2b*).

ii. Sequence changes in interferon-γ

Human interferon-γ also undergoes irreversible thermal inactivation according to the general mechanism in Scheme 2. The rate of inactivation increases dramatically over a short temperature range at or near the T_m of the protein

Ronald Wetzel

(*Figure 1*). Inactivation is accompanied by aggregation and light scattering, and inactive material can be reactivated by a denaturation/renaturation cycle (15). To stabilize the molecule against thermal inactivation, one might try two approaches: stabilize the protein against reversible thermal unfolding by making changes that improve the ΔG_{stab}, or alter the kinetic lifetime or solubility properties of a non-native state so that the protein is allowed to remain in solution (and is thus enabled to return to the native, folded state) upon cooling.

It is clear that the unfolding step and the aggregation step are separable, because pH affects the two processes differently (15). Human interferon-γ is most stable against reversible unfolding in the neutral pH range, as are most proteins. However, the protein is much more stable to irreversible thermal inactivation at pH 5 than at pH 7. One possible mechanism by which lower pH might stabilize an unfolded form is by increasing the net charge on the protein; however, interferon-γ has a pI above 10, and thus has a high net charge even at pH 7. On the other hand, there may exist a specific aggregation surface in the unfolded or partially folded state which undergoes an important change in net charge and aggregation propensity between pH 7 and 5. The histidine imidazole has a pK_a in this range, and two pieces of evidence support the role of His deprotonation in the sensitization of interferon-γ to thermal inactivation. First, species variants that do not contain His at the two positions occupied by His in the human sequence thermally unfold reversibly at pH 7 (as well as at pH 5). Secondly, specific chemical modification of His with ruthenium pentaamine converts the human protein to one that is highly stable against irreversible thermal inactivation/aggregation at temperatures above its T_m, in contrast to the wild type (15).

Thus, as with disulphide-crosslinked T4 lysozyme, His-modified interferon is stable to irreversible thermal inactivation because of a dramatic change in a kinetic or solubility property of a non-native state. It is too early to say how general an approach it will be to structurally alter proteins to suppress aggregation of non-native forms. As an alternative to rational design, genetics with screening might be used to identify suppressed aggregation mutants. If inclusion body formation is due to the poor solubility or extended kinetic lifetimes of folding intermediates *in vivo* (see Section 3), and if these intermediates resemble the aggregation-prone folding intermediates generated in stress-induced unfolding *in vitro*, then perhaps mutants that exhibit reduced inclusion body formation will also exhibit enhanced folding and/or stability *in vitro* due to reduced aggregation properties (12).

2.3 Aggregation accompanying protein folding *in vitro*

Protein folding *in vitro* not only continues to be of great theoretical interest, but, with the advent of recombinant DNA methods for the production of large amounts of protein in bacteria, it has also become of tremendous practical importance. As will be discussed in Section 3, many proteins that are expressed at high levels in foreign cells tend to form dense aggregates called inclusion bodies. These aggregates are generally resistant to solvation with native buffers, requiring for their recovery such conditions as extremes of pH, high concentrations of solute denaturants, or addition of detergents. Once dissolved under non-native conditions, the protein must be renatured. Sometimes refolding proceeds smoothly *in vitro*, but frequently refolding is not a straightforward matter of dilution or dialysis to return to native buffer conditions. Probably the most common, and the most serious, barrier to efficient refolding is the loss of material due to non-covalent aggregation and precipitation. The aggregation problem has been interpreted in the context of models such as that shown in Scheme 1, where, under the refolding conditions, the folding kinetics favour formation of a side-product that undergoes aggregation, and ultimately precipitation, to remove the protein effectively from the equilibrium. This kind of aggregation has been noted especially in multisubunit or multidomain proteins, where long-lived folding intermediates are more likely to be observed (35). It can also occur with proteins normally thought of as single-domain proteins; one well-characterized case is that of bovine growth hormone (bGH) (16). In principle, anything that suppresses the kinetics of aggregate formation while still encouraging formation of native protein structure should favour the desired folding reaction and improve yields. Some important parameters are listed below. More specific discussions on the recovery of recombinant polypeptide from inclusion bodies have been published recently (4, 36, 37).

2.3.1 Protein concentration

The correct folding of monomeric proteins should be of unimolecular kinetics and thus concentration independent. Since aggregation requires the encounter of at least two polypeptides, it may be enhanced at high protein concentration, depending on the rate-limiting step. Losses to aggregation might thus be suppressed by lowering protein concentration in refolding reactions; indeed, in some cases folding intermediates have been populated for characterization by doing the refolding at high protein concentrations. In one such experiment, a bGH folding intermediate was observed when partial refolding was done at 1.6 mg/ml, but was not observed at 16 µg/ml (38). This latter concentration is typical of those found to be optimal in *in vitro* protein-folding experiments.

2.3.2 Physical separation of folding molecules

The unfolding of trypsinogen by reduction of disulphides is usually irreversible, but when immobilized on a solid phase by covalent attachment, the protein is prohibited from undergoing encounters with other molecules, and it unfolds reversibly (39). Similarly, rhodanese can be thermally unfolded reversibly, without the normal loss of material due to aggregation, by immobilizing the protein before heating (40). Irreversible covalent attachment is impractical for refolding most proteins of interest, but the above results suggest that affixing unfolded proteins to a matrix, by non-covalent (41) or reversible covalent interactions, might increase refolding yields by discouraging aggregation.

2.3.3 Choice of denaturing solvent

Especially if the source of denatured protein is a bacterial inclusion body (IB), some effort must be expended in solubilizing the starting material. Inclusion bodies from different proteins can exhibit radically different solubilities in denaturants, and the proteins, once solubilized, may also refold with differing efficiencies, depending on denaturant. There are other practical considerations regarding choice of denaturant, some of which are listed below:

(a) 6 M guanidine hydrochloride is a strong denaturant, but its high ionic strength precludes ion-exchange clean-up of solubilized protein.

(b) 5 M guanidine thiocyanate is significantly better than the hydrochloride as a denaturant and a better solvent for at least some IBs, but little work has been reported on refolding proteins from this denaturant.

(c) 8 M urea is less denaturing than the guanidine salts, but is neutral and therefore useful in ion-exchange purifications. The chief disadvantage of urea is the potential for carbamylation of amino groups due to cyanate—this can be minimized by using *freshly* deionized urea, working in the cold at neutral or slightly acidic pHs, and/or use of low pK_a amine-containing buffer components like glycylglycine to scavenge cyanate. It should be noted that 'electrophoresis-grade' or other 'high purity' ureas, which are low in *cyanide*, are almost certainly not low in *cyanate*, which forms in solution (even frozen solutions) or as a solid by disproportionation; the degree of disproportionation products present in urea solution is conveniently assessed by conductivity measurements.

(d) The presence of reducing agent in the buffer will suppress the formation of disulphide-linked aggregates, and will break up and help dissolve those already present.

(e) Proteins have been refolded successfully after extraction from inclusion bodies at pH 10–11 (42) or pH 2–3 (43); however, proteins contain groups such as side-chain amides and disulphides which are sensitive to extremes of pH. Although these reactions may give only trace amounts of side-products, even low levels of chemically altered product might still be deleterious to its use—for example, if they are difficult to detect and remove but are highly immunogenic. Therefore, extreme pH conditions may prove useful only in cases where purity requirements are less stringent.

2.3.4 Details of denaturant removal

Solutes used to facilitate the dissolution of inclusion bodies or otherwise insoluble proteins, such as urea or guanidine hydrochloride, may be considered to have two aspects in the refolding reaction, acting both as *denaturants* to control the thermodynamics and kinetics of chain folding, and as *solvents* to maintain all the molecular forms, such as key folding intermediates, in solution. As the concentration of denaturant is reduced to initiate refolding, the protein can pass through a series of folding intermediates, each of which has characteristic kinetics of formation and decay (both productive decay to native protein, and unproductive decay, to aggregates and precipitates) at any denaturant concentration. Each successively lower level of denaturant can *differentially* influence the stability and solubility of these intermediates and aggregates.

Thus the challenge in such a refolding reaction is to identify a series of conditions (of denaturant concentrations, pH, other buffer components, etc.) each of which promotes further folding while at the same time retains all forms of the protein in solution. Conditions can fail either because denaturant is removed too slowly or too quickly. In a classic example illustrating the former possibility, tryptophan synthase was incubated in a series of concentrations of urea from 0 to 8 M, then dialysed to remove the urea. Active protein was recovered after incubation at both high and low urea concentrations, but incubation in 3 M urea led to poor recovery of activity. This was shown to be due to aggregation of a folding intermediate (11). Passing quickly through 3 M urea (as in the dialysis starting at 8 M urea), aggregation was avoided. Thus, lingering at a particular denaturant level can allow the accumulation of an aggregating intermediate which would not have caused a problem if the critical denaturant concentration had been more quickly passed over. In an illustration of the latter case, Brems and co-workers found that bGH diluted into native buffer from 2.8–5 M guanidine hydrochloride (where a folding intermediate is populated) forms irreversible precipitates. However, if a preliminary dilution is done from 2.8–5 M to 2 M denaturant and the resulting solution allowed to stand briefly, the poorly soluble intermediate breaks up and further dilution can be done to obtain the native protein in good yield (16).

2.3.5 Protein purity

Intuitively, one imagines that irreversible processes such as aggregation might be accelerated in the refolding of a protein contaminated with other proteins; this seems to be the case in the salting out of native proteins (8). Goldberg and co-workers detected no influence of either serum albumin or of crude *E. coli* extract on the refolding of tryptophan synthase, which is known to be sensitive to aggregation in its refolding (11). However, the presence of reduced, denatured serum albumin dramatically reduces the refolding yield of turkey lysozyme from the reduced, denatured protein (44). These differing results may have to do with differences in the way the experiments were done, or in differing responses of different proteins to serum albumin. In any case, the bacterial proteins that contaminate inclusion body preparations may have unusual, unique aggregation properties. It is possible that the presence of these proteins in particular, which one might presume are especially aggregation-prone by virtue of their association with inclusion bodies, might be disadvantageous in an *in vitro* refolding reaction. Co-contaminants can do mischief in other ways. In one well-documented case, efforts to refold creatine kinase from *E. coli* inclusion bodies failed until detergent washes of the IBs were introduced into the work-up protocol. Subsequently it was found that the detergents remove a proteolytic activity which reactivates during refolding reactions to destroy the creatine kinase before it has a chance to protect itself by folding (45). Methods for purifying IB-derived material before refolding have been reviewed (36).

A number of different detergent treatments have been used for removal of associated *E. coli* proteins and other molecules from IBs. They include Triton X-100, deoxycholate, and Nonidet (36). In our experience, Triton X-100 gives good recovery of IB protein, but incomplete removal of contaminating proteins, while deoxycholate gives improved purification but also partially solubilizes and removes the major, desired IB protein. Since the effectiveness of denaturants can depend on the specific, heterologously expressed protein in the IB (C. Jones, personal communication), it may be important to survey a number of detergent conditions.

2.3.6 Buffer components

Ionic strength and pH are obvious variables in a search for conditions that favour productive collapse of an intermediate over its aggregation. In analogy to the isoelectric precipitation of native, folded proteins, it is generally believed that one should avoid pHs near the pI of one's protein in refolding experiments. In addition, the choice of buffer ions and their concentrations may also be important. For example, anions are known to vary widely in their abilities both to stabilize the folded structure of proteins (46, 47), and to induce the aggregation of folded proteins ('salting out') (8). This trend is described in the Hofmeister series $\{SO_4^{2-} > HPO_4^{2-} > \text{acetate} > \text{citrate} > Cl^- > NO_3^- > I^- > ClO_4^- > SCN^-\}$, in which multiply charged anions are more effective in both phenomena compared with singly charged anions, due to the varying influences of these anions on the strength of the hydrophobic effect (9). In the thiol–disulphide exchange refolding of ribonuclease, the influence of neutral salts on folding rates or final yields showed the same trend as the effects of these salts on the stability of the native molecule (48). On the other hand, we have observed a decrease in folding efficiency with increasing strength in the Hofmeister series, due to the enhancement of aggregation and concomitant decrease in recovery of native product, both in thermal denaturation of T4 lysozyme (12) and in *in vitro* refolding reactions of other proteins (R. Wetzel, unpublished). Thus it may be important to survey different ions in exploring refolding conditions, possibly by picking representative ions from different positions in the Hofmeister series. As discussed for other variables, the effect of Hofmeister salts on a particular refolding reaction may depend on its relative influence on the rates of productive folding vs. aggregation of a folding intermediate.

Another possible variable is the addition of neutral solutes, such as glycerol or sucrose, known to stabilize the native states of proteins (49). Such additives can improve *in vitro* refolding yields in some cases (50). Polyethylene glycol has also been used to inhibit the aggregation and precipitation of an intermediate in the refolding of carbonic anhydrase (51).

2.3.7 Detergents

There are two ways in which added levels of detergent might influence refolding reactions. Some proteins, such as peripheral membrane proteins and independently expressed fragments of proteins, might be expected to exhibit limited solubility even if properly folded. The presence of a non-denaturing detergent in the refolding mixture might help to preserve the folded product in solution. For other proteins, detergents might be expected to mask hydrophobic surfaces on folding intermediates transiently, suppressing their aggregation and precipitation and thus facilitating completion of folding. 'Detergent-assisted refolding' has been observed in the refolding of rhodanese (52, 53). Rhodanese has been shown to refold inefficiently because of the aggregation of an intermediate (52). The generality of this method has not been explored; success will depend, in part, on the detergent being otherwise benign, both in the refolding reaction and in interactions with the folded product.

2.3.8 Temperature effects

Most protein-folding reactions are facilitated at higher temperatures, up to about 40°C (54). However, low reaction temperatures have also been observed to improve refolding yields, by suppressing aggregation in favour of native structure formation (55, 56). Hydrophobic interactions become stronger with higher temperature over the useful temperature range for proteins (9). Perhaps this effect is important in both trends. Where high temperature improves yields, perhaps aggregation is not an important side pathway, or, if it is, hydrophobic interactions play a bigger role in the transition state to N than they do in the transition state to I or Ag. The opposite may be true when low temperature improves folding yields. Once again, the important lesson is to realize that temperature is an important parameter to explore.

2.3.9 Accessory proteins

Recently it has become clear that at least some proteins benefit in their *in vivo* folding from their interaction with some members of the heat-shock family of

proteins termed chaperonins. Chaperonins are envisioned as facilitating folding by interacting with partially folded proteins. In some cases this interaction
can serve to delay folding until secretion can be accomplished. The interaction can also promote folding of non-secretory proteins, by inhibiting the
aggregation of intermediates until folding/assembly is complete (57). Might
these proteins be useful in providing the same role *in vitro*? There are several
published reports of the use of the *E. coli* heat-shock proteins GroEL and
GroES to facilitate *in vitro* folding of denatured proteins such as ribulose-1,5-
bisphosphate carboxylase (58), citrate synthase (59), dihydrofolate reductase
(60), and rhodanese (60). In a different kind of experiment, DnaK has been
used to suppress or reverse soluble aggregate formation in the thermal inactivation of RNA polymerase (61). In these experiments, molar excesses of heat-
shock protein were required to achieve their maximum effect; such requirements may limit the utility of these factors in the large-scale production of
proteins, unless reusable or continuous chaperone reactors can be developed.

The use of added proteins to facilitate folding by disfavouring aggregate
formation may extend beyond the use of the chaperonin class. Brems has
demonstrated that addition of a structured fragment of bGH prevents aggregation of an intermediate in bGH folding (62); the effect on rates of formation or
yield of N during refolding have not been reported. Added protein fragments
were reported to *inhibit* the rate and final extent of folding of *E. coli* dihydrofolate reductase (63). The value of added fragments may depend on their ability
to dissociate eventually from I, allowing folding to proceed to completion.

The above discussion concerns added molecules whose role is to disfavour
aggregation, effectively slowing the aggregation rates of species by changing
their properties. One can also imagine the use of added proteins for enhancing the forward rate of folding, which would have the same end-result of
increasing yield by decreasing losses due to aggregation. For example, enzymes such as peptidyl-prolyl *cis-trans* isomerase (PPIase) (64) and protein
disulphide isomerase (PDI) (65) might be expected to catalyse the productive
collapse of a folding intermediate.

One limitation to the use of accessory proteins to promote folding may be
their potential sensitivity to the partially denaturing conditions required in the
refolding process.

2.3.10 Thiols and thiol–disulphide equilibria

Aggregation by incorrect disulphide formation is a common problem in
protein-folding experiments. Some disulphide-mediated aggregation can
occur while the protein is in the denatured state, since under strongly
denaturing conditions there is no structural facilitation of correct disulphide
formation. In proteins containing cysteine residues but no disulphide bonds in
the native molecule, the cysteines should be maintained in a reduced state
during solubilization and refolding by the presence of reducing agent and
EDTA and/or low pH. Proteins that require disulphide bond formation to

achieve their native states should also be solubilized under conditions that generate and preserve the reduced sulphydryl forms of cysteine residues. Alternatively, such proteins should be converted to reversibly blocked cysteine forms such as S-sulphonates (66) or mixed disulphides with cysteine or glutathione. These groups prevent undesired intra- and intermolecular disulphide bond formation while the protein is in denaturant, and also can sometimes confer some solubility on to the unfolded protein, even after it is restored to native conditions. At the same time these groups are readily displaced in the early stages of thiol–disulphide exchange-mediated disulphide bond formation.

Protein folding by thiol–disulphide interchange is a process in which the reaction product distribution is determined by the kinetics and thermodynamics of interconversion of the various possible disulphide-bonded states of the protein. Were thermodynamics to dominate, and if all the states were soluble, then one would normally obtain a high yield of native product. Contrary to this simple expectation, however, the product of such refolding reactions is not always exclusively—or even predominantly—the native state. Significant material is often found in disulphide-linked aggregates which are incapable under reaction conditions of re-entering the productive folding pathway.

One convenient way to check for the formation of disulphide-linked aggregates is by non-reducing sodium dodecyl sulphate (SDS) polyacrylamide gel electrophoresis (PAGE). It is important that such an experiment be carried out correctly, however. Before adding an aliquot of the refolding reaction to SDS gel loading buffer, all traces of sulphydryl groups—both small molecular weight and protein-associated—must be removed from the sample. Failure to do this may lead to SDS-mediated facilitation of thiol–disulphide interchange in the gel sample, with loss of intermolecular disulphide bonds. A convenient way to do this is by 'capping' the sample with an excess of iodoacetamide or iodoacetate. A second mistake that is sometimes made is to run a control of a reduced sample in a lane adjacent to the 'capped' sample; the mercaptoethanol in the reducing lane can diffuse through the gel to the adjacent lane, reducing the 'capped' sample so that it runs as a monomer even though it may have been an oligomer. A non-reducing lane should separate the reducing lane from thiol-sensitive samples.

Other practical aspects of thiol–disulphide interchange-mediated refolding have been discussed elsewhere (54).

3. Formation of protein inclusion bodies (IBs) in *vivo*

3.1 Nature of inclusion bodies

Bacteria under certain growth circumstances can accumulate so much of a particular molecular species that it becomes localized in dense intracellular structures called inclusion bodies; these may or may not be surrounded by membrane (67). *Protein-rich* inclusion bodies have been observed when *E. coli* is grown in amino acid analogues, due to the synthesis of non-native analogue-containing protein (68). With the advent of recombinant DNA techniques, protein inclusion bodies are more commonly observed, in the biosynthesis of unnaturally high levels of homologous (69) or heterologous (70, 71) proteins.

IBs are most commonly observed in the cytoplasm, but they can be formed in the periplasm as well (72). Little structural characterization of IBs has been published; surrounding membranes are not observed (68). A crude fraction rich in IBs can be obtained by centrifugation of native lysates (lysozyme or mechanical lysis) and analysed by SDS PAGE of supernatants and pellets (68, 73). Since IBs are very dense, it is also possible to isolate them from the insoluble cellular debris by differential centrifugation (74, 75). Although such procedures can be difficult to carry out in good yield, they can also generate very clean samples for analytical characterization. When isolated, inclusion body preparations often contain some associated *E. coli* protein. Since detergent washes and density gradient centrifugation can sometimes selectively remove much of this *E. coli* material (10, 75), its *in vivo* significance is not clear. Deposition into inclusion bodies appears to allow stable accumulation of a poorly stable polypeptide chain (76). IB formation can also provide an effective initial purification from soluble *E. coli* protein (36). On the other hand, inclusion bodies can complicate recovery of heterologous protein by requiring sometimes difficult refolding steps. Clearly, general methods for controlling IB formation would be valuable.

One can imagine several mechanisms by which IBs might be formed (see Scheme 1):

(a) *Aggregation of the folded state, favoured by limited protein solubility and high expression levels.* It is reasonable to assume that aggregation of 'N' is the basis of IB formation in at least some cases, for example those involving unnatural sequences generated as fusion proteins or truncated proteins, or in some examples of expression of normally membrane-associated proteins, which may have poor solution properties even when in their most stable folded state.

(b) *Aggregation of unfolded states, populated in the cell due to low thermodynamic stability and/or high growth temperature.* Many disulphide-bonded proteins when made in *E. coli* are isolated as IBs, and more often than not the IBs are not intermolecularly disulphide cross-linked. Because the intracellular SH/SS balance may be highly reducing, proteins that require disulphide bonds for thermodynamic stability may exist in the cytoplasm predominantly in the reduced form, leading to the population of unfolded states I or U, which may be more prone to aggregation, at equilibrium.

(c) *Aggregation of folding intermediates, enhanced by long half-lives and/or low solubility of the intermediates.* Recent evidence suggests that, at least in some cases, IB formation must be due to the aggregation of transient, kinetically populated folding intermediates. King and co-workers have identified a series of mutations in the tailspike protein of *Salmonella* phage P22 which cause loss of active phage formation at a restrictive growth temperature of 42 °C. The mutations are not inherently inactivating, since the mutants produce viable phage when grown at a lower, permissive temperature. The defect is not a decrease in the thermal stability of the homo-trimeric tailspike protein, since the mutant proteins, like the wild type, melt near 90 °C. What is affected is the extent to which tailspike protein is deposited into IBs. Although some wild-type protein goes into IBs throughout the growth temperature range, in the mutants so much protein is lost to IBs at the restrictive temperature that there is no native, soluble trimer available to complete phage assembly. In both wild type and mutants the extent of IB formation increases with growth temperature; those mutants examined exhibit a very sharp temperature-dependence of IB formation, consistent with the involvement of a co-operative folding process. Since the data indicate a thermal effect far below the range of the T_m of native trimer, the evidence points to a thermal effect on a folding/assembly intermediate (77).

3.2 Detection and analysis of inclusion bodies

Many protein inclusion bodies are readily detected as refractile particles in phase-contrast microscopy of growing bacterial cells (71). They are also readily detected by electron microscopy (68, 71). IB protein can also be distinguished from other debris-associated protein by its greater density than normal cell debris in low-speed spins (74). In some cases heterologously expressed protein is found associated with the native lysis pellet even though refractile bodies are not observed in the cells by phase-contrast microscopy. This might occur if proteins are trapped in the membrane, but in such a case one might expect detergent washes to readily solubilize the protein. A class of protein IB which does not refract light but which is apparent in electron micrographs of cells has been described recently (10). A quantitative measure of the amount of gene product in IBs can be made by analysing the native lysis supernatants and pellets of the cell by SDS PAGE (68, 73). Although the crude lysis pellet contains membrane and non-specifically associated *E. coli* protein as well as IB protein, a detergent wash of the pellet (Section 2.3.5) can give a relatively clean preparation of IB protein (10).

Mutational effects on the partitioning of heterologous gene product into IBs have been observed in several systems (see Section 3.3.5). Although this suggests that random mutagenesis might be used to generate useful mutants, this requires an efficient selection or screen. Unlike the P22 phage experiments summarized in Section 3.1, most expression systems of interest involve proteins that produce no selectable phenotype. However, there are several ways in which moderate to high throughput screens can be devised using antibodies to the protein of interest. With a protein giving high levels of IBs, with low expression in a soluble form, it should, in principle, be possible to do immunoblots of cells or phage plaques to look for increases in levels of soluble expression. Such a method would be less useful in looking for mutations which increase levels of IBs, with concomitant decreases in levels of soluble protein; in this case immunoblots could not distinguish increased IB mutants from increased proteolysis mutants. An antibody-based screening protocol has been devised which allows identification of both classes of mutants (78). In outline, this method is as follows:

Protocol 2. Antibody-based screening protocol

1. Grow cells in 96-well microtitre plates. Include replicate wild-type cultures (see ref. 79).

2. Replicate transfer aliquots of cultures into two microtitre plates.

3. Lyse one replicate plate by lysozyme lysis.

4. Lyse the other replicate plate with 6–8 M guanidine hydrochloride, which also dissolves IBs. Dilute from guanidine hydrochloride into refolding buffer in another plate.

5. Do ELISA assay on appropriately diluted samples of both plates.

6. Normalize values of each plate by dividing by wild-type values.

7. Ratio of normalized values of guanidine hydrochloride plate to native plate gives IB distribution relative to wild type (>1, more IBs than wild type; <1, fewer IBs than wild type).

3.3 Influencing inclusion body formation

3.3.1 Temperature

The temperature effect on IB formation is a widely observed, but not universal, phenomenon. Besides the P22 tailspike protein, lower temperatures have been reported to favour soluble expression with interferon-β (80), creatine kinase (81), immunoglobulin Fab fragment (82), β-galactosidase fusions (83, 84), human interferon-γ (78, 85), subtilisin E (86), and glycogen phosphorylase (87). It may sometimes be necessary to monitor carefully the time course of expression and IB partitioning in order to observe a decrease in IB levels on decreasing the temperature (84). Reducing the growth temperature does not always improve soluble expression levels significantly. In some cases, it may be that low-enough temperatures were not tried, or were not possible.

For example, *E. coli* must be grown at 22°C to suppress significantly IB formation in the high-level expression of glycogen phosphorylase (87).

The temperature dependence of IB formation suggests a possible general drawback of temperature-induced promoter systems. Although high temperature can induce heat-shock proteins which may help suppress IB formation (see below), it can also promote IB formation.

Thus, soluble expression *in vivo* and soluble refolding *in vitro* (Section 2.3.8) can both be improved in some cases by decreasing the temperature. Of particular interest is the folding of immunoglobulin, where low-temperature suppression of aggregate formation has been observed both *in vivo* (82) and *in vitro* (56) with similar molecules.

3.3.2 Expression level

Increasing the rate or extent of expression of some homologous *E. coli* proteins leads to formation of inclusion bodies (72). Thus, in heterologous expression of proteins featuring high IB formation, it stands to reason that a reduction in the rate or extent of expression might favour soluble expression. However, there are few, if any, examples of any practical benefit clearly attributable to the reduction of expression rate or extent. At the same time, one possible explanation for the low-temperature suppression of IB formation described above is in its expected effect in decreasing the biosynthesis rate.

3.3.3 Growth medium

E. coli producing high levels of pre-β-lactamase deposit some of the unprocessed gene product in cytoplasmic IBs and some of the secreted and processed β-lactamase in periplasmic IBs (72). Growth of cells in broth containing about 0.5 M sucrose almost completely suppresses IB formation, leading to high levels of soluble periplasmic lactamase (88). Similar levels of sucrose are effective in improving the correct refolding of lactamase (at the expense of aggregate formation) *in vitro*. Both *in vivo* and *in vitro* effects may be related to the known stabilizing effects of sugars and other solvent additives on protein stability (49). This effect of sucrose is presumably only possible with secreted proteins, since the sugar has access only to the periplasm. It is not yet clear how general this method of suppressing IB formation in the periplasm will prove to be.

3.3.4 Strains

It is difficult to find conclusive evidence for the role of strain differences on extent of IB formation. Although significant anecdotal information exists in the literature (see, for example, ref. 89), in most cases other differences between experiments obscure the interpretation. In one controlled study, Browner *et al.* report significant differences between two strains in their ability to make soluble glycogen phosphorylase (87). In this comparison, growth temperature, growth medium, vector and promoter, and total expression level were equivalent between the two *E. coli* strains, but one strain produced about 50% of the phosphorylase as IB, while the other produced only a few per cent. The authors note, however, that the optimal strain for wild type is not always the optimal strain for mutants.

Besides existing strains, it might be possible to construct strains with especially good soluble expression properties. There is currently substantial interest in the expression or overexpression of some heat-shock proteins, the 'chaperonins', as potential aids of soluble expression of other recombinant proteins (see Section 2.3.9). In one well-documented case, overexpression of the *E. coli* hsp GroEL allowed soluble expression of the major plant protein ribulose bisphosphate carboxylase, while in the normal background only insoluble, inactive material was made (90). Interestingly, a number of classic heat-sensitive mutations in *E. coli* proteins can be suppressed by overexpression of GroEL (91). Another class of folding-related proteins of current interest are the PPIases (64). These enzymes catalyse the isomerization of peptidyl-prolyl bonds, a class of peptide bond implicated in slow folding steps in several proteins in *in vitro* studies. According to Scheme 1, enhancement of the forward rate of maturation of intermediates to the N state should suppress aggregation side-reactions. It may be that PPIases can play such a role in at least some cases *in vivo*.

3.3.5 Mutation

The work of the King group was the first evidence that single amino acid replacements could significantly influence the partitioning of protein between IBs and the soluble cytoplasm. In the meantime, similar effects have been reported with the *E. coli* proteins Che B and Che Y (92), with human interferon-γ (78), and with human interleukin-1β (93). The sequence of the linker has also been reported to influence the degree of IB formation in expression of a fusion protein (84).

The amino acid sequences of the tailspike protein mutants are consistent with turn *loci* in this protein, whose structure, however, has not been determined. Inclusion body mutations do not occur exclusively at turns, however; considering only the interferon and interleukin examples, mutations that affect IB formation can be located in α-helix, β-sheet, ordered loops, and disordered structure (93). The possible ways in which mutation may be acting to influence inclusion body formation have been discussed elsewhere (2, 93).

In most cases, mutations have been observed to increase IB formation compared to wild type, but in at least two cases the opposite effect has been observed. Second-site revertants have been identified by the King group, which, when introduced into a wild-type background, yield lower IB formation in P22 tailspike protein compared to the wild type (94). Human interferon-γ is made almost exclusively in IBs when the wild type is expressed in *E. coli*, but mutations have been isolated in which most of the protein is soluble in the cytoplasm (78). These examples raise the possibility that mutagenesis might be used to search for useful 'production variants' with improved expression properties, such as decreased IB formation; such variants might be expected to have other useful properties, such as improved stability to irreversible thermal inactivation *in vitro* (12). Such an approach may not be feasible in cases where increased immunogenicity would be undesirable, but may be useful, for example, in the production of industrial enzymes.

Acknowledgements

I acknowledge gratefully the efforts and insights of my co-workers and collaborators who contributed to work described here. In addition, I thank Terry Porter, Chris Jones, Mark Hurle, Boris Chrunyk, and Jim Callaway for useful suggestions and critical reading of the manuscript.

References

1. Volkin, D. B. and Klibanov, A. M. (1989). In *Protein function: a practical approach*, (ed. T. E. Creighton), pp. 1–24. IRL Press, Oxford.
2. Wetzel, R. (1992). In *Stability of protein pharmaceuticals:* in vivo *pathways of degradation and strategies for protein stabilization*, (ed. T. J. Ahern and M. C. Manning). Plenum Press, New York, in press.
3. Georgiou, G. and De Bernardez-Clark, E. (ed.) (1991). *Protein refolding*, ACS Symposium Series. American Chemical Society, Washington, DC.
4. Buchner, J. and Rudolph, R. (1991). In *Current opinion in biotechnology*, (ed. R. Freedman and R. Wetzel), pp. 532–8. Current Biology Ltd, London.
5. Kim, P. S. and Baldwin, R. L. (1990). *Ann. Rev. Biochem.*, **59**, 631–60.
6. Mitraki, A. and King, J. (1989). *Bio/tech.*, **7**, 690–7.
7. Schulz, G. E. and Schirmer, R. H. (1979). In *Principles of protein structure*, (ed. C. Cantor), Springer Advanced Texts in Chemistry. Springer-Verlag, Berlin.
8. Scopes, R. K. (1987). In *Protein purification: principles and practice*, (ed. C. R. Cantor), Springer Advanced Texts in Chemistry, pp. 41–54. Springer-Verlag, Berlin.
9. Creighton, T. E. (1984). *Proteins: structures and molecular properties*, pp. 145–52. W. H. Freeman, New York.
10. Hart, R. A., Rinas, U., and Bailey, J. E. (1990). *J. Biol. Chem.*, **265**, 12728–33.
11. London, J., Skrzynia, C., and Goldberg, M. E. (1974). *Eur. J. Biochem.*, **47**, 409–15.
12. Wetzel, R., Perry, L. J., Mulkerrin, M. G., and Randall, M. (1990). In *Protein design and the development of new therapeutics and vaccines; Proceedings of the Sixth Annual Smith, Kline and French Research Symposium*, (ed. G. Poste and J. B. Hook), pp. 79–115. Plenum, New York.
13. Nikkel, H. J. and Foster, J. F. (1971). *Biochemistry*, **10**, 4479–86.
14. Campbell, I. D. and Dwek, R. A. (1984). *Biological spectroscopy*, pp. 217–37. Benjamin/Cummings, Menlo Park, CA.

15. Mulkerrin, M. G. and Wetzel, R. (1989). *Biochemistry,* **28,** 6556–61.
16. Brems, D. N. (1988). *Biochemistry,* **27,** 4541–6.
17. Vicik, S. and De Bernardez-Clark, E. (1991). In *Protein refolding*, (ed. G. Georgiou and E. De Bernardez-Clark), pp. 180–96. American Chemical Society, Washington, DC.
18. Cleland, J. L. and Wang, D. I. C. (1990). *Biochemistry,* **29,** 11072–8.
19. Bradford, M. M. (1976). *Anal. Biochem.,* **72,** 248–54.
20. Wetzel, R., Perry, L. J., Baase, W. A., and Becktel, W. J. (1988). *Proc. Natl Acad. Sci. USA,* **855,** 401–5.
21. Matthews, B. W., Nicholson, H., and Becktel, W. J. (1987). *Proc. Natl Acad. Sci. USA,* **84,** 6663–7.
22. Matsumura, M., Becktel, W. J., and Matthews, B. W. (1988). *Nature,* **334,** 406–10.
23. Nicholson, H., Becktel, W. J., and Matthews, B. W. (1988). *Nature,* **336,** 651–6.
24. Zhang, X.-J., Baase, W. A., and Matthews, B. W. (1991). *Biochemistry,* **30,** 2012–17.
25. Liao, H., McKenzie, T., and Hageman, R. (1986). *Proc. Natl Acad. Sci. USA,* **83,** 576–80.
26. Matsumura, M. and Aiba, S. (1985). *J. Biol. Chem.,* **260,** 15298–303.
27. Alber, T. and Wozniak, J. A. (1985). *Proc. Natl Acad. Sci. USA,* **82,** 747–50.
28. Matsumura, M., Yasumura, S., and Aiba, S. (1986). *Nature,* **323,** 356–8.
29. Matsumura, M., Yahanda, S., Yasumura, S., Yutani, K., and Aiba, S. (1988). *Eur. J. Biochem.,* **171,** 715–20.
30. Goldenberg, D. (1988). *Ann. Rev. Biophys. Biophys. Chem.,* **17,** 481–507.
31. Wells, J. A. (1990). *Biochemistry,* **29,** 8509–17.
32. Wetzel, R. (1987). *TIBS,* **12,** 478–82.
33. Matsumura, M., Becktel, W. J., Levitt, M., and Matthews, B. W. (1989). *Proc. Natl Acad. Sci. USA,* **86,** 6562–6.
34. Anderson, W. D., Fink, A. L., Perry, L. J., and Wetzel, R. (1990). *Biochemistry,* **29,** 3331–7.
35. Jaenicke, R. (1984). *Angewandte Chemie,* **23,** 395–466.
36. Marston, F. A. O. and Hartley, D. L. (1990). In *Methods in enzymology* (ed. M. P. Deutscher), Volume 182, pp. 264–76. Academic Press, San Diego.
37. Schein, C. H. (1989). *Bio/tech.,* **7,** 1141–9.
38. Brems, D. N., Plaisted, S. M., Kauffman, E. W., and Havel, H. A. (1986). *Biochemistry,* **25,** 6539–43.
39. Light, A. (1985). *Bio/tech.,* **3,** 298–306.
40. Horowitz, P. and Bowman, S. (1987). *J. Biol. Chem.,* **262,** 5587–91.
41. Creighton, T. A. (1986). Patent, WO 86/05809.
42. Marston, F. A. O., Lowe, P. A., Doel, M. T., Schoemaker, J. M., White, S., and Angal, S. (1984). *Bio/tech.,* **2,** 800–4.
43. Kronheim, S. R. (1986). *Bio/tech.,* **4,** 1078–82.
44. Goldberg, M. E., Rudolph, R., and Jaenicke, R. (1991). *Biochemistry,* **30,** 2790–7.
45. Babbitt, P. C., West, B. L., Buechter, D. D., Kuntz, I. D., and Kenyon, G. L. (1990). *Bio/tech.,* **8,** 945–9.
46. von Hippel, P. H. and Wong, K.-Y. (1964). *Science,* **145,** 577–80.
47. Washabaugh, M. W. and Collins, K. D. (1986). *J. Biol. Chem.,* **261,** 12477–85.
48. Schaffer, S. W., Ahmed, A. K., and Wetlaufer, D. B. (1975). *J. Biol. Chem.,* **250,** 8483–6.

49. Timasheff, S. N. and Arakawa, T. (1989). In *Protein structure: a practical approach*, (ed. T. E. Creighton), pp. 331–45. IRL Press, Oxford.
50. Valax, P. and Georgiou, G. (1991). In *Protein refolding*, (ed. G. Georgiou and E. De Bernardez-Clark), pp. 97–109. American Chemical Society, Washington DC.
51. Cleland, J. L. and Wang, D. I. C. (1990). *Bio/tech.*, **8**, 1274–8.
52. Tandon, S. and Horowitz, P. M. (1986). *J. Biol. Chem.*, **261**, 15615–18.
53. Tandon, S. and Horowitz, P. M. (1987). *J. Biol. Chem.*, **262**, 4486–91.
54. Jaenicke, R. and Rudolph, R. (1989). In *Protein structure: a practical approach* (ed. T. E. Creighton), pp. 191–223. IRL Press, Oxford.
55. Mitraki, A., Betton, J.-M., Desmadril, M., and Yon, J. (1987). *Eur. J. Biochem.*, **163**, 29–34.
56. Buchner, J. and Rudolph, R. (1990). *Bio/tech.*, **1**, 157–62.
57. Pelham, H. R. B. (1986). *Cell*, **46**, 959–61.
58. Goloubinoff, P., Christeller, J. T., Gatenby, A. A., and Lorimer, G. H. (1989). *Science*, **342**, 884–9.
59. Buchner, J. (1991). *Biochemistry*, **30**, 1586–91.
60. Martin, J., Langer, T., Boteva, R., Schramel, A., Horwich, A. L., and Hartl, F.-U. (1991). *Nature*, **352**, 36–42.
61. Skowyra, D., Georgopoulos, C., and Zylicz, M. (1990). *Cell*, **62**, 939–44.
62. Brems, D. N. (1990). In *Protein folding: deciphering the second half of the genetic code*, (ed. L. M. Gierasch and J. King), pp. 129–35. American Association for the Advancement of Science, Washington DC.
63. Hall, J. G. and Frieden, C. (1989). *Proc. Natl Acad. Sci. USA*, **86**, 3060–4.
64. Lang, K., Schmidt, F. X., and Fischer, G. (1987). *Nature*, **329**, 268–70.
65. Freedman, R. B. (1989). *Cell*, **57**, 1069–72.
66. Morehead, H., Johnston, P. D., and Wetzel, R. (1984). *Biochemistry*, **23**, 2500–7.
67. Shively, J. M. (1974). *Ann. Rev. Microbiol.*, **28**, 167–87.
68. Prouty, W. F., Karnovsky, M. J., and Goldberg, A. L. (1975). *J. Biol. Chem.*, **250**, 1112–22.
69. Gribskov, M. and Burgess, R. R. (1983). *Gene*, **26**, 109–18.
70. Williams, D. C., van Frank, R. M., Muth, W. L., and Burnett, J. P. (1982). *Science*, **215**, 687–9.
71. Wetzel, R. and Goeddel, D. V. (1983). In *The peptides: analysis, synthesis, biology*, (ed. J. Meienhofer and E. Gross), pp. 1–64. Academic Press, New York.
72. Georgiou, G., Telford, J. N., Shuler, M. L., and Wilson, D. B. (1986). *Appl. Environ. Microbiol.*, **52**, 1157–61.
73. Haase-Pettingell, C. A. and King, J. (1988). *J. Biol. Chem.*, **263**, 4977–83.
74. Fish, N. M. and Hoare, M. (1988). *Biochem. Soc. Trans.*, **16**, 102–4.
75. Bowden, G. A., Paredes, A. M., and Georgiou, G. (1991). *Bio/tech.*, **9**, 725–30.
76. Shortle, D. and Meeker, A. K. (1989). *Biochemistry*, **28**, 936–44.
77. King, J., Fane, B., Haase-Pettingell, C., Mitraki, A., Villafane, R., and Yu, M.-H. (1990). In *Protein folding: deciphering the second half of the genetic code*, (ed. L. M. Gierasch and J. King), pp. 225–40. American Association for the Advancement of Science, Washington, DC.
78. Wetzel, R., Perry, L. J., and Veilleux, C. (1991). *Bio/tech.*, **9**, 731–7.
79. Wetzel, R., Perry, L. J., Veilleux, C., and Chang, G. (1990). *Prot. Engng*, **3**, 611–23.
80. Mizukami, T., Komatsu, Y., Hosoi, N., Hoh, S., and Oka, T. (1986). *Biotech. Lett.*, **8**, 605–10.

81. Chen, L. H., Babbitt, P. C., and Kenyon, G. L. (1990). *FASEB J.*, **4**, A2119.
82. Cabilly, S. (1989). *Gene*, **85**, 553–7.
83. Lee, S. C., Choi, Y. C., and Yu, M.-H. (1990). *Eur. J. Biochem.*, **187**, 417–24.
84. Strandberg, L. and Enfors, S.-O. (1991). *Appl. Environ. Microbiol.*, **57**, 1669–74.
85. Schein, C. H. and Noteborn, M. H. M. (1988). *Bio/tech.*, **6**, 291–4.
86. Takagi, H., Morinaga, Y., Tsuchiya, M., Ikemura, H., and Inouye, M. (1988). *Bio/tech.*, **6**, 948–50.
87. Browner, M. F., Rasor, P., Tugendreich, S., and Fletterick, R. J. (1990). *Prot. Engng*, **4**, 351–7.
88. Bowden, G. A. and Georgiou, G. (1988). *Biotechnol. Prog.*, **4**, 97–101.
89. Tucker, J., Sczakiel, G., Feuerstein, J., John, J., Goody, R. S., and Wittinghofer, A. (1986). *EMBO J.*, **5**, 1351–8.
90. Goloubinoff, P., Gatenby, A. A., and Lorimer, G. H. (1989), *Nature*, **337**, 44–7.
91. van Dyk, T. K., Gatenby, A. A., and LaRossa, R. A. (1989). *Nature*, **342**, 451–3.
92. Krueger, J. K., Stock, A. M., Schutt, C. E., and Stock, J. B. (1990). In *Protein folding: deciphering the second half of the genetic code*, (ed. L. M. Gierasch and J. King), pp. 136–42. American Association for the Advancement of Science, Washington, DC.
93. Wetzel, R. and Chrunyk, B. A. (1992). In *Biocatalys Design for Stability and Specificity* (ed. M. Himmel and G. Georgiou) ACS Symposium series, American Chemical Society Books, Washington, DC. In press.
94. Mitraki, A., Fane, B., Haase-Pettingell, C., Sturtevant, J., and King, J. (1991). *Science*, **253**, 54–8.

9

Principles of protein turnover: possible manipulations

STEFAN JENTSCH and ANDREAS BACHMAIR

1. Introduction

Protein degradation is an essential biological activity of living cells. In several cases protein turnover has a regulatory function, e.g. in reducing the levels of key enzymes of metabolic pathways and in the elimination of cellular regulators. In particular, some transcription factors, repressors, certain products of viral and cellular oncogenes, and crucial cell-cycle regulators, such as the cyclins, are short-lived *in vivo*. Turnover rates for individual proteins can vary considerably depending on the cell type, nutritional and other influences, and the position within the cell cycle. Another function of intracellular protein degradation is the elimination of misfolded, misassembled, mislocalized, damaged, or otherwise abnormal proteins.

A striking feature of intracellular protein degradation is its remarkable selectivity. Substrate-specific protein degradation requires structural features on proteolytic substrates which target these proteins for destruction. These determinants must be recognized either directly by specific proteases or, alternatively, by specific recognition components of more complex proteolytic systems. Thus, a half-life of a given protein in a given cellular environment is probably a function of the affinity of recognition components to available target sites on proteins and of the cellular concentration of these components or other limiting components of a proteolytic system. Experimental evidence suggests that many proteins bear several distinct proteolytic determinants for different recognition components or proteolytic pathways (1). Furthermore, long-lived proteins may be turned into proteins with short half-lives by structural alterations which expose normally cryptic signals. This might be the case for the degradation of abnormal proteins and the regulated degradation of proteins triggered, for instance, by substrate phosphorylation or other modifications which may influence protein structure. In addition to amino acid sequences that stimulate protein degradation, structural determinants have been described which confer enhanced protein stability (2, 3). These sequences, functioning at the N-terminus and C-terminus of a protein, may either influence existing proteolytic determinants or may block degradation by proteases which degrade proteins progressively from the N- or C-terminus, respectively.

A frequent consequence of protein engineering is the rapid degradation of these proteins when expressed in prokaryotic and eukaryotic cells. Unlike natural proteins (including short-lived ones: refs 4, 5), engineered proteins are often thermodynamically unstable. Thermodynamic instability is, in many cases, correlated with proteolytic susceptibility of the engineered protein (6). Given the current unsatisfactory success in predicting protein structure from sequence, estimates about the *in vivo* turnover of proteins are, similarly, difficult to make. Some recent experiments, however, gave the first insights into the nature of the signals that may control protein degradation *in vivo* (7, 8). Although the overall picture is rather incomplete, some of these results, which are discussed in this chapter (summarized in *Table 1*), may provide some hints for designing proteins with a controlled half-life.

2. Protein turnover in prokaryotes

2.1 Degradation pathways

In *Escherichia coli*, selective protein degradation is mediated by a large set of different proteases (for a review see ref. 9). Degradation of most unstable engineered proteins is thought to be mediated by the proteases La (the *lon* gene product) and Ti (the *clp* gene products). Both enzymes are involved in the ATP-dependent degradation of abnormal and natural short-lived proteins (10, 11, 12). The *lon* gene is activated by heat shock and other stresses, including stresses induced by the overexpression of abnormal and foreign gene products (13, 14). *lon* mutants are known to stabilize some short-lived regulative proteins, such as SulA, RscA, and λN and are widely used for gene expression studies (9). The usefulness of mutants should be established for each individual case because in some mutant backgrounds (e.g. *lon* mutants) some proteins (e.g. λcII) are *destabilized* (10). The reason for this phenomenon may be that other proteases with different substrate specificities are induced (10). Protein degradation of many (abnormal) proteins in *E. coli* appears to depend on the activities of major heat-shock (stress) proteins. Stress proteins (which are also expressed at basal levels in unstressed cells) are thought to participate in protein folding and may hold abnormal proteins in a conformation suitable for proteolytic recognition or degradation. Mutants in the genes *dnaK* (hsp70), *dnaJ*, *groEL* (hsp60), *grpE*, and in the stress-specific sigma factor (*htpR*) exhibit severe deficiencies in protein degradation (15, 16). In particular, *lon htpR* double mutants appear to be well suited for the expression of a variety of otherwise unstable proteins (16). Furthermore, certain proteins exhibit prolonged half-lives in mutants in the *hflA* locus which is required for the degradation of λcII protein (17), and in

Table 1. Examples of protein stabilizations and destabilizations

Organism/host	Protein	Type of alteration	References
Stabilizations			
E. coli	Proinsulin	N-terminal extension	2
E. coli	Arc repressor phage P22	C-terminal extension	3
E. coli	Fragment of cI repressor phage λ	C-terminal extension	8
CHO cells	Ornithine decarboxylase	C-terminal truncation	36
COS cells	T-cell antigen receptor (TCRα)	C-terminal truncation, transmembrane domain exchange	24
Xenopus oocytes	Cyclin	N-terminal truncation	37, 45
Destabilizations			
E. coli	Fragment of cI repressor phage λ	C-terminal extension	8
Yeast	β-galactosidase fusion	N-terminal extension	1
Yeast	β-galactosidase fusion	Single amino acid (N-end rule)	7, 31
Yeast	Dihydrofolate reductase	Single amino acid, N-terminal extension (N-end rule)	31

degP mutants. The latter mutants have been shown to stabilize some fusion proteins in the periplasm (18).

2.2 Structural determinants for protein stability

Recent systematic approaches helped to decipher some proteolytic determinants and gave important insights into the mechanisms that control protein stability in *E. coli*. In experiments from one laboratory, the amino-terminal domain of λ repressor was used to determine signals for proteolytic susceptibility (8, 19). Deletion analysis and random site-directed mutagenesis revealed that as few as five non-polar amino acid residues at the very C-terminus of a protein may lead to enhanced intracellular proteolysis without affecting the thermodynamic stability of the protein. Interestingly, degradation of proteins bearing non-polar extensions is not ATP-dependent and occurs in *lon* and *hptR* mutants. This suggests that these proteins are degraded by a pathway distinct from those known to degrade abnormal proteins. Additional studies from the same laboratory (3, 8) showed that charged amino acids at the C-terminus of a protein may have stabilizing influences. In particular, a single charged residue (Asp) has the greatest stabilizing effect when it is at the very C-terminal position and its influence diminishes as its distance from the C-terminus is increased. Remarkably, addition of such sequences to an unrelated metabolically and thermodynamically unstable protein strongly increased the half-life of that protein (3). Similar experiments with one protein (proinsulin) suggest that N-terminal additions of short homopolymers of 6–7 amino acids (Ala, Asn, Cys, Gln, His, Ser, Thr) positively affect the *in vivo* half-life of that protein in *E. coli* (2).

3. Protein turnover in eukaryotes

3.1 Degradation pathways

Eukaryotes have evolved different degradative pathways which can be distinguished by their cellular localization and their substrate selectivity.

Protein degradation via lysosomes is strongly induced under starvation conditions and appears to be rather unselective. Some long-lived cytosolic proteins can be degraded by this system if they bear a specific sequence motif (e.g. Lys–Phe–Glu–Arg–Gln; 20). These sequences are recognized by a member of the hsp70 stress-protein family which may function in the translocation of proteins across lysosomal membranes (21). In the yeast system, a number of mutants deficient in vacuolar (lysosomal) proteases are in use. These mutants have been employed successfully to avoid degradation problems during purification, e.g. after lysis of cells. These mutants, however, probably have no influence on half-lives of short-lived cytoplasmic and abnormal proteins, and are described elsewhere (22).

Recently, a novel proteolytic pathway has been described, which functions in the endoplasmic reticulum (23, 24). This 'ER-degradative pathway' appears to be responsible for the elimination of newly synthesized unassembled or misassembled protein complexes of the secretory pathway but not of cytosolic proteins.

Selective turnover of short-lived and abnormal proteins in eukaryotes is mostly cytoplasmic and ATP-dependent, and is suggested to be mediated largely by a ubiquitin-dependent pathway (see refs 25 and 26 for reviews). Protein degradation by this system involves ATP-dependent covalent attachment of ubiquitin to proteolytic substrates and their subsequent degradation by specific ATP-dependent protease complexes (multicatalytic proteinase complex/proteasome). Regulation and selectivity are mediated predominantly at the level of substrate recognition by the ubiquitin–protein ligase system. Ubiquitin conjugation is a multistep process and requires the activities of a ubiquitin-activating enzyme, E1, and members of a family of ubiquitin-conjugating enzymes, E2s (27). Certain ubiquitin-conjugation reactions need additional protein factors, E3s, for activity. These E3 proteins (or 'ubiquitin-ligases') are thought to recognize proteolytic signals directly and trigger E2-catalysed ubiquitin–protein conjugation (28). From *Saccharomyces cerevisiae* the key E2 enzymes responsible for protein degradation have recently been identified and are encoded by the genes *UBC1*, *UBC4*, and *UBC5* (29, 30). These enzymes are functionally overlapping and are together essential for cell viability. Similar to the stress-inducible *E. coli* protease La, the ubiquitin-dependent proteolysis pathway is induced by heat shock and other stresses. In yeast, ubiquitin (encoded by the yeast gene *UBI4*) and the enzymes UBC4 and UBC5 are stress proteins and essential components of the stress response (29, 31). *UBC4* and *UBC4 UBC5* double mutants have slow growth properties and show strong defects in the degradation of specifically abnormal and otherwise short-lived proteins (29). Moreover, in these mutants certain short-lived engineered fusion proteins are stabilized (32).

3.2 Structural determinants for protein stability

3.2.1 The N-end rule

Systematic approaches in yeast indicated that one simple determinant for proteolytic susceptibility is the first (N-terminal) residue (7, 33, 34). Certain test proteins, where the first amino acid is unblocked (unacetylated) and has a bulky side-chain (Met excluded; the basic: Arg, Lys, His; the hydrophobic: Trp, Leu, Phe, Tyr; the acidic and polar: Asp, Glu, Asn, Gln) are short-lived both *in vivo* in yeast and *in vitro* in reticulocyte extracts. The aforementioned basic and hydrophobic N-terminal amino acid residues are so-called *primary* destabilizing residues (34). As part of a more complex N-end rule pathway, proteins with N-terminal Asp, Glu, Asn, and Gln residues are converted *in vivo* into proteins with short half-lives only after they become modified by cellular enzymes (34). N-terminal Asn and Gln residues (*tertiary* destabilizing) are first converted by a deamidase into Asp and Glu. Catalysed by an aminoacyl-tRNA-protein transferase, N-terminal Asp and Glu residues (*secondary* destabilizing) are further modified by the N-terminal addition of a destabilizing Arg residue (34). Recently, the recognition component of the N-end rule pathway has been identified as an E3 protein of the ubiquitin–protein ligase system (35, 36). This protein has a high affinity to proteins with N-terminal primary destabilizing residues and a low affinity to proteins with N-terminal stabilizing residues.

Further studies revealed that in addition to a destabilizing N-terminal residue, protein substrates for the N-end rule pathway of the ubiquitin system must have an exposed internal Lys residue in close (steric) proximity to the N-terminus (33). This Lys residue serves as an acceptor site for E2-enzyme catalysed ubiquitin conjugation. Notably, a 30-amino-acid fragment with an N-terminal destabilizing residue and an exposed Lys confers a short half-life when N-terminally fused to β-galactosidase, and also confers a short half-life to an unrelated protein, dihydrofolate reductase (33).

Proteins with defined N-terminal amino acid residues can be constructed experimentally. Since the enzyme involved in the generation of termini of most cellular proteins, methionine aminopeptidase, does not remove the initiator Met if the second residue has a bulky side-group, other ways have to be exploited. Probably the best way to generate proteins with defined N-terminal residues in eukaryotic cells is to express proteins as protein fusions with an N-terminal ubiquitin moiety (7, 33). Yeast, and probably all eukaryotes, possess enzymes (ubiquitin C-terminal hydrolases) which precisely cleave ubiquitin from ubiquitin precursors and ubiquitin conjugates. Since these enzymes also work for a variety of synthetic ubiquitin–protein fusions, such gene fusions can be expressed in eukaryotic cells, resulting in a precise removal of the ubiquitin moiety (7). This method has further promising applications and has been used recently for a high-efficiency gene expression system (see below).

3.2.2 The PEST-hypothesis

Recently, it has been suggested that proteins bearing stretches of amino acids rich in Pro, Glu, Ser, and Thr ('PEST'-sequences) are short-lived *in vivo* (37). Computer analysis revealed that these sequences are one characteristic feature of many short-lived proteins, including E1A, c-myc, c-fos, and p53. It has been speculated that PEST-sequences serve as calcium-binding sites which may trigger proteolysis by calcium-dependent calpains (37, 38). However, since experiments testing the influence of PEST-regions on protein stability gave occasionally ambiguous results (39), a PEST-mediated pathway requires further confirmation.

4. Further comments

Regions controlling protein degradation are usually distinct from those directly involved in protein function. Removal of degradation signals can be done on a trial and error basis by introducing simple truncations or internal in-frame deletions. Such manipulations are experimentally easy to perform and quite often lead to dramatic stabilizations of otherwise unstable proteins (40).

To optimize gene expression and facilitate protein purification, proteins are often designed and expressed as fusion proteins. Frequently, however, synthetic fusions are degraded rapidly *in vitro*. In some cases, high-expression yields can be obtained by fusing intact open reading frames directly to a start codon of a high-expression 'ATG-vector' (e.g. by the polymerase chain reaction; 41).

Recently, expression of proteins as ubiquitin–protein fusions has been reported to lead to high levels of protein production in both *E. coli* and yeast (42, 43, 44). More importantly, proteins in native conformation, e.g. in enzymatically active form, were obtained in cases where the unfused protein could not be produced to the same level and quality. Ubiquitin fusions can also be used for high-efficiency expression of short peptides in *E. coli* cells (45). The mechanisms underlying these effects are unclear but may be related to the chaperone-like function of ubiquitin (46). Bacteria apparently lack the ubiquitin system so that, unlike the situation in eukaryotes, ubiquitin fusions are not cleaved *in vivo*. In those cases where removal of the 8 kDa ubiquitin moiety from the N-terminus of the desired protein is not necessary, this method is certainly very promising. If necessary, ubiquitin–protein fusions can be recovered from bacterial extracts and processed *in vitro* by ubiquitin C-terminal hydrolases. Unfortunately, these enzymes are not yet commercially available and experiments suggest that different C-terminal hydrolases exist with distinct ubiquitin–protein fusion substrate specificities (47). However, standard extracts, such as reticulocyte lysates, apparently contain all the necessary activities and could be used instead.

In recent years major research interests have focused on determining protein structure–function relationships and the identification of signals and determinants controlling the fate and destination of proteins. Systematic genetic, biochemical, and biophysical studies have improved our current knowledge in this field of research greatly, and considerably more progress can be anticipated in the near future. However, given our still poor understanding of the mechanisms and principles underlying the recognition and degradation of proteolytic substrates, the design of proteins with predetermined half-lives remains a combination of science and luck.

References

1. Hochstrasser, M. and Varshavsky, A. (1990). *Cell*, **61**, 697.
2. Sung, W. L., Yao, F.-L., Zahab, D. M., and Narang, S. A. (1986). *Proc. Natl Acad. Sci. USA*, **83**, 561.
3. Bowie, J. U. and Sauer, R. T. (1989). *J. Biol. Chem.*, **264**, 7596.
4. Rogers, S. W. and Rechsteiner, M. (1988). *J. Biol. Chem.*, **263**, 19833.
5. Rogers, S. W. and Rechsteiner, M. (1988). *J. Biol. Chem.*, **263**, 19843.
6. Parsell, D. A. and Sauer, R. T. (1989). *J. Biol. Chem.*, **264**, 7590.
7. Bachmair, A., Finley, D., and Varshavsky, A. (1986). *Science*, **234**, 179.
8. Parsell, D. A., Silber, K. R., and Sauer, R. T. (1990). *Genes Dev.*, **4**, 277.
9. Miller, C. G. (1987). In Escherichia coli *and* Salmonella typhimurium, (ed. F. C. Neidhardt), Vol. 1, pp. 680–96. American Society for Microbiology, Washington DC.
10. Gottesman, S., Gottesman, M., Shaw, J. E., and Pearson, M. L. (1981). *Cell*, **24**, 225.
11. Katayama, Y., Gottesman, S., Pumphrey, J., Rudikoff, S., Clark, W. P., and Maurizi, M. R. (1988). *J. Biol. Chem.*, **263**, 15226.
12. Gottesman, S., Clark, W. P., and Maurizi, M. R. (1990). *J. Biol. Chem.*, **265**, 7886.
13. Phillips, T. A., Van Bogelen, R. A., and Neidhardt, F. C. (1984). *J. Bacteriol.*, **159**, 283.
14. Goff, S. A. and Goldberg, A. L. (1985). *Cell*, **41**, 587.
15. Straus, D. B., Walter, W. A., and Gross, C. A. (1988). *Genes Dev.*, **2**, 1851.
16. Baker, T. A., Grossman, A. D., and Gross, C. A. (1984). *Proc. Natl Acad. Sci.*, **81**, 6779.

17. Cheng, H. H., Muhlrad, P. J., Hoyt, M. A., and Echols, H. (1988). *Proc. Natl Acad. Sci.*, **85**, 7882.
18. Strauch, K. L. and Beckwith, J. (1988). *Proc. Natl Acad. Sci.*, **85**, 1576.
19. Pakula, A. A. and Sauer, R. T. (1989). *Ann. Rev. Genet.*, **23**, 289.
20. Chiang, H.-L. and Dice, J. F. (1988). *J. Biol. Chem.*, **263**, 6797.
21. Chiang, H.-L., Terleckey, S. R., Plant, C. P., and Dice, J. F. (1989). *Science*, **246**, 382.
22. Hirsch, H. H., Rendueles, P. S., and Wolf, D. H. (1989). In *Molecular and cell biology of yeasts*, (ed. E. F. Walton and G. T. Yarranton), pp. 134–200. Blackie, London.
23. Lippincott-Schwartz, J., Bonifacino, J. S., Yuan, L. C., and Klausner, R. D. (1988). *Cell*, **54**, 209.
24. Bonifacino, J. S., Suzuki, C. K., and Klausner, R. D. (1990). *Science*, **247**, 79.
25. Finley, D. and Chau, V. (1991). *Ann. Rev. Cell Biol.*, **7**, 25.
26. Hershko, A. (1991). *TIBS*, **16**, 265.
27. Jentsch, S., Seufert, W., Sommer, T., and Reins, H.-A. (1990). *TIBS*, **15**, 195.
28. Ciechanover, A. and Schwartz, A. L. (1989). *TIBS*, **14**, 483.
29. Seufert, W. and Jentsch, S. (1990). *EMBO J.*, **9**, 543.
30. Seufert, W., McGrath, J. P., and Jentsch, S. (1990). *EMBO J.*, **9**, 4535.
31. Finley, D., Özkaynak, E., and Varshavsky, A. (1987). *Cell*, **48**, 1035.
32. Johnson, E. S., Bartel, B., Seufert, W., and Varshavsky, A. (1992). *EMBO J.*, **11**, 497.
33. Bachmair, A. and Varshavsky, A. (1989). *Cell*, **56**, 1019.
34. Gonda, D. K., Bachmair, A., Wünning, I., Tobias, J. W., Lane, W. S., and Varshavsky, A. (1989). *J. Biol. Chem.*, **264**, 16700.
35. Reiss, Y. and Hershko, A. (1990). *J. Biol. Chem.*, **265**, 3685.
36. Bartel, B., Wünning, I., and Varshavsky, A. (1990). *EMBO J.*, **9**, 3179.
37. Rogers, S., Wells, R., and Rechsteiner, M. (1986). *Science*, **234**, 364.
38. Wang, K. K. W., Villalobo, A., and Roufogalis, B. D. (1989). *Biochem. J.*, **262**, 693.
39. Ghoda, L., van Daalen Wetters, T., Macrae, M., Ascherman, D., and Coffino, P. (1989). *Science*, **243**, 1493.
40. Murray, A. W., Solomon, M. J., and Kirschner, M. W. (1989). *Nature*, **339**, 280.
41. MacFerrin, K. D., Terranova, M. P., Schreiber, S. L., and Verdine, G. L. (1990). *Proc. Natl Acad. Sci. USA*, **87**, 1937.
42. Butt, T. R., *et al.* (1989). *Proc. Natl Acad. Sci. USA*, **86**, 2540.
43. Ecker, D. J., *et al.* (1989). *J. Biol. Chem.*, **264**, 7715.
44. Sabin, E. A., Lee-Ng, C. T., Shuster, J. R., and Barr, P. J. (1989). *Biotech.*, **7**, 705.
45. Yoo, Y., Rote, K., and Rechsteiner, M. (1989). *J. Biol. Chem.*, **264**, 17078.
46. Finley, D., Bartel, B., and Varshavsky, A. (1989). *Nature*, **338**, 394.
47. Miller, H. I., Henzel, W. J., Ridgway, J. B., Kuang, W.-J., Chisholm, V., and Liu, C.-C. (1989). *Biotech.*, **7**, 698.
48. Glotzer, M., Murray, A. W., and Kirschner, M. W. (1991). *Nature*, **349**, 132.

PART III

Modification and expression of proteins

10

Chemical approaches to protein engineering

ROBIN E. OFFORD

1. Introduction

This chapter exemplifies the chemical approach to protein engineering. We shall see that such an approach permits the preparation of, among other things, human insulin from porcine insulin, a super-active version of the same hormone, an antibody adapted for tumour location by a regio-specific substitution, proteins with high-abundance isotopic substitution of single atoms (radioactive and stable), specific C-terminal amidation of peptides produced by recombinant means, and various types of protein chimeras. Some of these applications are admittedly quite specialized, but it is clear that the methodology in general has much to offer, not only to research workers concerned with a particular system, but also to those who produce large quantities of recombinant polypeptides for some practical purpose, while encountering some limitation in the capacity of the normal biological expression system to introduce some particular structural feature.

The methods used in the present chapter are chemical, not recombinant, since the latter methods are covered elsewhere in this book. Recombinant methods are often simpler and tend to be convertible to new systems without too much special development; they are adapted to the needs of the 'kit generation' of scientists, and are none the worse for that. Chemical methods can score over recombinant ones if the circumstances happen to be particularly favourable (semisynthetic human insulin took less time and effort to develop as a bulk pharmaceutical than did its recombinant counterpart) and while predictions (e.g. ref. 1) that recombinant methods could readily introduce non-coded substitutions have at last come true (e.g., ref. 2 and references cited therein), chemical methods can be used to reach a much wider range of goals. Also, if circumstances are favourable, they can comfortably be used at a large scale. In any case there should be no conflict, since the most creative developments lie at the interface between the two types of methods, chemical and recombinant, and the publications of Markussen (3) offer an excellent example.

Chemical methods of protein engineering often have as their goal the production of a semisynthetic structure, in which a natural polypeptide is brought into association with an artificial (or chemically modified) one. We have essentially four choices as regards bringing and keeping these components together. These options are non-covalent association, disulphide bridging, peptide-bond formation, and the production of unnatural types of covalent links. All these options are discussed in Section 3 of this chapter, but beforehand we should look at means for specific modification of a protein's functional groups: these methods are useful things in themselves as well as constituting essential initial steps towards the production of complete semi-synthetic proteins.

2. Specific modification of functional groups

Of the 20 natural types of amino acid side-chain, about half can be made to accept chemical substitution under conditions sufficiently mild as to be compatible with the survival of the peptide bond. Of these, the amino, thiol, and carboxyl groups are particularly readily, and usefully, substituted. Many good reviews of this subject exist (e.g. ref. 4).

Since any given type of amino acid residue will tend to occur in a protein in more than one place, chemical methods normally substitute them all. Also, as far as amino and carboxyl groups are concerned, it is rare for the α-amino or α-carboxyl groups of the peptide chain to be chemically distinguishable from their side-chain counterparts, notwithstanding the differences in pK that exist between side-chain and terminal groups. Some of the methods for the general substitution of all available groups are mentioned in Section 2.1. In addition, a very few chemical methods exist that do discriminate, and these are discussed in Section 2.2, but on the whole this discrimination remains one of the most intractable problems for the protein chemist.

Distinction between α- and side-chain groups is a worthwhile goal. For instance, many clinically important peptides are C-terminally amidated but, if produced by recombinant methods, would normally have the C-terminus in the free state. The oxidative enzyme system that carries out such amidations in nature (5) seems to be difficult to apply in practice, and we are left to look for a route that gives C-terminal amidation without converting aspartyl and glutamyl residues to asparaginyl and glutaminyl ones. We shall see in Section 2.2 that it is largely to the preference shown by proteases to act at the α-carboxyl of amino acid residues that we must look for a solution to this problem.

This same distinction, between α- and side-chain groups, is important if we wish to construct protein chimeras through main-chain bonds; moreover, the fact that each chain has only one α-amino group and one α-carboxyl group automatically provides unique locations for site-specific substitution if only one can make use of them (Section 2.2).

2.1 Multiple substitutions

2.1.1 Conventional amino protection

Amino groups can be modified by substituents drawn from the reversible protection strategies of conventional peptide synthesis. The *t*-butyloxycarbonyl (Boc) group is a typical aid-labile substitution and the methanesulphonylethyloxycarbonyl (Msc) group is complementary to it in that it is a base-labile one. Full experimental details for the use of these groups with natural polypeptides are given in refs 1, 6, and the references quoted in ref. 7. These groups are most often used as temporary protection in order to direct elsewhere the action of some subsequent reagents that would otherwise have attacked amino groups. The ionization of the groups is suppressed and, to a different extent depending on the group used, such substitutions therefore alter the balance of hydrophilicity and hydrophobicity in favour of the latter: this may one day prove useful in an industrial context. The conditions for the removal of these groups look severe (anhydrous trifluoroacetic acid for the Boc group, a brief exposure to strong base for the Msc group), but it is astonishing how well many polypeptides and small proteins survive them.

2.1.2 The acetimidyl group

We have already said that most methods of amino substitution do not discriminate in any reliable way between α- and ε-amino groups, in spite of the difference in their pK values. On the other hand, the acetimidylation reaction has long been recognized (8) as an exception to this rule. While discrimination is not perfect, the α-amino group tends to be left free to a substantial and sometimes useful extent (9, 10), even when the ε-group are fully substituted.

The acetimidyl group is also unusual in that the substituted amino group is still ionizable,

$$\text{R--NH--}\overset{\overset{\text{NH}}{\|}}{\text{C}}\text{--CH}_3 + \text{H}^+ \rightleftharpoons \text{R--NH--}\overset{\overset{\text{NH}_2{}^+}{|}}{\text{C}}\text{--CH}_3 \quad (1)$$

though with a slightly higher pK. Proteins that are fully acetimidylated retain their solubility in aqueous media and usually have either virtually completely authentic behaviour (10) or at least a reasonable approximation to it. It is normally only in the limited number of cases in which a lysine side-chain is directly involved in the mechanism of action of the protein that biological activity is lost on acetimidylation (e.g. ref. 11). For this reason, the group is frequently left on the protein, once it has been applied. It can, however, be removed if really necessary (8), although under strongly basic conditions that are not always resisted by the protein.

To take proper advantage of the acetimidylation reaction it is necessary to carry it out with care. The most straightforward mechanism for the acetimidylation reaction is not the only one possible. As pointed out by Browne and Kent (12) another mechanism becomes more significant the lower the pH, which involves the formation, not of the N^ε-acetimidylated lysyl residue, but an N-alkylimidate, e.g.

$$
\underset{\displaystyle |}{\overset{\displaystyle CH_3}{}}
$$
$$
-(CH_2)_4-NH^+=C-OCH_3
$$

which, worse still, can react further to give stable cross-links between lysine side-chains. Wallace and Harris (13) took a variety of conditions, both standard ones and those designed explicitly to diminish or to exaggerate the side-reaction, and showed that the problem can be quite serious unless special precautions, which they fully describe, are taken. Their method, of which the key feature is pH control, represents a compromise between a desire to minimize side-reactions, to preserve biological activity, and to avoid too rapid loss of reagents by hydrolysis. Their procedure could usefully be compared with those of Inman *et al.* (14) in which the retention of biological activity is slightly less of a preoccupation.

As discussed previously (7), several groups active in this branch of protein chemistry have found acetimidylation a rewarding operation, and it deserves to be exploited further.

2.1.3 Other side-chain substitutions

Methyl esterification of carboxyl groups is a simple operation (1). The conditions (e.g. exposure for a few minutes at room temperature to dry methanol in which a known concentration of HCl has been generated by prior addition of the calculated amount of acetyl chloride) seem severe, but polypeptides and small proteins often survive them well. The chances of retaining activity, or of being able subsequently to restore it, can perhaps be increased if the product is not dried down, but diluted in ice-water, and the HCl (normally 0.1–0.5 M) neutralized by careful addition of base. Such exposure to strong acid can induce the reaction known as the N → O acyl shift, but this can often be reversed by a mild basic treatment (e.g. ref. 15).

Cysteine occurs less frequently in proteins than most other types of residue, and the thiol group has a quite distinctive type of chemical reactivity relative to other functional groups. Consequently, cysteine residues are preferred targets for those wishing to introduce a relatively small number of substitutions into defined places, and some applications of this principle are very elegant indeed (e.g. ref. 16).

When pairs of cysteine residues are linked through a disulphide bridge, it is sometimes necessary to break them. This can be done by reduction, preferably with dithiothreitol (17) or, oxidatively, by sulphitylation (18).

$$-CH_2S-SCH_2- + 2SO_3^- \rightarrow -CH_2S-SO_3^- + -CH_2S-SO_3^- \quad (2)$$

This modification (for which full experimental details appropriate to the present topic have been published; ref. 19) has the advantage over simple reduction in that the resulting cysteine sulphonyl residues are stable to oxidation, whereas the free cysteine side-chains will readily re-oxidize to form disulphide bridges, often in the wrong places or at the wrong times. On the other hand, the cysteine sulphonyl groups can be immediately converted to the free —SH form by reduction with a thiol.

2.2 Single, or restricted, substitution: chemical versus enzymic approaches

It is unreasonable to expect a chemical reagent directed at a particular kind of functional group to strike consistently at only one of these in the protein, and leave all the others untouched. A few reactions (e.g. metal-catalysed trans-amination of the α-amino group, with glyoxylate as the acceptor) work by means of stereochemical mechanisms that make use of the specific juxtaposition of atoms of the peptide bond to those of terminal groups, and are, as a consequence, specific for the latter. Some of these reactions have been reviewed with experimental detail (20), but are not exemplified here because, while they certainly work in favourable cases, they are somewhat difficult to control when first applied to new systems.

One or two reagents seem to be relatively specific for the α-amino group: what decides the matter is not just a question of relative values of pK but of the value of Brønsted's β as well (e.g. refs 21, 22) and an example is given in Section 2.2.1.

On the other hand, enzymes are perfectly adapted to distinguish between closely similar groups, and their use revolutionizes the approach to this problem. Proteases normally function in the natural state to catalyse the hydrolysis of the peptide bond:

$$-NH—CH(R_n)—CO—NH—CH(R_{n+1})—CO—$$

$$\downarrow H_2O \qquad\qquad (3)$$

$$-NH—CH(R_n)—COOH + H_2N—CH(R_{n+1})—CO—$$

Many do so by means of a mechanism in which the enzyme's active site is first esterified by the carboxyl group coming from the peptide bond undergoing cleavage, and in which the resulting acyl-enzyme intermediate is then decomposed by nucleophilic attack by water. As always, the same mechanism must be able to work in reverse, and so, at the very least, the theoretical possibility exists of attack on the acyl-enzyme intermediate by the amino group of a peptide or by some other, non-peptidic nucleophile. Reactions of the first type, peptide ligations, are of great academic and practical importance, since the side-chain protection schemes of conventional peptide synthesis are avoided. One enzymic ligation at least is practised at the 100 kg scale in the pharmaceutical industry, and we will discuss ligations further in Section 3.3.

Reactions of the second type, specific C-terminal substitutions.

$$—NH—CH(R_n)—CO—NH—CH(R_{n+1})—CO—$$

$$\downarrow NH_2—R \tag{4}$$

$$—NH—CH(R_n)—CO—NH—R + H_2N—CH(R_{n+1})—CO—$$

can call on a very wide range of non-peptidic nucleophilic components indeed, and examples are given later in the present section.

Reactions catalysed by proteases in ordinary aqueous media proceed overwhelmingly in the direction of hydrolysis, but this is not because of the favourable value for $\Delta G^{o\prime}$ for the process as written in Equation 2. One decisive factor that pushes towards hydrolysis is the high activity of water in such media, and another the tendency, over a considerable pH range, of the newly liberated carboxyl and amino groups to ionize. If we wish to go in the other direction in either Equation 3 or Equation 4, then we will usually have to pay the very considerable energy costs of suppressing these charges. It has been known for a long time that this energy cost can be paid off if, for example, the product precipitates, since mass action will then push the reaction over in the direction of synthesis. But a usefully large solubility difference between substrate and product is rare when the product is a large molecule, and the ingenious idea of trapping the product by some biologically specific affinity process (e.g. ref. 23) has so far been little exploited in practice.

Another stratagem exists which provides an attractive way round the energy problem. This measure, introduced into this field by Laskowski Jr (24), consists of the addition to the reaction mixture of an uncharged but water-miscible solvent, normally termed a 'cosolvent'. Not only does this reduce the effective concentration of water but, if enough of the cosolvent is added, the dielectric constant is very significantly lowered. As a result the pK values of the ionization processes of the α-amino and carboxyl groups are brought closer together. Because it involves a separation of opposed charges, the carboxyl ionization is the most affected. It transpires that, at an apparent pH value midway between the two new pK values, there is a very significant relaxation of the ionization-energy obstacle to coupling.

In practice, the cosolvents most often used include butane-1,4-diol, glycerol, dimethylacetamide, dimethylformamide, and dimethylsulphoxide. The denaturing tendency of these solvents increases roughly as we go along the list and thus, since we must preserve the catalytic activity of the enzyme, it is among the solvents mentioned first that we would look if we wished to reduce the water content to a strict minimum. For instance, a water content as low as 4% is tolerated by trypsin (25), although very high coupling yields can be obtained with as much as 20% water (e.g. ref. 26) and even 50% water allows considerable product formation in many cases. The reader might wish to consult the chapter on protease-catalysed ligations written by Kasche in another volume in the *Practical Approach* series (27). In addition, a thorough review of the theory of all such methods has been published (28) together with an account of the many uses to which they have been put. Also, these two publications, or the original papers that they cite, can be consulted for a complete description of the second, kinetic approach, which does not depend on the use of cosolvents. The kinetic approach is not reviewed here because, while it can be very effective, it requires an extensive programme of preliminary experiments before yields can be optimized, and it is difficult to give an example of general significance. None the less, the literature shows that if the target polypeptide or protein molecule is small enough to permit reverse-phase HPLC as an analytical method, the optimization can be very effective indeed and high yields obtained.

Two further points need to be made. First, it is not necessary for successful ligation or substitution that the cleavage of the original peptide bond and the subsequent coupling occur as a single, concentrated process. Indeed, the synthetic reactions are many times quicker if the cleavage has already taken place.

$$\text{—NH—CH(R}_n\text{)—COOH} + \text{H}_2\text{N—R} \rightarrow$$
$$\text{—NH—CH(R}_n\text{)—CO—NH—R} + \text{H}_2\text{O} \tag{5}$$

It is therefore often simplest, and best, to allow cleavage to take place under normal, hydrolytic conditions, which are then adjusted for synthesis in the same reaction vessel by adding a large excess of the amino component (for ligation) or other nucleophilic compound (for C-terminal substitution) and the cosolvent, changing the pH to the expected midpoint of the new pK values and, usually, adding enough extra enzyme to compensate for any general slowing down of catalytic action due to the pH changes and the presence of cosolvent. All this makes for a quite easy 'one-pot' reaction (53).

The second point is that the approach does not require the peptide substrates to be free of other side-chains that satisfy the specificity requirements of the protease. Such residues can, in fact, be tolerated at internal as well as terminal sites. While the danger is not completely absent in such cases that, while reacting at the C-terminus, one might be cleaving (and perhaps reacting also) at others, the severity of this problem is often overestimated. In particular, one can exploit the greater rapidity of reaction with pre-existing α-carboxyl groups just mentioned. Furthermore, steric and sequential factors can severely limit the number of sensitive sites along the chain even when many potential cleavage sites are present.

However, before looking more closely at these enzymically directed processes, let us examine a case in which purely chemical factors dictate a high degree of specificity.

2.2.1 Selectivity for α-amino groups: α-phenylisothiocyanate

This reaction

$$C_6H_5NCS + H_2N\text{---}CH(R_1)\text{---}CO\text{---}NH\text{---}$$
$$\downarrow$$
$$C_6H_5\text{---}NH\text{---}CS\text{---}NH\text{---}CH(R_1)\text{---}CO\text{---}NH\text{---} \tag{6}$$

is not absolutely specific for the α- as opposed to the ε-amino group, but reaction with limiting, low concentrations of reagent gives some measure of specificity, particularly if the pH is kept at a value low enough to maximize the difference in the degree of ionization of α- and ε-groups. Reference 29 is typical, and a reasonable degree of selectivity can be obtained. The reaction does not constitute, strictly speaking, a reversible protection, but of course the acid treatment which constitutes the second phase of the Edman reaction removes both the group and the N-terminal residue to which it was attached. The removal of this residue sometimes forms part of the desired reaction scheme in any case.

2.2.2 Enzymically directed substitution at the C-terminus: protein active nucleophiles

Here, R—NH$_2$ in Equation 5 is H$_2$N—NH—CO—NH—NH$_2$, i.e. 1,1'-carbodihydrazide. This compound, introduced into this field by K. Rose (30, 31), has a number of very considerable advantages. First, as a hydrazide, it has a pK of the —NH$_2$ groups several units lower than the values encountered for the ordinary aliphatic amino group. In a sense, the pK values of the —COOH and —NH$_2$ components are thus already close enough together that, with carbohydrazide, a cosolvent is no longer essential (although the use of one will increase the yield of the wanted product still further, see *Protocol 1*, footnote *c*). Secondly, the compound is symmetrical, which means that the effective concentration of —NH—NH$_2$ groups is twice that which would be calculated on the basis of milligrams per millilitre. What is more, while not as soluble as its apparent homologue, urea, the carbohydrazide will easily give a 2.5 M solution at room temperature (i.e. 5 M in hydrazide groups). The third advantage is that the product,

$$\text{Protein---NH---CH(R}_n)\text{---CO---NH---NH---CO---NH---NH}_2$$

possesses, regio-specifically, a hydrazide group. This strong nucleophile has a chemical reactivity quite unlike that of any natural functional group of a protein, and this reactivity can be exploited under mild aqueous conditions, to permit specific reaction by a variety of reagents that do not otherwise touch the protein (see Section 2.3).

Protocol 1 describes the reaction with insulin as a small protein test substrate, recommended as an initial step in gaining familiarity with these processes. The same procedure works well with much larger proteins, including antibody fragments (30, 53) and intact antibodies (B. Dufour, I. Fisch, and R. E. Offord, unpublished work).

Protocol 1. Production of a protein specifically modified at the C-terminus with carbohydrazide

1. Prepare *des*-Ala^{B30}-insulin by incubating 100 mg zinc-free insulin[a] with 0.5 mg lysyl endopeptidase from *Achromobacter lyticus* (Wako Pure Chemical Co.) in 20 ml of freshly made 0.1 M ammonium bicarbonate solution for 30 h at 37°C, followed by lyophilization of the digest.[b]

2. Make a 2.5 M aqueous[c] solution of carbodihydrazide and adjust it to pH 5.5[d] (glass electrode, uncorrected) with glacial acetic acid.

3. Dissolve 20 mg of the *des*-Ala^{B30}-insulin in 1 ml of the carbohydrazide solution at room temperature.

4. Add 2 mg of the *Achromobacter* protease[e] in 0.2 ml water.[c]

5. Incubate at room temperature for 6 h.[f]

6. Stop the reaction by acidification to an apparent pH of 3 (glacial acetic acid) or by adding a tenfold excess over enzyme (w/w) of Trasylol[g,h] (Bayer) as a 78 mg/ml aqueous solution, or by HPLC.[i]

---

[a] Insulin can be freed of zinc by exhaustive dialysis of a freshly made solution of 2.5 mg/ml zinc insulin (porcine, Novo) in fresh 8 M urea, against 1% acetic acid.

[b] Here, the solution is lyophilized for the sake of convenience. As stated in the text, protease digests, particularly when the target molecule is subject to denaturation, can just as well be adjusted to the final coupling conditions by the direct addition of the various components mentioned in steps **2–4**. If this is to be done, we prefer to adjust the protein concentration before digestion to at least 10 mg/ml, using a Centricon-10 membrane concentrator cell (Amicon). When calculating the quantities to add, it is necessary to take account of the increase in volume (about 10%–15%) caused by the addition of carbohydrazide to a final concentration of 2.5 M. In some cases, Tris–HCl gives better results than NH$_4$HCO$_3$.

[c] All the solutions mentioned in steps **2–4** can be made in dimethylsulphoxide–water 1:1 (v/v). This increases the conversion yield from roughly 70–80% to almost quantitative.

[d] At pH values lower than 5.5, the reaction is slower, although it tends to the same end-point. As one goes above pH 5.5, the reaction becomes progressively more rapid, but so, in general, does any tendency to side-reactions involving cleavage. The optimum pH must be determined for each target system, but it is very often within the range pH 6 ± 0.5.

[e] This enzyme acts only on lysyl residues. An alternative is to use the same volume of a 10 mg/ml solution of trypsin (porcine). It is probably a good general principle to use TPCK-treated trypsin but in the present instance it makes no difference. Conversion yields are about the same as for the *Achromobacter* protease. After 4 h tryptic incubation in the presence of dimethylsulphoxide only about 1% of the protein has been cleaved at Arg^{B22} (even less in the absence of dimethylsulphoxide) in spite of the susceptibility of this residue to trypsin.

[f] In the dimethylsulphoxide mixtures, reaction is nearly complete after 15 min, whereas, in the absence of dimethylsulphoxide, the reaction goes about one-quarter of the way to its end-point during the first 15 min.

[g] Porcine trypsin's active range extends surprisingly far down in the pH scale.

[h] The inhibitor is the best choice for proteins that cannot, unlike insulin, withstand acidification. It is active against both trypsin and the *Achromobacter* protease.

Protocol 1. *Continued*

[i] For analytical, as opposed to preparative runs, direct application for HPLC in any normal, acidic system will do. Full details for the two systems that we use are given in ref. 31. However, even for analytical runs, direct application is only possible when the pH is, as in the present case, at 5.5 or below: the buffering power of small samples will then tend to be overcome at once by the HPLC eluant, and the pH will not be above 3 for any significant length of time. On the other hand, if the reaction pH is too high, or if the sample is present in preparative quantities, there will be a significant degree of buffering power with respect to the HPLC system. If the pH rises even a little, there is so much protease present that the aqueous dilution of the reaction mixture is followed very swiftly by reversal of the coupling reaction and loss of most of the product.

2.2.3 Enzymically directed substitutions at the C-terminus: protein aldehydes

The aldehyde group is another that answers to the twin criteria of having a reactivity unlike that of natural functional groups, and reactivity that can be exploited under mild aqueous conditions. Reactions 7 and 8 taken together indicate one method of introducing it, site-specifically, into a protein (30, 31, 32).

$$—NH—CH(R_n)COOH + H_2N—CH_2—CHOH—CH_2NH_2$$
$$\downarrow$$
$$—NH—CH(R_n)CO—NH—CH_2—CHOH—CH_2NH_2 \qquad (7)$$

$$—NH—CH(R_n)CO—NH—CH_2—CHOH—CH_2NH_2$$
$$\downarrow$$
$$—NH—CH(R_n)CO—NH—CH_2—CHO \qquad (8)$$

The choice of nucleophile for Equation 7 was governed by the fact that periodate oxidation, well known for vicinal diols ($—CHOH—CHOH—$) is even more effective for 1-amino, 2-hydroxy compounds. The reaction in Equation 8 is 1000–2000 times faster than the oxidation of a diol. This means that only a very slight excess of periodate is required and that, surprisingly, the reaction can even take place in a solution quite rich in ethylene glycol. The 1-amino, 2-hydroxy compound has first call on the periodate, and the glycol then takes what is left. Thus, the chances are greatly reduced of damaging side-reactions with the protein, such as will be discussed in Section 2.2.5.

The chosen nucleophile in Equation 7, $H_2NCH_2—CHOH—CH_2NH_2$, is once again symmetrical and its effective working concentration is thereby doubled. But, as a normal primary amine it has, unlike carbohydrazide, an ordinary pK and requires a co-solvent (for practical information, see *Protocol 1* footnote *c*, and *Protocol 2*).

As was true in Section 2.2.2, we have again produced, in a site-specific way, a protein with a chemical reactivity not found in nature; one that, as we shall once again see in Section 2.3, can be exploited for further specific reactions under mild, dilute conditions.

2.2.4 Protein and peptide α-amides by reverse proteolysis

If R—NH$_2$ in Equation 5 is NH$_3$, we might hope to obtain a specifically C-terminally amidated product, and this reaction has been shown to work in practice, though not in particularly high yield (34). If R—NH$_2$ in Equation 5 is an amino acid amide, the result is once again a specifically amidated product, but the yields are now much higher (33, 34; A. S. Calne, K. Rose, and R. Offord, in preparation). This approach is a promising alternative to the natural amidating system (5), particularly for large-scale applications.

2.2.5 Protein aldehydes: chemical oxidation of sugars

This method is applicable only to glycoproteins, and relies on the more familiar reaction between periodate and vicinal diols. However, it should be given serious consideration whenever a glycoprotein is the subject of interest (and many such molecules are important), and particularly when there is a single sugar moiety or when sugars occur at only a few sites. Naturally, if there are too many glycosylated regions, then oxidation may produce too many aldehydic sites.

The idea of producing protein aldehydes in this way for subsequent specific reaction with other molecules is due to Murayama *et al.* (35) and was subsequently revived to good effect, notably with reference to antibodies, by Rodwell (36).

Antibodies frequently have only one major glycosylated site (although this rule is less universal than is often supposed). While there are normally several sugar residues at this position, and thus multiple aldehydic sites are generated, the fact that they are concentrated in one region of space, far from the antigen-binding sites, often minimizes any disadvantage that might otherwise ensue.

More serious is the need, given that most sugar diols do not have the extreme reactivity to periodate that is shown by the 1-amino, 2-hydroxy compounds mentioned previously, to use higher concentrations of periodate and the impossibility of having glycol present as an *in situ* scavenger. Under these conditions, sensitive residues, particularly cysteine, methionine, and tryptophan, can be expected to undergo oxidative attack to some extent, and this may be the reason for the total loss by some antibodies of binding activity after periodate oxidation, even where the latter process is attempted under the very mildest conditions compatible with a reasonable yield of aldehyde.

The reader should be aware of these potential complications, but the procedure is simple enough (e.g. ref. 36) to make it worthwhile to try it. Section 2.3 seeks to reinforce the point that it is often very valuable to have easy access to protein aldehydes.

2.3 Second substitutions

Methods were discussed in Section 2.2 that give us, with site- or regio-specificity, protein active nucleophiles and protein aldehydes. These derivatives permit the formation, under conditions that are normally compatible with the retention of biological activity, of very useful conjugates with molecules having appropriate complementary reactivity.

2.3.1 Coupling to protein aldehydes: nucleophilic substituents

Amino-oxy compounds (i.e. derivatives of hydroxylamine) readily form oxime linkages with protein aldehydes.

$$\text{(Protein chain)}-\text{CHO} + \text{H}_2\text{N}-\text{O}-\text{CH}_2-\text{R}$$
$$\downarrow$$

$$\text{(Protein chain)}-\text{CH}=\text{N}-\text{O}-\text{CH}_2-\text{R} \qquad (9)$$

In principle, the lower the pH, the less stable the linkage, but the degree of acidity needed to cleave the bond is normally more extreme than most proteins can stand, and little loss is seen even at pH 2. Oxime linkages become even more stable at high pH values, but they then form much more slowly: pH 4.5 is a good compromise between rate and position of equilibrium (37) and we find this a useful value when, for example, conjugating a radio-active chelator to an antibody (30, 38). This value is also often acceptable from the point of view of protein stability, but the reaction can often be used productively at higher or lower pH values.

We find aminooxy compounds to be extremely useful: ref. 38 gives typical instructions for making such compounds.

As has been said, the oxime linkage can be regarded as essentially a stable one under physiological or near-physiological conditions. The reductive stabilization of this bond is therefore seldom necessary. On the other hand, the hydrazone linkage is slowly but certainly destroyed, and has therefore been envisaged for certain clinical applications that require slow release of the conjugated molecule (39).

To give complete stability, the hydrazone linkage can be stabilized by cyanoborohydride reduction

$$\text{(Protein chain)}-\text{NH}-\text{NH}_2 + \text{HCO}-\text{R} \rightleftarrows \text{(Protein chain)}-\text{NH}-\text{N}=\text{CH}-\text{R}$$
$$\downarrow$$

$$\text{(Protein chain)}-\text{NH}-\text{NH}-\text{CH}_2-\text{R} \qquad (10)$$

but the product should be checked for success or failure by chromatography after some appropriate strong acid treatment, since cases are known of inexplicable failures. Several literature descriptions of purportedly successful reductions of this kind give no data from such tests and, since the unreduced hydrazone linkage decomposes only slowly at physiological pH, the successful isolation of a conjugate cannot in such cases be taken as proof that it represents the reduced form.

2.3.2 Coupling to protein active nucleophiles: aldehydic substituents

It is well known that seryl and threonyl compounds with a free α-amino group are extremely rapidly oxidized by periodate (40) and they thus give compounds ready for coupling to the protein nucleophiles described above. This idea can be applied, not only to seryl or threonyl derivatives of small, non-protein molecules (30) but to whole proteins, or fragments, in which serine or threonine is N-terminal. Such residues, when within the chain, are not, of course, affected.

Alternatively, a pre-existing amino-oxy compound can be converted into an aldehydic one (31). We find this a very useful approach, since it permits us to keep stocks of amino-oxy compounds and adapt them to the aldehydic form at will.

$$\text{HCO}-m-\text{C}_6\text{H}_4-\text{CHO} + \text{H}_2\text{N}-\text{O}-\text{R} \underset{\leftarrow}{\rightarrow}$$
$$\text{HCO}-m-\text{C}_6\text{H}_4-\text{CH}=\text{N}-\text{O}-\text{R} \tag{11}$$

Alternatively, any other appropriate method (i.e. acylation of an amino group by acetal-protected aldehydic carboxylic acids) could be adapted to this purpose almost as easily.

3. Working with protein fragments

Semisynthetic operations, like their counterparts in total synthesis, involve building up a molecule by bringing smaller segments together. This can be done by exploiting non-covalent interactions, disulphide bridging, ordinary peptide bonds (made chemically or enzymatically), or other, non-peptide covalent bonds.

3.1 Chimeric proteins: non-covalently associating systems

The three-dimensional structure of a protein depends so much on the non-covalent interaction between the side-chains that a single nick somewhere in the polypeptide chain does not necessary permit the two parts of the chain to fall apart. (In rare cases, more than one nick can be tolerated.) Frequently, even biological activity is retained. If the interactions are broken in some denaturing system, and the two fragments separated, those same non-covalent forces will often, if the fragments are once again mixed together in a non-denaturing medium, bring them back together in the original configuration, with consequent restoration of activity. The implications of these points for the production of semisynthetic analogues have been known for many years, and have been reviewed (1, 7).

3.1.1 Discovery of non-covalent associating systems

The information obtained with non-covalently associated analogues has often been of great value, but the procedures involved are normally so specific to the particular system being studied that they are probably not appropriate for detailed description here. On the other hand, it is highly likely that many more such complementary systems remain to be discovered, including other new proteins, or new proteolytic cleavages of old ones. References 41 and 42, and the papers cited therein, give useful indications as to how to discover such complexes.

3.1.2 Separation and recombination of non-covalently associating fragments

Reference 41 gives details which are typical of procedures for the separation and controlled recombination of associating fragments, and these principles are readily adaptable to other systems.

3.2 Chimeric proteins: disulphide bridges

Another quick route to semisynthetic analogues is provided by those systems in which separate segments of polypeptide chains are linked, not non-covalently, but by disulphide bridges. If these bridges are broken, as described in Section 2.1.3, and the fragments separated, one of them can be mixed with an appropriate modified or synthetic partner and the bridge reformed. This method has given us much valuable information on insulin, and has also contributed to knowledge of antibody structure and function. Reference 43, notable for the fact that the semisynthetic end-product has five times the biological activity of the natural, parent molecule, describes the re-formation of the disulphide bridges under carefully chosen, optimized conditions.

3.3 Chimeric proteins: formation of peptide bonds through chemical activation

Many chemical methods of forming peptide bonds exist and have been reviewed, both in general (44) and in the specific context of semisynthesis (1, 7). We restrict ourselves here to the use of the so-called active esters for the addition of single amino acid residues to the N-terminus of a peptide chain.

3.3.1 Chemical coupling with active esters

The chemical process involved is

$$-NH-CH(R_n)-CO-O-R + H_2N-CH(R_{n+_1})-CO-$$
$$\downarrow$$
$$-NH-CH(R_n)-CO-NH-CH(R_{n+1})-CO- \tag{12}$$

It is, of course, necessary to protect all amino groups present, with the exception of the α-amino group of the chain that is to be extended. This protection should normally be reversible. Full experimental details typical of the many successful examples of this approach are given in ref. 45 (the production of an insulin specifically tritiated at one residue only) and ref. 46 (the production of analogues of cytochrome *c*).

3.4 Chimeric proteins: formation of peptide bonds through enzymic ligation

The ability of proteases to work usefully in reverse has already been explained and exemplified in Section 2.2. In that section, we were more concerned with incoming nucleophiles that were not amino acid or peptide in nature; we will now examine the very productive use of the same methodology for the formation of true peptide bonds.

3.4.1 Human insulin from porcine insulin

The process carried out in *Protocol 2* is indicated in the equation:

$$\tag{13}$$

$$\text{Lys–Ala+Thr(Bu}^t\text{)–OBu}^t \rightleftharpoons \text{Lys–Thr(Bu}^t\text{)–OBu}^t\text{+Ala}$$

See ref. 47 for some remarks on the mechanism. The resulting *tert*-butyl derivative of human insulin is readily converted to the free hormone by acid treatment. *Protocol 2* constitutes a convenient and convincing practical introduction to the field of reversed proteolysis for those contemplating an application of their own. It might be found instructive to follow the reaction as a time course by analytical HPLC. Even when, as described in *Protocol 2*, insulin is the substrate, the half-time is only about 10 min; when *des*-Ala^{B30}-insulin is taken (cf. *Protocol 1*), the half-time is nearer 2 min.

The same practical approach as that described in *Protocol 2* can be readily adapted to the use of esters and amides of many other amino acids. It can equally be adapted to other protein targets, if need be with a range of different proteolytic enzymes. It is, of course, necessary that the target protein's solubility, enzyme susceptibility, and tendency to denaturation in co-solvent mixtures are satisfactory for the purpose. The general factors affecting the choice of certain of the reaction conditions were discussed in ref. 25.

It is noteworthy that trypsin, the enzyme that carries out the removal of alanine-B30 and its replacement by the threonine derivative, ought in principle also to cut at arginine-B22. That it does not do so is an illustration of the point made earlier, i.e. enzymes that ought to attack multiple sites in the protein target can none the less be effectively used at only one of these sites.

This is more easily explained when the enzyme has only to ligate (that is, in the present case, if alanine-B30 has been removed beforehand) since we have seen that ligation is much faster than intrachain cleavage. On the other hand, the successful removal in some procedures of residue B30 by trypsin with the virtually complete avoidance of cleavage at arginine-B22 was, when first discovered (26), quite surprising.

An analogous procedure to *Protocol 2* is currently used (26) to supply a significant fraction of the world's diabetics with human insulin made by semi-synthetic means. Through their simplicity, such procedures present no apparent obstacles to scaling up to the 100 kg level.

If *Protocol 2* is carried out using $H_2{}^{18}O$, it will introduce ^{18}O into the peptide bond B29–B30 (48). This label becomes non-exchangeable as soon as the trypsin is removed or inactivated. Alternatively, tritium can be introduced, site-specifically at position B30, if a ^3H-labelled threonine derivative is used (45).

Protocol 2. The enzymic conversion of pig insulin into human insulin

1. Mix 231 mg threonine (Bu^t)–OBu^t (i.e. threonine protected by tertiary butylation of both the side-chain —OH and the α—COOH, obtainable from Bachem AG) with 710 μl of butane-1,4-diol and 40 μl of glacial acetic acid.

2. Add 10 mg of porcine insulin (zinc-free, see *Protocol 1*) to 150 μl of the above mixture, followed by 5 μl H_2O.[a] Stopper the tube, incubate at 37°C for 10 min and agitate the mixture from time to time: the insulin disperses but does not go fully into solution.

3. While the above incubation is taking place, dissolve 5 mg of porcine[b] trypsin (previously treated with TPCK, 1-chloro-4-phenyl-3-tosylamido-butane-2-one, Worthington) in 40 μl of water.[a] Use gentle agitation with a fine stirring-rod.[c]

4. Add 5 μl of the trypsin solution to the insulin suspension with immediate and rapid mixing.[d]

5. Stopper the tube once again and incubate at 37°C with occasional mixing. The remaining insulin soon goes into solution.

6. After 2 h, cool to 0°C and acidify the mixture with an equal volume of glacial acetic acid.

7. Collect the insulin fraction by gel filtration (Sephadex G50 fine, 60 cm × 1 cm diam. in 1% acetic acid[c] at 50 ml/h) followed by lyophilization.[e]

8. Remove the tertiary butyl groups by dissolving the product in 400 μl of trifluoroacetic acid (Pierce, or redistilled) at 22°C for 45 min. Then blow off the majority of the acid in a stream of dry filtered air or nitrogen, and

precipitate the protein with 3 ml of sodium-dried diethyl ether. Centrifuge, wash the pellet in a further 3 ml of diethyl ether, centrifuge, and dry in a vacuum desiccator.

---

[a] Reference 48 describes the use of $H_2^{18}O$ for the production of an isotopically labelled insulin. In the interests of economy, the very minimum quantity of the trypsin solution would then be made up for step **3**. On the other hand, it is worthwhile to dry the proteins down from a little $H_2^{18}O$ beforehand, as this displaces bound $H_2^{16}O$, with a slight but measurable increase in the isotopic abundance of the final product.

[b] The porcine form of trypsin is much more resistant to solvent denaturation than is the bovine form. It also continues to act at remarkably low pH values: hence the rather severe acidification in step **6** (see also footnotes *h* and *i* to *Protocol 1*).

[c] This is a very concentrated solution of protein and there is a risk of the enzyme's denaturing with the formation of a gel. This denaturation is promoted by acid-washed glass surfaces, and so glassware should be cleaned by pyrolysis. By far the most effectively controlled method for such cleaning is to install in the laboratory an ordinary domestic oven equipped for pyrolytic self-cleaning (which will cost a great deal less than most items of scientific apparatus). Simply rinse glassware with distilled water and place it in the oven for a normal pyrolysis cycle.

[d] When larger quantities are used (the protocol will scale up without any difficulty by a very large factor indeed) the amount of heat released by dilution of the organic solvent is considerable, and a temperature rise must be prevented. The heat released can, for example, denature the enzyme very rapidly, since even porcine trypsin is not all that stable when there is so little water present.

[e] If the gel-filtration step were to be omitted, or if (particularly with a protein substrate of higher molecular weight than insulin) the gel filtration did not completely remove the trypsin, then this enzyme would still be present in the ether precipitate. Although it would have been largely inactivated by this time, so much was initially present that even a relatively small proportion remaining active could rapidly destroy the product. Should this danger present itself, all aqueous solutions of the precipitate should be kept below pH 3 until some appropriate separating system can be used to eliminate the remaining proteolytic activity. Possibilities are HPLC with an acidic solvent system, as mentioned in *Protocol 1*, or ion-exchange chromatography in 8 M urea (in which trypsin is not active, see ref. 25). The use of an appropriate inhibitor (see *Protocol 1*, footnote *h*) is also worth considering.

3.4.2 Fragment condensation

The incoming nucleophile does not have to be as small as a single amino acid derivative, and larger peptides can be used (49). However, efficiency suffers, as larger and larger peptides are chosen. This is less from increasing steric hindrance than from problems related to the concentration of reactants. Even if the larger fragments are readily soluble, the quite high molar concentrations required to drive the process as far as possible are increasingly difficult to reach as the molecular weight increases. It is not easy, when working with a decapeptide as the amino component in the coupling, to keep the molar concentration the same as that in *Protocol 2*, since the concentration in mg/ml will need to be more or less 10 times that of the derivative of a single amino acid. None the less ref. 49, and others like it, show that such couplings will work.

3.4.3 Enzymic coupling to active esters

Despite the similarity of the title of this section to that of Section 3.3.1, the process involved is quite different here: the incoming nucleophile is an active ester *without* protection of its α-amino group

$$—NH—CH(R_n)—COOH + H_2N—CH(R_{n+1})—CO—O—C_6H_4Cl_2$$

$$\downarrow \text{ (enzymic coupling)}$$

$$—NH—CH(R_n)—CO—NH—CH(R_{n+1})—CO—O—C_6H_4Cl_2$$

$$H_2N—CH(R_{n+2})CO— \qquad (14)$$

$$\downarrow \text{ (chemical coupling)}$$

$$—NH—CH(R_n)—CO—NH—CH(R_{n+1})—CO—NH—CH(R_{n+2})—CO—$$

Rose *et al.* (50) selected the dichlorophenyl ester as being one that, at an apparent pH of 5, had a sufficiently slow reactivity to be protected from self-polymerization by the protonation of the amino group, and was sufficiently stable to hydrolysis to permit the isolation of the peptide active ester product of Equation 14. Hydroxysuccinimido esters were, for example, too reactive to give satisfactory results. On the other hand, some other esters were found to have been like the dichlorophenyl one in permitting the successful formation and isolation of the peptide active ester, but they did not confer sufficient reactivity on the product to enable it to participate efficiently in the second stage of Equation 14.

This procedure, merging as it does the enzymic and chemical approaches, has several practical applications. For example, while having some of the convenience of an enzymic ligation scheme, it can be used to make bonds that would be difficult to form by enzymic methods alone: e.g. if residue (n+2) in Equation 14 were proline, or a D-amino acid. Among the practical examples illustrated in the literature are those in refs 15 and 52. Wallace (51) wished to test the effect of excising the so-called 'bottom' or 'Ω' loop of the 104-residue protein cytochrome *c*. The beginning and end of the segment of sequence that constitutes the loop (residues 40–55) are close together in space, and could be bridged by peptide bonds if the loop was excised and a single residue inserted in its place. This was done with remarkable efficiency. The dichlorophenyl ester of alanine was first added to the C-terminus of the proteolytic fragment (1–38) of cytochrome *c* (cf. the upper part of Equation 14). When mixed with proteolytic fragment (56–104) the resulting peptide active ester coupled in excellent yield, doubtless assisted by the non-covalent forces discussed in Section 3.1. The benefit of this conformational assistance was retained even when the same principle was used to introduce sequence variations (coded and non-coded) at or near the bridging residue (52).

This method has probably still much to offer.

3.5 Chimeric proteins: non-peptide bonds

If we are released from the constraint of using natural (i.e. disulphide or peptide) bonds to link our fragments, the possibilities are endless. The catalogue of the Pierce Chemical Company lists a great number of bifunctional agents that can be used to link proteins together, most often by placing the ends of the linker on to lysine side chains of the two molecules to be joined, but the preceding sections should have given a picture of how protein–protein conjugation can be accomplished in a more controlled, site-specific way. It would, for instance be possible to make a tail-to-tail dimer with a main-chain oxime link in which the two reactants are those discussed in Sections 2.2.2 and 2.2.3. This tail-to-tail type of dimer should be adaptable to many purposes: for instance if the monomers were suitably activated Fab fragments of antibodies, we could presumably prepare stable, heterofunctional $F(ab)'_2$ analogues. R. E. Offord and B. Dufour have also made head-to-tail chimeras, between a C-terminal activated protein nucleophile prepared as described above and one of the specific N-terminal aldehydic derivatives mentioned in Section 2.2. Such an operation could be useful, for example, as a pilot study to see if some favourable biological property made the production of a recombinant version (but with true peptide bond linkers, naturally) worthwhile.

4. Conclusion

To have given detailed protocols for all the operations described in this chapter would have made it so long as to have distorted the balance of the whole book. It is hoped that interested readers might try the two model operations on insulin given in *Protocols 1* and *2*. With luck, this should whet the appetite for more adventurous applications of the same general principles which, as stated at the beginning of this chapter, can in favourable cases give valuable results that cannot be obtained by any other type of methodology.

Acknowledgements

I thank my past and present colleagues, particularly Keith Rose, for many original contributions to the work described above, and my friends outside my own group for a free and productive exchange of ideas over the years. The academic work of our own laboratory is supported by the Fonds National de la Recherche Suisse, the Ligue Suisse contre le Cancer, and the Krebsliga der Zentralschweiz, to whom I offer my sincere thanks.

References

1. Offord, R. E. (1980). *Semisynthetic proteins*. John Wiley, Chichester.
2. Bain, J. D., *et al.* (1991). *Biochemistry*, **30**, 5411–21.
3. Markussen, J., Hougaard, P., Ribel, U., Sorensen, A. R., and Sorensen, E. (1987). *Prot. Engng*, **1**, 205–33.
4. Means, G. E. and Feeney, R. E. (1990). *Bioconjugate Chem.*, **1**, 2–12.
5. Bradbury, A. F., Finnie, M. D. A., and Smyth, D. G. (1982). *Nature*, **298**, 686–8.

6. Geiger, R., Obermeier, R., and Tesser, G. I. (1975). *Chem. Ber.,* **108,** 2758–63.
7. Offord, R. E. (1990). In *Protein design and the development of new therapeutics and vaccines*, (ed. J. B. Hook and G. Poste), pp. 253–82. Plenum, New York.
8. Hunter, M. L. and Ludwig, M. J. (1962). *J. Am. Chem. Soc.,* **84,** 3491–504.
9. Mahrenholz, A. M., Flanders, K. C., Hoosein, N. M., Gurd, F. R. N., and Gurd, R. S. (1987). *Arch. Biochem. Biophys.,* **257,** 379–86.
10. Nureddin, A. and Inagami, T. (1975). *Biochem. J.,* **147,** 71–81.
11. Mahoney, P. J. (1982). Ph.D. thesis. University of Oxford.
12. Browne, D. T. and Kent, S. B. H. (1975). *Biochem. Biophys. Res. Commun.,* **67,** 126–32.
13. Wallace, C. J. A. and Harris, D. E. (1984). *Biochem. J.,* **217,** 589–94.
14. Inman, J. K., Perham, R. N., Dubois, G. C., and Appella, E. (1983). In *Methods in enzymology* (ed. C. H. W. Hirs), Vol. 91, pp. 559–69. Academic Press, New York.
15. Saunders, D. J. and Offord, R. E. (1977). *Biochem. J.,* **165,** 479–86.
16. Kaiser, E. T. and Lawrence, D. S. (1984). *Science,* **226,** 505–11.
17. Cleland, W. W. (1964). *Biochemistry,* **3,** 480–2.
18. Swan, J. M. (1957). *Nature,* **180,** 643–5.
19. Rees, A. R. and Offord, R. E. (1976). *Biochem. J.,* **159,** 467–79.
20. Dixon, H. B. F. and Fields, R. (1972). In *Methods in enzymology* (ed. C. H. W. Hirs and S. N. Timasheff), Vol. 25, pp. 409–19. Academic Press, New York.
21. Lande, S. (1971). *J. Org. Chem.,* **36,** 1267–70.
22. Lande, S. and Burton, J. (1971). In *Peptides 1979*, (ed. E. Scoffone), pp. 109–12. North-Holland Publishing Company, Amsterdam.
23. Nyberg, F. (1988). *J. Mol. Recognit.,* **1,** 59–62.
24. Laskowsky, M., Jr (1978). In *Semisynthetic peptides and proteins*, (ed. R. E. Offord and C. Di Bello), pp. 255–62. Academic Press, London.
25. Rose, K., DePury, H., and Offord, R. E. (1983). *Biochem. J.,* **211,** 671–6.
26. Markussen, J. (1980). German Patent DE 31 04 949.
27. Kasche, V. (1989). In *Proteolytic enzymes, a practical approach*, (ed. R. J. Beynon and J. S. Bond), pp. 125–43. IRL Press, Oxford.
28. Kullmann, W. (1987). *Enzymatic peptide synthesis.* CRC Press, Boca Raton.
29. Jones, R. M. L., Rose, K., and Offord, R. E. (1987). *Biochem. J.,* **247,** 785–8.
30. Rose, K. and Offord, R. E. (1987). European Patent Application, EP 243929.
31. Rose, K., *et al.* (1991). *Bioconjugate Chem.,* **2,** 154–9.
32. Offord, R. E. and Rose, K. (1986). In *Protides of the biological fluids XXXIV*, (ed. H. Peeters), pp. 35–8. Pergamon Press, Oxford.
33. Breddam, K., Widmer, F., and Meldal, M. (1991). *Int. J. Pep. Prot. Res.,* **37,** 153–60.
34. Rose, K., *et al.* (1992). In *Innovation and perspectives in solid phase synthesis and related technologies,* (ed. R. Epton). SPCC, Birmingham, in press.
35. Murayama, A., Shimada, K., and Yamamoto, T. (1978). *Immunochem.,* **15,** 523–8.
36. Rodwell, J. D., *et al.* (1986). *Proc. Natl Acad. Sci. USA,* **83,** 2632–6.
37. King, T. P., Zhao, S. W., and Lam, T. (1986). *Biochemistry,* **25,** 5774–9.
38. Pochon, S., *et al.* (1989). *Int. J. Cancer,* **43,** 1188–94.
39. Ghose, T. and Blair, A. H. (1987). *CRC Crit. Rev. Ther. Drug Carrier Syst.,* **3,** 263–359.

40. Dixon, H. B. F. and Weitkamp, L. R. (1962). *Biochem. J.*, **84**, 462–8.
41. Harris, D. E. and Offord, R. E. (1977). *Biochem. J.*, **161**, 21–5.
42. Li, C. H., Blaker, J., and Hayashida, T. (1978). *Biochem. Biophys. Res. Commun.*, **82**, 217–22.
43. Schwartz, G. P., Burke, G. T., and Katsoyannis, P. G. (1987). *Proc. Natl Acad. Sci. USA*, **84**, 6408–11.
44. Bodanszky, M. and Bodanszky, A. (1984). *The practice of peptide synthesis*. Springer-Verlag, Berlin.
45. Davies, J. G. and Offord, R. E. (1985). *Biochem. J.*, **231**, 389–92.
46. Wallace, C. J. A. and Corthésy, B. E. (1986). *Prot. Engng*, **1**, 23–7.
47. Rose, K., *et al.* (1991). *Prot. Engng*, **4**, 409–12.
48. Rose, K., Gladstone, J., and Offord, R. E. (1984). *Biochem. J.*, **220**, 189–96.
49. De Filippis, V. and Fontana, A. (1990). *Int. J. Pep. Prot. Res.*, **35**, 219–27.
50. Rose, K., Herrero, C., Proudfoot, A. E. I., Offord, R. E., and Wallace, C. J. A. (1988). *Biochem. J.*, **249**, 83–8.
51. Wallace, C. J. A. (1987). *J. Biol. Chem.*, **262**, 16767–70.
52. Proudfoot, A. E. I., Wallace, C. J. A., Harris, D. E., and Offord, R. E. (1986). *Biochem. J.*, **239**, 333–7.
53. Fisch, I., Kunzi, G., Rose, K. and Offord, R. E. (1992). *Bioconjugate Chemistry*, **3**, 147–53.

11

Protein engineering of antibody combining sites

KATE L. HILYARD, DAVID STAUNTON, ALISON E. JONES, and ANTHONY R. REES

1. Introduction

The immunoglobulin G (IgG) molecule consists of a tetramer of two identical 25 kDa polypeptides (the light chain, L) and two identical 50 kDa polypeptides (the heavy chain, H). Crystal structures of IgG and IgG fragments have shown the antibody combining site (ACS) to be formed by the juxtaposition of six hypervariable loops or *complementarity determining regions* (CDRs), three from the light chain variable domain and three from the heavy chain variable domain. The CDRs of each chain are supported on a *framework* region which consists of conserved β-strands that fold to form a β-sandwich. When a light and heavy chain come together, one surface of each sandwich associates to form a β-barrel structure (*Figure 1a, b*). Supported on this β-barrel scaffold the six CDRs, pack together in the tertiary structure to form a relatively flat platform with a surface area of about 700 Å^2 (1). This *variable domain*, which contains all the determinants of antigen recognition, is one of six domains in the IgG molecule, the remaining five being Constant or C-type domains. The relative positions of these domains in the characteristic Y-shaped IgG and their nomenclature are shown in *Figure 1c*.

One of the functions of the antibody molecule is to bind specifically to its antigen, forming a high-affinity complex. The enormous diversity of the antibody repertoire and the high specificity of each antibody within this repertoire for its cognate antigen, has created a paradigm for molecular recognition. To understand fully the principles governing this molecular recognition process it is necessary to dissect the two interacting faces, the *epitope* on the antigen and the *paratope* on the antibody. The recent publication of the crystal structures of five antiprotein antibody–antigen complexes (HyHEL-5-lysozyme, HyHEL-10-lysozyme, D1.3-lysozyme, NC41-neuraminidase, and NC10-neuraminidase) and two antipeptide antibody–antigen complexes (anti-myohemerythrin and anti-haemaglutinin, F17/9) has given

(a)

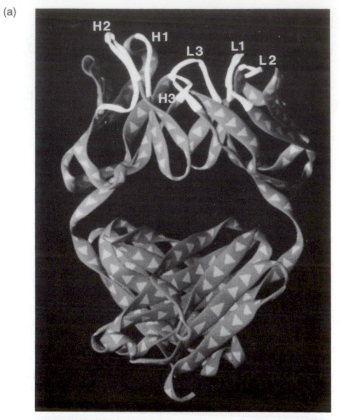

(b)

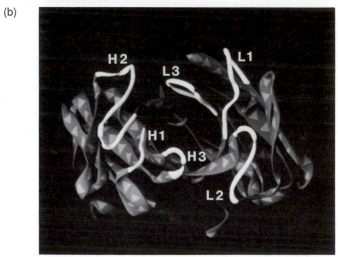

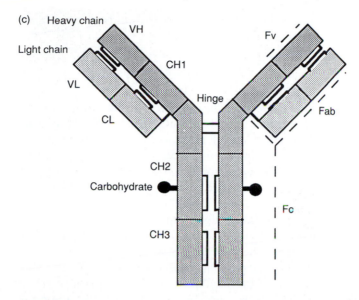

Figure 1. Side (a) and plan (b) views of the variable domain (Fv) of an immunoglobulin. The Cå coordinates from the anti-lysozyme antibody, Gloop2 (11) are plotted to show framework residue positions (open bonds) and CDR residue positions (filled bonds). The CDRs of the heavy chain are indicated as H1, H2, and H3; the light chain CDRs are indicated as L1, L2, and L3. (c) Relative positions of the variant domains of the light (VL) and heavy (VH) chains, which together make an Fv region, and the constant domains for an immunoglobulin G (IgG). Intra- and interchain disulphide bonds are shown in heavy lines.

an insight into the types of interactions that occur between antibody and protein or peptide antigens (for a review see ref. 2).

However, even with the information provided by structural studies, the processes that occur on antigen recognition are not completely explained, such as the effects on binding affinity and specificity caused by somatic point mutations, a natural process in antibody maturation. Additional information on the fine points of the molecular interaction can be provided by protein engineering studies. With this knowledge it may be possible to exploit the enormous natural repertoire of antibody specificities (estimated to be 10^9 in the primary repertoire) and, in addition, to produce engineered antibodies containing novel functions, such as enzymatic activity, allosteric binding, or redox properties.

This chapter describes methods used by our group to study the molecular interactions within an antibody–antigen complex in order to understand the specificity and affinity of the interaction, and to explore the potential of the antibody molecule as a template for the introduction of novel functions.

To begin a protein engineering study of the antibody combining site (ACS) the following are required:

(a) *A cDNA clone for the antibody Fv, Fab, or complete IgG molecule*. This can now be generated relatively easily from a myeloma or hybridoma cell using the polymerase chain reaction (3).

(b) *Structural information for the antibody*. Structural information is crucial to guide the rational design of mutations. Ideally, the structure of the Fab–antigen complex or of the uncomplexed Fab would be available. Where there is no experimentally determined structure for the antibody, a model of the variable region can be generated using suitable computer programs that operate on the amino acid sequence of the light and heavy chain variable region. The modelling procedures involve prediction of CDR conformations using algorithms that combine knowledge-based and conformational search methods. The framework region conformation is also generated by the procedure, resulting in a model for the entire antibody variable region. This can now be carried out with an accuracy approaching that of the medium-resolution X-ray structure, so that engineering studies can be designed by reference to a reasonably accurate structure of the combining site. The most comprehensive algorithm for antibody modelling (*Figure 2* and described in detail in refs 4–6) is now available as the program CAMAL through Oxford Molecular Ltd, Oxford, UK.

(c) *The antibody epitope on the antigen*. Numerous methods have been developed for epitope mapping, for example serological mapping (7), synthetic peptides (9), or NMR studies (10). If the structure of the antigen is available, a docked model of the complex can be attempted, although docking algorithms are less well advanced than those described in (b).

2. Analysis of antibody binding affinity and specificity

In this chapter, the procedures for engineering the combining site of an antibody are described by reference to two antibodies that have been the subject of such studies in our own laboratory. For the point mutation studies we have used Gloop2 (7, 8), an antibody that was raised against a peptide fragment of hen egg lysozyme (HEL, amino acid residues 57–84) and binds to that peptide with an affinity of 10^8 M^{-1} and to HEL with an affinity of 10^7 M^{-1}. The native Gloop2 Fab structure has been solved by X-ray crystallography (11). Initial protein engineering studies were guided by a three-dimensional model of the Gloop2–HEL complex (12). For the CDR grafting experiments Gloop2 was used as a framework while the CDRs were transferred from another anti-HEL antibody, HyHEL-5 (13).

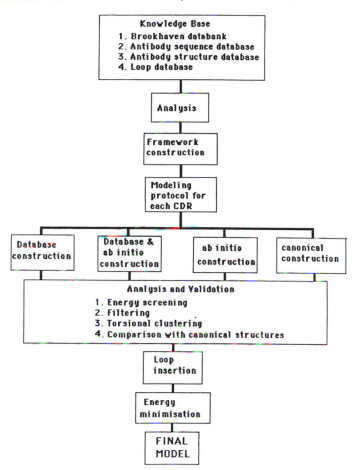

Figure 2. Flow diagram of the antibody modelling algorithm, CAMAL. Details of the algorithm can be found in refs 4–6 and of the program from Oxford Molecular Ltd, UK. The authors are happy to receive any enquiries.

2.1 Design of the substitutions

Protein engineering can be used to generate residue insertion, deletion, or substitution. It has been observed that the crucial determinants of CDR conformation are backbone length (12, 14) and the presence of 'key' residues at critical positions within the CDR (15, 16). Therefore, when contemplating engineering of the ACS it should be noted that residue insertion or deletion within the CDR loop may have a major effect on the overall CDR conformation and hence of the rest of the ACS. In this study we have limited our descriptions and procedures to the analysis of residue substitution only.

To study the role of amino acid residues in the interaction with the antigen, single or multiple substitutions can be made by site-directed mutagenesis.

Mutations can be systematic; for example, each amino acid can be substituted by alanine (17), or designed according to some structural rationale, such as for example the removal of the OH group of Thr by substitution to Val. However, the design approach should not attempt to be too 'clever' since important information can be obtained from both conservative (that is, retaining approximate volume and charge characteristics) and non-conservative substitutions. In Gloop2, substitution of Glu28 (CDR1 of the light chain) to Ser, Ile, and Arg caused an increase in the affinity for the protein antigen, but had no major effect on the peptide affinity (18, 19). It was concluded that the higher binding affinity to the protein was due to the increased hydrophobic character of the side-chain (the arginine side-chain can have both hydrophilic interactions via the guanido group and hydrophobic interactions via the methylene groups; see ref. 20). These results would not have been observed if only conservative substitutions had been made. On the other hand, where mutiple changes are planned it is important to preface a suite of non-conservative changes by *the most conservative* mutation. If this is not done, vital information may be missed. For example, in mutation of Tyr94 to Phe in CDR L3 of Gloop2 (the most conservative mutation for Tyr) binding to peptide was abolished while binding to HEL was unchanged (21).

The results of our studies have shown that some positions in the CDRs appear to be more tolerant to substitution than others. For example, the non-conservative substitution Tyr32 to Asp in CDR1 of the light chain of Gloop2 reduced the peptide affinity by only tenfold and had no effect on protein affinity (21). This tolerance to change is probably an inherent feature of certain positions in the CDRs as reflected in their high level of sequence variation *in vivo*. There are, however, other positions where change cannot be tolerated, either because the residue is critical for the antigen interaction or for maintaining the structure of the ACS. For example, the apparently conservative substitution Glu99 to Asp in CDR3 of the heavy chain of Gloop2 abolished detectable binding of both peptide and protein antigen (21). In this situation it is not possible to say whether the residue concerned is a critical antigen contact or performs a structural role. To lessen the chances of mutating a structurally important residue, each CDR should be compared with the 'canonical structure' for that CDR, as detailed by Chothia and Lesk *et al.* (15, 16). Each canonical structure describes a preferred conformation for a CDR, the preference being determined by key residues whose presence controls a particular main-chain conformation. Clearly, mutagenesis of 'key' residues should be avoided where possible.

In summary, CDR residues can be classified in three groups for the purpose of protein engineering studies:

(a) Residues that interact with the antigen directly via salt bridges, hydrogen bonds, or van der Waals interactions. Substitution of these residues would be expected to affect antigen binding directly.

(b) Residues that are critical for maintenance of the ACS structure; for example residues involved in CDR/CDR or CDR/framework contact, or structure determining (key) residues. Substitution of these residues would be expected to affect antigen binding indirectly.

(c) Residues that are neither involved in direct interaction with the antigen nor in maintenance of the ACS structure. Substitution of these residues should have no effect on antigen binding, though this can only be known after their mutagenesis.

2.2 Generation of the mutant antibodies

For the generation of point mutations, relatively small oligonucleotides (20mers) and standard site-directed mutagenesis (SDM) techniques can be used. Details of the SDM methods will not be discussed as they are covered in Chapters 2 and 3 of the Practical Approach volume, *Directed mutagenesis* (22).

2.3 Production of the mutant antibodies

Expression systems for antibodies can be either prokaryotic or eukaryotic, and transient or stable. Here, we describe a small-scale transient system for preliminary binding analysis, and a prokaryotic secretion system for larger-scale production. For transient expression we have used co-translation of heavy and light chain synthetic RNA in *Xenopus laevis* oocytes. This is a rapid secretion expression system producing microgram amounts of antibody 24 h after injection of the oocyte. The IgG molecule produced is correctly processed and has the same affinity for the antigen as the hybridoma-produced IgG (23). The yield of secreted IgG is low compared to other expression systems, but it can be used directly in radioimmunoassays without purification (23).

The heavy and light chain RNA is synthesized *in vitro* using an SP6 transcription vector containing the 5' and 3' flanking regions of the efficiently transcribed *X. laevis* β-globin mRNA (*Figure 3*). The mutant antibody cDNA

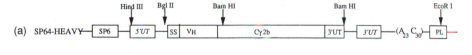

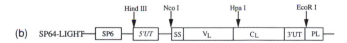

Figure 3. Details of the SP6 transcription vector showing (a) heavy chain construct (HG201) and (b) light chain construct (LG211). SS, immunoglobulin signal peptide; V, variable region; C, constant region; L, light chain; H, heavy chain (gamma2b) UT, untranslated region of the *Xenopus* β-globin gene; PL, polylinker.

(with the initiation codon and immunoglobulin signal sequence) is subcloned into the vector via a unique restriction site. The resultant SP6 polymerase transcription products contain the coding region flanked by the 5' and 3' untranslated regions. Transcripts with a 5' methyl cap structure (required for efficient translation in the oocyte) can be generated by including a synthetic cap dinucleotide in the synthesis reaction. The optimum conditions for RNA synthesis have been described previously (24) and will not be covered here. The preparation and microinjection of the *X. laevis* oocytes are described in *Protocol 1*.

Protocol 1. Preparation and microinjection of *X. laevis* oocytes

Specialized equipment:
- dissection kit
- stereomicroscope, micromanipulator, microsyringe, and fibreoptic lamp
- hard glass capillary tubes (10 cm) and micropipette puller
- 96-well tissue culture plates

Method

1. Dissect the oocytes in small clumps from the abdomen of the *X. laevis* frog and wash in modified Barth's saline (MBS; see *Protocol 2*). The oocytes can be stored in this form for several days in MBS at 22°C.

2. Strip the oocytes from the connective tissue using forceps and sort according to size and condition. Use only those oocytes of a large size and uniform pigment. Incubate the oocytes in fresh MBS for at least 4 h before injection, to allow any non-viable oocytes to be identified and removed.

3. Mix an aliquot of the heavy and light chain RNA to give a final concentration of 250 ng/µl for both chains. Load the micropipette needle with 1 µl of the RNA mix. Store the remaining mix at 4°C.

4. Inject 40 nl of RNA mix (10 ng of each chain) into each oocyte. A microscope slide onto which a small piece of plastic with a serrated edge has been glued can be used to immobilize a number of the oocytes and facilitate injection. Alternatively, a commercially available nylon mesh on the bottom of a Petri dish can be used to the same effect. The injection should be into the equatorial region or vegetal pole (the light pigment) to avoid injection of the nucleus which lies under the animal pole (the dark pigment).

5. Return the injected oocytes to fresh MBS and leave at 22°C for 1–3 h before sorting and removing any damaged oocytes. Incubate the oocytes in 96-well tissue culture plates previously blocked with BSA (0.5 µg/ml) in MBS for 30 min at 37°C and washed twice in MBS. Incubate five oocytes per well in 60 µl of incubation medium (500 µM L-methionine; 5% v/v fetal calf serum (dialysed against MBS) in MBS) for 15–20 h at 22°C. For

the synthesis of ^{35}S-labelled transcripts, incubate the oocytes in labelling medium (45 pmole L-[^{35}S]methionine; 40 pmole cold L-methionine (0.8 μM); 5% v/v fetal calf serum (dialysed against MBS) in MBS).

6. Remove the supernatant from those wells where no cell lysis has occurred. Pool the supernatants and centrifuge for 10 min at 4°C to remove cell debris. Store at −70°C.

Protocol 2. Preparation of modified Barth's saline

1. Prepare the following stock solutions:
   - solution A: 128 g NaCl, 2 g KCl, 5 g NaHCO$_3$, 45 g Tris base: dissolve in water, adjust pH to 7.6 with concentrated HCl, and make up to a final volume of 1 litre. Store as 40 ml aliquots at −20°C
   - solution B: 1.9 g CaNO$_3$:4H$_2$O, 1.54 g CaCl$_2$:6H$_2$O: dissolve in 1 litre of water and store as 40 ml aliquots at −20°C
   - solution C: 5 g MgSO$_4$:7H$_2$O: dissolve in 1 litre of water and store as 40 ml aliquots at −20°C
   - solution D: sodium penicillin and streptomycin sulphate each at 10 mg/ml: store as 1 ml aliquots at −20°C

2. Prepare MBS by mixing 40 ml of solutions A, B, and C, and 1 ml of D in a total volume of 1 litre.

3. The final composition of MBS is:
   - 88.0 mM NaCl
   - 1.0 mM KCl
   - 2.4 mM NaHCO$_3$
   - 15.0 mM Tris–HCl, pH 7.6
   - 0.3 mM CaNO$_3$:4H$_2$O
   - 0.41 mM CaCl$_2$:6H$_2$O
   - 0.82 mM MgSO$_4$:7H$_2$O
   - 10 μg/ml sodium penicillin
   - 10 μg/ml streptomycin sulphate

2.4 Analysis of antigen binding by the mutant antibodies

2.4.1 Quantitation of the IgG concentration in the oocyte supernatant

In order to estimate the concentration of functional IgG in the oocyte supernatant an equilibrium is established between the mutant IgG bound to the microtitre plate via goat anti-mouse polyclonal antibodies (GAM) and ^{125}I-labelled antigen (*Protocol 3*). The assay should indicate the optimum dilution of the oocyte supernatant that retains maximum binding of the labelled antigen.

Protocol 3. Quantitation of the IgG concentration in the oocyte supernatant

Specialized equipment:
- flexible PVC V-bottom microtitre plates
- gamma-counter

Method
1. Coat the microtitre plates with 50 μl of goat anti-mouse polyclonal anti-bodies (GAM) (affinity purified at 50 μg/ml in phosphate buffered saline (PBS)) at 4 °C overnight. Wash and block in PBS containing 0.05% v/v Tween:20 (PBS/Tween) for 20 min at 22 °C.
2. Serially dilute 50 μl of the oocyte supernatant containing the mutant IgG into PBS/Tween. Incubate at 22 °C for 4–6 h. Wash three times in PBS/Tween.
3. Add 50 μl of ^{125}I-labelled antigen (at approximately 20×10^3 c.p.m./50 μl in PBS/Tween) to each well and incubate at 4 °C overnight. Wash the wells for 3×1 min in ice-cold PBS/Tween. Cut out the individual wells and count on a gamma-counter for 1 min.

2.4.2 Inhibition radioimmunoassay

The solid phase radioimmunoassay used to determine the affinity constant for the mutant antibodies is a modification of that developed by Darsley and Rees (7) and is described in *Protocol 4*. The oocyte supernatant containing the mutant antibody is used at the optimum dilution determined in *Protocol 3*.

Protocol 4. Solid phase inhibition radioimmunoassay

Specialized equipment: as for *Protocol 3*.
1. Coat microtitre plates with GAM (50 μg/ml in PBS) and block with PBS/Tween as in *Protocol 3*.
2. Add 50 μl of the test supernatant diluted in PBS/Tween (and a control wild-type antibody). Incubate at 22 °C for 4–6 h. Wash three times in PBS/Tween.
3. Add 25 μl of the inhibitor solution at varying concentrations, plus 25 μl of ^{125}I-labelled antigen (at approximately 20×10^3 c.p.m./25 μl in PBS/Tween) to each well and incubate overnight at 4 °C. Wash the wells for 3×1 min in ice-cold PBS/Tween. Cut out the individual wells and count on a gamma-counter for 1 min.

The binding data from the inhibition assays can be analysed using the Scatchard method or non-linear regression. There are a number of algorithms available that automatically perform such an analysis, e.g. Ligand (25). The apparent affinity constants generated may vary slightly from assay to assay due to small variations in the reaction conditions, for example the specific activity of the labelled antigen or the efficiency of plate coating. For this reason a control assay with the wild-type antibody should always be performed in parallel with the experimental assays.

3. Production of binding determinants by CDR grafting

The sequence of the six CDRs determines the structure of a given antibody combining site (ACS) and hence the antigen specificity and affinity. The transfer of these CDRs to a similar framework structure will reconstruct the ACS and the binding specificity of the original antibody. The substitution of amino acid residues in the ACS has already been discussed, and CDR grafting is an extension of this technique, with entire CDRs being replaced by site-directed mutagenesis. This transfer can either be from one rodent antibody to another or, in the special case of 'humanization', involves the transfer of rodent CDRs to a human framework (26).

Frequently, the transfer of CDRs alone is insufficient to reproduce the ACS, and further mutations are required to restore the original antigen affinity. This reflects differences between the donor and recipient framework structures, some of which must be corrected, particularly those that form critical interactions with residues at the base of the CDRs or are involved in antigen contact. The antigen contact residues can only be corrected, of course, if the structure of the Fab–antigen complex is known. Generally, the structures of the donating and receiving antibodies will be unknown and so a modelling procedure must be employed prior to mutagenesis. The procedure uses the algorithm described in Section 1(b) and should observe the following steps.

(a) If the X-ray structures of the donor and recipient Fabs are known, see the procedure for the example described in Section 3.1 below.

(b) If only one or neither of the Fab structures are known:

   i. model the variable region of one or both antibodies by the method described in Section 1(b)

   ii. overlay the two variable regions using a multiple structure fitting program such as MOLFIT (27)

   iii. identify those differences between the structures that involve residues in contact across the framework-CDR boundaries, particularly where the differences result in changes in side-chain packing

While the above procedure is not given in detail, since there are too many different grafting situations that would need to be described, it will provide a sufficiently useful starting point for the identification of essential sites for mutation. The molecular biologist will be advised to consult an antibody structure expert for refinement of the mutation suite before embarking on a long and costly list of mutations, many of which may be unnecessary for restoring antigen binding activity.

3.1 Creation of HyHEL-5 cDNA by CDR grafting

HyHEL-5 is a murine antibody against HEL whose complex with its antigen has been determined to high resolution by X-ray crystallography (13). The structural information makes it an ideal candidate for protein engineering of the ACS. However, since cDNA clones were not available from the hybridoma cell, the antibody was constructed by grafting the HyHEL-5 CDRs on to the framework of Gloop2, another murine antibody for which the cDNAs were available. Although this example concerns the transfer of CDRs from one mouse framework to another, the principles will be the same for the humanization of rodent antibodies.

(a) The backbones of the framework structures of HyHEL-5 and Gloop2 were overlaid by the multiple structure fitting program MOLFIT to ensure that the 'takeoff' positions of the CDRs (the positions of the polypeptide chain at the framework–CDR junctions) were the same in Gloop2 as in HyHEL-5. Both structures showed good agreement, except for the end of the third heavy chain framework region where the HyHEL-5 backbone deviates from that of Gloop2 at positions 93 and 94 and results in a shift in the takeoff position between the two structures of 1.5 Å. The residues at these positions in HyHEL-5 are Leu and His, respectively, unlike the majority of murine antibodies, including Gloop2, where they are Ala and Arg. These substitutions cause the backbone, and hence CDR H3, to deviate from the plane that it normally occupies. To recreate the HyHEL-5 H3 structure these mutations within the framework region must be included.

(b) The CDRs can be defined by sequence alignment (28) and the residues to be transferred identified. Although these residues will contribute the majority of interactions, residues classified as framework may also interact with the antigen. Trp47 of the HyHEL-5 heavy chain framework region interacts with HEL in the complex, but is also present in the Gloop2 framework.

(c) The topology of the ACS involves not only the CDRs but also discontinuous framework residues interacting with CDR residues. Examination of the ACS structure or model will identify these residues. For example, in HyHEL-5, Ile2 of the light chain amino terminus packs into CDR L1 (16). This packing is retained in Gloop2 and no mutation was necessary. Again, in HyHEL-5, Gln89 of CDR L3 forms a hydrogen bond with

Tyr36 of the second framework region of the light chain. However, in Gloop2 residue 36 is a leucine. To restore this interaction, Leu36 of the Gloop2 light chain was mutated to a tyrosine.

Six mutagenic oligonucleotides were synthesized for each CDR, consisting of the HyHEL-5 CDR sequence flanked by arms (12 to 15 nucleotides long) complementary to the Gloop2 framework sequence, necessary to anchor the mutagenic oligonucleotide in position on the Gloop2 single-strand cDNA template. Mutations outside the CDRs were also incorporated into the oligonucleotides and the arms extended in length to compensate.

These oligonucleotides can range in size from 40 to 60 nucleotides in length. The phosphodiester method of oligonucleotide synthesis has a coupling efficiency of 97% for each cycle. For small oligonucleotides (20 nucleotides in length) this coupling efficiency results in the majority of the product being of the correct sequence and length. However, with a 60-residue-long oligonucleotide the required product will represent only 20% of the total yield and so purification is essential to prevent random deletions appearing in the CDRs of the grafted antibody.

Protocol 5. Purification of oligonucleotides for CDR grafting

Specialized equipment:
- 20×40 cm gel plates with 1.5 mm spacers and comb
- long-wave UV lamp
- fluorescent TLC plate

Method
1. Dry down deprotected oligonucleotide, dissolve in formamide dye buffer (80% formamide, 20 mM EDTA, 0.1% (v/v) bromophenol blue), and heat at 70°C for 5 min.
2. Load samples on to 10% 7 M urea PAGE. Run at 45 mA.
3. When the bromophenol blue reaches bottom of the gel (approx. 3–4 h), stop electrophoresis and place the gel between two layers of clingfilm.
4. Visualize DNA by laying the clingfilm-covered gel on to a TLC plate and shining long-wave UV light on the gel. The oligonucleotide appears as shadows on a fluorescent background.
5. Mark the position of the slowest migrating bands and cut out. Transfer gel slices to a dialysis bag containing 1 ml of Milli-Q water. Dialyse overnight at 4°C against 4 litres of water.
6. Filter the contents of the bag through a 0.45 μm filter and dry down.
7. Resuspend the oligonucleotides in Milli-Q water and quantify by UV absorption at 260 nm.

The oligonucleotides are utilized in the same manner as small oligonucleotides for site-directed mutagenesis while maintaining the molar ratio of oligonucleotide to template. For each chain all three oligonucleotides are employed in one round of mutagenesis to save time and resources. Mutations containing all three grafted CDRs are identified by colony hybridization.

Protocol 6. Identification of grafted chains by colony hybridization

1. For each heavy and light chain, make four replica plates of mutant colonies, by transferring individual colonies/turbid plaques of transformed bacteria with a sterile toothpick, and retain one as the reference plate. Include the original mutagenesis template in *Escherichia coli* as a negative control.

2. Transfer three of the plates to filters (nitrocellulose or 541 ashless), lyse and fix (29).

3. Create probes by kinasing 15 pmoles CDR grafting oligonucleotides with [^{32}P] γ-ATP.

4. Prehybridize filters in 6 × SSC (a 20× SSC stock solution consists of 175.3 g NaCl and 88.2 g sodium citrate, pH 7), 0.5% (v/v) Nonidet P-40, 0.6 mg/ml yeast tRNA at 65°C for 2 h.

5. Hybridize with probe in fresh prehybridization buffer at 65°C overnight.

6. Wash filters three times in 6× SSC at room temperature, then 0.1 × SSC at 65°C for 20 min.

7. Autoradiograph for 2 h, and from the autoradiograph identify colonies positive for all three CDRs.

8. From the replica plate, restreak or plaque purify those colonies that are positive for all three CDRs and sequence mutants.

Once the sequences of the CDR-grafted heavy and light chains have been confirmed, they can be expressed and characterized as described in Section 2. The epitope should be identified by competition between the CDR-grafted antibody and the original antibody from which the CDR sequences were obtained.

4. Production of antibody molecules in *Escherichia coli*

The *X. laevis* oocyte expression system described in Section 2.3 produces sufficient quantities of antibody for determination of antigen binding properties. For additional characterization by physical methods, such as NMR, X-ray crystallography, or fluorescence, larger amounts (>10 mg) of pure antibody are required. Since the antigen binding function of an antibody resides in the variable region, the smallest viable fragment containing the ACS, such as an Fab or Fv fragment, can be expressed. There are now numerous examples of the production of such fragments in *E. coli* using both intracellular and secretion expression systems. Two such systems, for intra-cellular expression or secretion of Fv fragments in *E. coli* on the 1 litre scale, are described in detail in Chapter 13 and will not be further elaborated here. Large-scale (20 litre) fermentation methods are described in Chapter 16. In this section we describe how to establish a relatively inexpensive laboratory-scale fermentation system for Fv production.

4.1 The secretion vector

Figure 4 shows a secretion vector that has been successfully used to express several different Fvs (21, 30). The vector is based on pUC19 and utilizes the *lacZ* promoter. The leader sequence from the pectate lyase gene directs the VH and VL polypeptide chains to the periplasm. Because the Fv product is toxic to *E. coli*, expression has to be induced when the host cell density is sufficiently high. Generally, after induction the cell density increases for a few hours, and then decreases as cell death occurs. The risk of plasmid loss favours the use of intracellular antibiotic selection, e.g. kanamycin. However, ampicillin resistance is adequate for plasmid selection if sufficient concentrations are used (100 µg/ml). Other precautions that can be used to reduce plasmid loss are to use fresh transformants and to grow the cells at 30°C (31). The *E. coli* cell strains that are often used for expression are W3110 and BMH71.18.

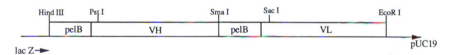

Figure 4. The pelB-Fv expression construct used in the Fv production methods described in the text. pelB, pectate lyase leader peptide; VH and VL, coding regions for the variable domains of the heavy and light chain, respectively.

4.2 Laboratory fermentation of *E. coli* for Fv production

This section describes the use of a small, 5–7 litre, benchtop fermenter that is suitable for laboratory-scale production of Fv fragments. The method used in the authors' laboratory employs a fed-batch technique with a defined minimal medium. This enables high cell densities to be reached and makes for easier downstream processing of the Fv. To minimize plasmid loss, and enable high biomass to be achieved without exceeding the oxygen transfer rate of the vessel, the growth rate of the cells is kept low by using glycerol as a carbon source and a growth temperature of 30°C. Under these temperature conditions the

Fv is secreted into the medium in soluble form and does not build up as insoluble material in the cells or the periplasm.

A defined minimal medium composed of water, salts, trace elements, thiamine, and a carbon source is used for fermentation rather than a complex medium, such as Luria broth (L-broth), for the following reasons:

(a) L-broth contains a number of different proteins, which complicates downstream processing.

(b) The growth rate of the cells in L-broth may vary as cells switch from metabolizing one nutrient to the next. Using defined media, growth is controlled by addition of a single carbon source.

(c) It is not necessary to use catabolite repression of promoters by adding glucose to the medium since the levels of uninduced expression are negligible.

Protocol 7. Preparation of minimal medium

Method for making 100 ml of minimal medium for shaker flasks.

1. Stock components of analytical (AR) grade should be autoclaved or filter sterilized as appropriate.

2. Prepare the following solutions: trace element stock (*Table 1*), media base (*Table 2*).

3. To avoid precipitation, the correct amount of each component is added using sterile technique in the order shown in *Table 3*. The composition of the completion medium for a 6 litre fermentation is shown in *Table 4*.

4. To prepare minimal agar plates: prepare 100 ml of sterile molten noble agar made in water with 4 g of agar per 100 ml. Mix 100 ml of minimal medium with the agar and pour the plates as normal.

Table 1. Composition of trace element stock

Trace element	Concentration in media (g/l)	Concentration of stock (g/100 ml)
$MnSO_4.4H_2O$	0.026	2.6
$ZnSO_4.7H_2O$	0.01	1.0
H_3BO_3	0.0005	0.05
$CoCl_2.6H_2O$	0.005	0.5
$CuSO_4.5H_2O$	0.002	0.2
$AlCl_3$	0.01	1.0
$Na_2Mo_4.2H_2O$	0.002	0.2

The trace element stock is made in 2 M hydrochloric acid and filter sterilized.

Table 2. Composition of media base

Basal media	Concentration in media (g/l)	Concentration for 10× stock (g/l)
$(NH_4)_2SO_4$	5	50
Na_2NPO_4	12	120
KH_2PO_4	6	60

A 10× stock concentrate of media base is used for making media in flasks. For fermenter runs, use working concentrations.

Table 3. Composition of minimum medium

Component	Concentration in media (g/l)	Concentration of stock	Volume of stock addition (ml)
H_2O	–	–	85
EDTA	0.4	40 g/l	1.0
Trace elements	Table 1	Table 1	0.1
$FeCl_3.6H_2O$ [a]	0.08	80 g/l	0.1
$CaCl_2.2H_2O$	0.04	40 g/l	0.1
$MgSO_4.7H_2O$	2.0	200 g/l	1.0
10× Media base	Table 2	Table 2	10.0
80% Glycerol	–	–	2.5
Thiamine	0.01	1 mg/ml	1.0
Antibiotic	as required	as required	0.1

[a] $FeCl_3.6H_2O$ stock solution is acidified with 2 ml concentration hydrochloric acid per 100 ml.

Table 4. Composition of completion medium for 6 litre fermentation

Component	Volume of stock (ml) [a,b]
H_2O	300
EDTA	60
Trace elements	6
$FeCl_3.6H_2O$	6
$CaCl_2.2H_2O$	6
$MgSO_4.7H_2O$	60
80% Glycerol	48
Thiamine	60
Antibiotic	as required

[a] The stock solutions used are the same as for minimal media.
[b] Add the correct amount of stock to water in the order shown, mixing thoroughly after each addition.

4.3 Fermentation equipment

The control unit should have PID (*P*roportional, *I*ntegrating and *D*ifferentiating; see ref. 32) control and be fully automatic for measuring and controlling pH, temperature, and dissolved oxygen. The fermenter should be equipped with probes to measure these parameters and have a high stirrer speed. It must also have a cooling device (for example a 'cold finger') for maintaining a temperature of 30°C during exponential growth when more heat is given out by the cells than can be lost by the vessel. A suitable low-cost set-up is shown in *Figure 5*, consisting of an 'Anglicon Microlab-0' control box from Brighton Systems, and an 'Electrolab' fermenter from Electrolab Ltd.

4.4 Preparation of inoculum for fermentation

W3110 *E. coli* cells are suitable for fermentation expression of antibody Fv fragments. Minimal medium is a poor buffer and will not sustain bacterial growth without external pH control. Therefore, when growing cultures in the shaking incubator start with a small volume of culture and use this to inoculate successively larger volumes of media.

Protocol 8. Adaptation of transformed *E. coli* to minimal medium

1. Streak *E. coli* W3110 (or other suitable cell line) onto a minimal agar plate.
2. Incubate at 37°C for up to 48 h to produce good-sized colonies.
3. Prepare competent cells in L-broth and transform as normal.
4. Plate out transformed culture on to L-agar plates and incubate at 30°C for 24 h. *Note*: transformation on to minimal plates is found to be inefficient.
5. Restreak transformed colonies on to minimal/antibiotic plates.
6. Incubate for up to 48 h to obtain large colonies.
7. Inoculate 10 ml of minimal medium with a large colony from the restreaked plate.
8. Incubate at 30°C, 250 r.p.m., for 24 h.
9. Inoculate 100 ml of minimal medium with the 10 ml culture from above.[a]
10. Incubate at 30°C, 250 r.p.m., until the culture is in the late exponential phase.
11. Make 12 ml glycerol stocks from this culture, consisting of 6 ml culture and 6 ml 80% (v/v) glycerol solution, using sterile technique. Mix and store at −70°C.

[a] A high-viability inoculum is essential for successful fermentation. The volume of the inoculum should be ~10% of the volume of the fermentation to be carried out.

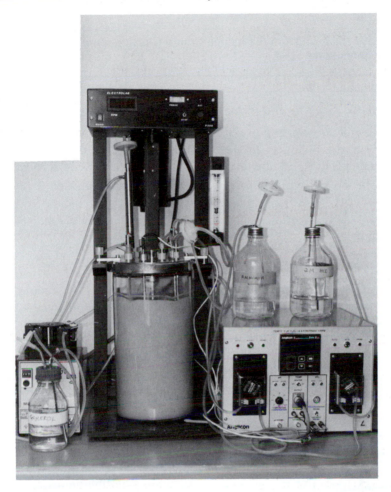

Figure 5. Laboratory-scale (6 litre) fermenter assembly in the authors' laboratory, consisting of an Anglicon Microlab-0 control unit and an Electrolab fermenter with a 6 litre capacity glass vessel. The experiment shown is an Fv production in *E. coli* (strain W3110) and the cell density at the time of the photograph was about 20 OD units.

Protocol 9. Preparation of inoculum for a 6 litre fermentation

1. Inoculate 100 ml of minimal medium (without added glycerol) with a 12 ml glycerol stock of transformed cells. The glycerol present in the stock is sufficient for this stage of growth.

2. Incubate at 30°C, 250 r.p.m., for 24 h or until the culture is in the late exponential phase.

3. Use this 100 ml culture to inoculate 500 ml of minimal medium and incubate as above.

Protocol 9. *Continued*

4. Remove the flask from the incubator and transfer, using sterile technique, to an inoculation flask.

Protocol 10. Growth and induction of *E. coli* in fermenter

1. Set up and autoclave the vessel containing the basal medium.
2. Calibrate the dissolved oxygen (DO) probe and set the control unit to maintain pH at 6.8, temperature at 30°C, and DO between 20% and 40%.
3. Control the pH automatically using ammonia solution and 2 M hydrochloric acid.
4. Prepare the completion medium and add to the fermenter, followed by the inoculum.[a]
5. Until the DO falls below 40% air flow is not needed and stirrer speed should be kept low.
6. Growth is continued to as high a density a possible, depending on the plasmid stability of the culture. The culture should be induced when the optical density at 550 nm is greater than 10.0. A suitable antifoam such as poly(propylene glycol) 2000 (Aldrich), is added when required. 1–2 ml of neat antifoam is usually sufficient for a 6 litre fermentation.
7. The initial glycerol concentration (8 g/l) in the medium will become exhausted when the culture has reached an optical density 550 nm of 10 or 11. This is signalled by a sharp rise in the dissolved oxygen as the cells stop growing. Start the glycerol feed (24% (v/v) glycerol), initially at approximately 8 ml/h and increase the rate linearly in stages in parallel with the bacterial growth. This linear feed process is a simple but effective way of increasing cell density.
8. Induce with 1 mM IPTG, dissolved in water and filter sterilized.
9. Decrease the feed rate so that the growth rate is just above stationary level, and continue growth for at least another 12 h.

[a] If ampicillin is the antibiotic used, fresh filter-sterilized ampicillin should be added every 12 h to a final concentration of 100 µg/ml.

4.5 Purification of Fv from fermentation media

Although the purification procedures are essentially the same as those used for benchtop-scale production (0.5–1 litre, see Chapter 13), the higher cell densities and volumes in a fermentation run require different techniques for clarification and concentration. Affinity chromatography is the preferred method for purification of Fv from the culture supernatant. This method selects for antigen binding function and enables large volumes of crude supernatant to be concentrated to only a few millilitres of pure protein. Due to the intrinsic high affinity of many antibody–antigen interactions, relatively harsh conditions (pH 2.0 or pH 12.0) are often required to dissociate the complex. Fortunately, the Fv domain is fairly resistant to these extremes of pH, provided that the exposure is brief, and antigen binding is usually recovered after the pH is returned to neutral. Unfortunately, the harsh eluting conditions may hydrolyse the linkage between the ligand and matrix and so it is advisable to check the stability of the affinity material to the pH conditions being employed. If in doubt, do not reuse the matrix. An advantage of the high affinity ligand/Fv association is that stringent washing conditions can be used to remove other protein contaminants, e.g. 1.5 M NaCl in pH 8.5 buffer.

Protocol 11. Purification of Fv from a 6 litre fermentation

Specialized equipment:
- stacked disc separator (Alfa-Laval LAB 102B-05)
- tangential-flow filtration units (e.g. Millitan from Millipore) with 0.22 μm and 10 000 Da NMW (nominal molecular weight) cut-off filters
- HiLoad 16/60 Superdex 200 prep. grade column

Method
1. Remove cells by passing the medium directly from the fermentation vessel through a stacked disc separator. Adjust flow for maximum clarification and monitor towards the end of the separation to prevent overflow of cells from the separator vessel into the clarified broth.
2. Remove the remaining cells and cell debris by passing through a 0.22 μm tangential-flow filtration unit, e.g. Pellicon system with two 0.22 μm cassettes.
3. Concentrate the filtered media by tangential ultrafiltration through 10 000 Da NMW cut-off membranes. A Millitan 8 filter plate assembly offers a small hold-up volume with efficient ultrafiltration capable of reducing 5 litres to less than 1 litre in 3–4 h. Place the inlet and retentate tubing into the media reservoir and the filtrate to waste. Allow the media to recirculate at medium speed and approximately 0.5 bar g back pressure to minimize protein denaturation and fouling of filters. Tangential ultrafiltration is carried out at room temperature to maintain a manageable media viscosity, and a bacteriocide, such as sodium azide, is added to the media to prevent bacterial growth.
4. When the volume has been concentrated to below 1 litre, the phosphate-based media can be replaced by an alternative buffer if desired (phosphate sometimes interferes with antibody–antigen binding). In the authors' laboratory, Tris buffer is used when a lysozyme affinity column is employed.

Protocol 11. *Continued*

5. Recirculate concentrated medium through a 5 ml affinity column at 4°C overnight, wash and elute.

6. Identify Fv-containing fractions (e.g. by PHAST gel electrophoresis) and run on a 16/60 Superdex 200 gel permeation chromatography column. The Fv dimer elutes with a partition coefficient (K) value of 0.5. If there is a high proportion of denatured Fv, the affinity column elution conditions should be reassessed.

7. Assay for binding activity and antigen epitope specificity.

Acknowledgements

We sincerely thank Dr Nigel Woods of British Biotechnology Ltd for advice on fermentation techniques and for providing the formula for the minimal medium. We also acknowledge financial support from SERC, British Biotechnology Ltd, and The Wellcome Foundation.

References

1. Colman, P. M. (1988). *Adv. Immunol.,* **43,** 99–132.
2. Davies, D. R., Padlan, E. A., and Sheriff, S. (1990). *Ann. Rev. Biochem.,* **59,** 439–73.
3. Orlandi, R., Güssow, D. H., Jones, P. T., and Winter, G. (1989). *Proc. Natl Acad. Sci. USA,* **86,** 3833–7.
4. Martin, A. C. R., Cheetham, J. C., and Rees, A. R. (1989). *Proc. Natl Acad. Sci. USA,* **86,** 9268–72.
5. Martin, A. C. R., Cheetham, J. C., and Rees, A. R. (1991). In *Methods in enzymology* (ed. J. J. Langone), Vol. 203, pp. 121–53. Academic Press, San Diego.
6. Pedersen, J., *et al.* (1992). In preparation.
7. Darsley, M. J. and Rees, A. R. (1985). *EMBO J.,* **4,** 383–92.
8. Darsley, M. J. and Rees, A. R. (1985). *EMBO J.,* **4,** 393–8.
9. Geysen, H. M., Rodda, S. J., Mason, T. J., Tribbick, G., and Scherefs, P. G. (1987). *J. Immun. Methods,* **102,** 259–74.
10. Cheetham, J. C., Raleigh, D. P., Redfield, C.R., Dobson, C. M., and Rees, A. R. (1991). *Proc. Natl Acad. Sci. USA,* **88,** 7968–72.
11. Jeffrey, P. (1989). D.Phil. thesis. University of Oxford.
12. de la Paz, P., Sutton, B., Darsley, M. J., and Rees, A. R. (1986). *EMBO J.,* **5,** 415–25.
13. Sheriff, S., *et al.* (1987). *Proc. Natl Acad. Sci. USA,* **84,** 8075–9.
14. Darsley, M. J., de la Paz, P., Phillips, D. C., Rees, A. R., and Sutton, B. J. (1985). In *Methodological surveys in biochemistry and analysis,* (ed. E. Reid, G. M. W. Cook, and D. J. Morre), Vol. 15, pp. 63–8. Plenum Press, New York.
15. Chothia, C. and Lesk, A. M. (1987). *J. Mol. Biol.,* **196,** 901–17.

16. Chothia, C., *et al.* (1989). *Nature,* **342,** 877–83.
17. Cunningham, B. C. and Wells, J. A. (1989). *Science,* **244,** 1081–5.
18. Roberts, S., Hilyard, K. L., McKeowen, S., and Rees, A. R. (1991). In preparation.
19. Roberts, S., Cheetham, J. C., and Rees, A. R. (1987). *Nature,* **328,** 731–4.
20. Rees, A. R., Martin, A. C. R., Roberts, S., and Cheetham, J. C. (1990). *UCLA Symposia on Molecular and Cellular Biology 'Protein and pharmaceutical engineering',* (ed. C. S. Craik, R. Fletterick, C. R. Matthews, and J. Wells), Vol. 110, pp. 35–54. Wiley-Liss, New York.
21. Hilyard, K. L. (1991). D.Phil. thesis. University of Oxford.
22. McPherson, M. J. (ed.) (1991). *Directed mutagenesis: a practical approach.* IRL Press, Oxford.
23. Roberts, S. and Rees, A. R. (1986). *Prot. Engng,* **1,** 59–65.
24. Melton, D. A., Kreig, P. A., Rebagliati, M. R., Zinn, K., and Green, M. R. (1984). *Nucl. Acids Res.,* **12,** 7035–56.
25. Munson, P. J. (1983). In *Methods in enzymology* (ed. J. J. Langone and H. van Vunakis), Vol. 92, pp. 543–76. Academic Press, San Diego.
26. Verhoyen, M., Milstein, C., and Winter, G. (1988). *Science,* **239,** 1534–6.
27. McLachlan, A. D. (1982). *Acta Cryst.,* **A38,** 871–93.
28. Kabat, E. A., Wu, T. T., Reid-Miller, M., Perry, H. M., and Gottesmen, K. S. (1987). *Sequences of proteins of immunological interest,* (4th edn). US Department of Health and Human Services.
29. Sambrook, J., Fritsch, E. F., and Maniatis, T. (1989). *Molecular cloning: a laboratory manual,* (2nd edn). Cold Spring Harbor Press, Cold Spring Harbor, New York.
30. Ward, E. S., Güssow, D., Griffiths, A. D., Jones, P. T., and Winter, G. (1989). *Nature,* **341,** 544–6.
31. Schein, C. H. and Noteborn, M. H. M. (1988). *Bio/tech.,* **6,** 291–4.
32. McNeil, B. and Harvey, L. M. (ed.) (1989). *Fermentation: a practical approach.* IRL Press, Oxford.

Selection of variants of antibodies and other protein molecules using display on the surface of bacteriophage fd

RONALD H. JACKSON, JOHN McCAFFERTY, KEVIN S. JOHNSON, ANTHONY R. POPE, ANDREW J. ROBERTS, DAVID J. CHISWELL, TIMOTHY P. CLACKSON, ANDREW D. GRIFFITHS, HENNIE R. HOOGENBOOM, and GREG WINTER

1. Introduction

In recent years, recombinant DNA technology has allowed the generation of cloned genes for large numbers of closely related proteins. The researcher then faces the problem of selecting, from a population, a gene encoding a protein with desired properties. An example of wide interest is the isolation of specific antibodies from libraries of cloned antibody genes, derived using the polymerase chain reaction (PCR), from immunized or non-immunized animals. Also, random mutagenesis techniques can be used to make large numbers of derivatives of a cloned protein, from which a variant with desired improved characteristics needs to be selected. Display of protein molecules on the surface of bacteriophage (1) provides a powerful method of selecting a desired protein, from a mixture of closely related proteins, together with the gene encoding it. This ability to co-select proteins and their genes has been exploited to enable the isolation of high-affinity antigen-specific antibodies derived from an immunized mouse (2).

Bacteriophage fd is a filamentous, single-stranded DNA phage which infects male *Escherichia coli* cells. Adsorption to the host sex pilus is mediated by the gene 3 protein (g3p) displayed at the tip of the virion. The amino-terminal domains of the three g3p molecules on each virion form knob-like structures that are responsible for binding the phage to the F-pilus while the C-terminal domain is anchored in the phage coat (3, 4). Peptides have been displayed at the surface of phage by fusion to the N-terminus of g3p (5) and phage with binding activities isolated from random peptide libraries after repeated rounds of growth and selection for phage with desired binding characteristics (6–8).

We have extended the range of molecules displayed as gene 3 fusions to folded proteins (1). Antibodies have been displayed as functional binding molecules in the form of single-chain Fv fragments (1), Fab fragments (9), and VH fragments (Jackson and McCafferty, unpublished). This has allowed the selection of phage from a mixed population according to their binding characteristics, for example single-chain Fv antibody fragments with a high affinity for 2-phenyl-5-oxazolone have been selected from a library of antibodies derived from an immunized mouse (2). The enzymes alkaline phosphatase from *E. coli* (10) and *Staphylococcus* nuclease (Hoogenboom and Winter, unpublished) have catalytic activity when displayed on bacteriophage fd. Further, phage displaying the human receptor molecules, CD4 and platelet-derived growth factor BB receptor (Jackson and Hoogenboom, unpublished) specifically bind the appropriate ligand. Bass *et al.* (11) have displayed functional human growth hormone on the surface of the closely related bacteriophage M13. A recent report describes functional expression of an antibody fragment as a gene 8 protein fusion in M13 (12) but this system is, at present, not as well characterized as that using gene 3.

In this chapter we describe in detail procedures for the cloning and selection of antibody genes using display on bacteriophage as g3p fusions. Many of these techniques will be directly applicable to work with other proteins.

2. Vectors for the display of proteins on the surface of bacteriophage fd

2.1 Phage vectors

Phage vectors for the display of proteins on the surface of bacteriophage fd have been derived from the vector fd-tet (13). *Escherichia coli* cells containing these vectors grow as colonies on agar plates. The vector fdCAT1 (*Figure 1*) was designed to allow the insertion of single-chain Fv sequences as fusions at the N-terminus of the gene 3 protein and to be compatible with the PCR primers used by Orlandi *et al.* (14) to amplify antibody variable regions. The vector, therefore, contains codons for five amino acids derived from the N-terminus of the antibody heavy chain, together with five amino acids derived from the C-terminus of the light chain. The *Pst*I and *Xho*I restriction sites included in these amino acid sequences allow the insertion of single-chain Fv antibodies (15, 16). In these derivatives, the VH and VL domains of the antibody are linked by a peptide $(Gly_4Ser)_3$ to provide a single-chain molecule. For example, scFvD1.3 (anti-lysozyme) and scFvNQ11 (anti-2-phenyl-5-oxazolone) have been inserted into fdCAT1 and demonstrated to bind antigen. This vector, however, has the disadvantage that many antibody genes

(a)   Phage fd-tet-DOG1 cloning sites

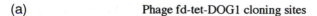

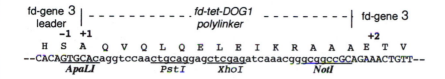

(b)   Phage fd-CAT1 cloning sites

fd gene 3 | — — — — — fd-CAT1 polylinker — — — — | fd gene 3
leader   +2
  -1   +1
   H   S   A   Q   V   Q   L   Q   L   E   I   K   R   E   T   V
   CACTCCGCTCAGGTCCAACTGCAggagcTCGAGATCAAACGGGAAACTGTT

              PstI   *XhoI*

Figure 1. Cloning sites for (a) fd-tet-DOG1; (b) fd-CAT1. The lower-case letters indicate in (a) the bases replaced when cloning *ApaL*I–*Not*I fragments into fd-tet-DOG1; in (b) the bases replaced when cloning *Pst*I–*Xho*I fragments into fd-CAT1.

contain *Pst*I sites (17), Also amino acid residues from the N-terminus and the C-terminus of antibody variable domains are incorporated into any protein expressed from genes cloned into this site.

For this reason, the vector fd-tet-DOG1 (*Figure 1*) was constructed. It is a derivative of fdCAT1 in which *ApaL*I and *Not*I restriction sites have been incorporated for cloning antibody genes as fusions to the N-terminus of g3p. The antibody sequence inserted between the *ApaL*I and *Not*I sites is excised and replaced by the sequence of DNA that is to be cloned. The restriction enzymes *ApaL*I and *Not*I have been selected because they cut only rarely in VH and VL domains. *ApaL*I does, however, cut in the CH1 domain of IgG heavy chains from mice and humans. For making libraries of heavy chains for display of Fab fragments, the phagemid vector pCANTAB-4 (*Figure 3*), which has a polylinker containing a *Sac*II site, is more suitable.

Protocol 1. Precipitation of bacteriophage with polyethylene glycol

1. Grow clones in fdCAT1 or fd-tet-DOG1 in 2xTY*[a]* containing 15 µg/ml tetracycline for 16–24 h. For clones in phagemids, rescue with helper phage (see *Protocol 2*). Typical phage titres are 10^{10}–10^{11} transducing units per ml.

2. Centrifuge at 10 000 *g* for 10 min. Collect supernatant.

Protocol 1. *Continued*

3. Add 1/5 volume PEG/NaCl (20% polyethylene glycol 8000, 2.5 M NaCl). Incubate at on ice for 1 h. Centrifuge at 10 000 *g* for 15 min at 4°C.

4. Retain the phage pellet. Remove as much supernatant as possible by decantation and allow the tube to drain on to a tissue paper. To remove residual traces of PEG, the tube may be respun for 1–2 min, although this is not usually necessary.

5. Resuspend the pellet in 10 mM Tris, 0.1 mM EDTA, pH 8.0 (or other appropriate buffer). Spin at 11 600 *g* for 5 min in a microcentrifuge to remove bacterial debris. Retain the supernatant.[b]

[a] 2xTY is 16 g tryptone, 10 g yeast extract, 5 g NaCl per litre H_2O.
[b] Phage can be titred if desired by infection of male *E. coli* strains such as TG1. Phage expressing fusion proteins appear to show reduced infectivity.

To prepare vector DNA for cloning, phage are grown in 2xTY medium containing 15 µg/ml tetracycline for 16–24 h. Vector RF DNA is isolated from the bacterial pellet by the plasmid alkaline lysis method described in ref. 18, purified using caesium chloride gradient centrifugation and digested with restriction enzymes according to the manufacturer's instructions. For enzymes such as *ApaL*I and *Not*I, which have different buffer requirements, DNA is ethanol-precipitated after the first enzyme digest. The final product is purified on a 0.7–1% agarose gel and the DNA is extracted using a Geneclean II kit and dissolved in TE buffer ready for ligation with the DNA insert.

2.2 Phagemid vectors

Phage vectors have a low efficiency of transformation when compared to plasmids such as the pUC series of vectors. Phagemid vectors (which contain an origin of replication for filamentous bacteriophage) allow 100-fold higher efficiencies of transformation to be obtained compared to phage vectors. Gene 3 from bacteriophage fd has therefore been inserted into phagemid vectors with restriction sites allowing insertion of foreign DNA sequences. Superinfection with helper phage enables packaging of this phagemid DNA into a phage particle which displays the protein encoded by the insert as an N-terminal fusion with the gene 3 protein. The higher efficiency of transformation permitted by the use of phagemids is important when large libraries are made, for example repertoires of antibody genes (where a mouse would be expected to produce 10^7 different specificities).

Cloning into phagemids allows the number of fusion proteins displayed on the phage particle to be varied. When the g3p fusion is expressed using a phage vector, all copies of gene 3 will be expressed as fusions. Expression in phagemids following superinfection with M13K07 generates phage displaying a mixture of native g3p and the g3p fusion. The ratio of this mixture can be varied by the induction of the g3p fusion using different concentrations of isopropylthiogalactoside (IPTG), which induces expression from the *lac* promoter (*Figure 2*). As discussed below, it may be useful when selecting for high-affinity antibodies expressed on bacteriophage to be able to express no more than a single antibody molecule on each virion. This would prevent the co-selection of phage expressing multiple copies of a lower-affinity antibody on their surface, which have a similar avidity to phage expressing the higher-affinity antibody as a single copy.

The phagemids pCANTAB-3 and pCANTAB-4 (*Figure 3*) allowing cloning of genes as *ApaLI–Not*I or *SacII–Not*I fragments at the N-terminus of gene 3 have been derived from pUC119. Export of g3p fusions to the periplasm is directed using the native g3p leader sequence. The vector pHEN1 (*Figure 3*; ref. 9) allows the cloning of genes as *SfiI–Not*I fragments. Export to the periplasm is directed by a *pel*B leader sequence. pHEN1 contains an amber codon inserted at the start of the gene 3 segment, allowing expression of the inserted protein from the gene 3 fusion as a soluble fragment in a non-suppressing *E. coli* strain such as HB2151, without recloning the gene. Further, pHEN1 introduces a c-myc tag sequence (19) at the C-terminus of the soluble antibody to facilitate detection by ELISA or purification by affinity chromatography. The vectors pCANTAB-3 and pCANTAB-4 have been designed for flexibility in the introduction of such features. For example,

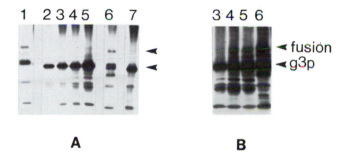

Figure 2. Western blot of PEG-precipitated phage used in ELISA probed with anti-g3p. Free g3p and the g3p-scFvD1.3 fusion bands are arrowed. Sample 1, fd-scFvD1.3; sample 2, pCANTAB-3 vector; sample 3, pCANTAB-3 scFvD1.3 rescued with M13KO7, no IPTG; sample 4, pCANTAB-3 scFvD1.3 rescued with M13KO7, 50 μM IPTG; sample 5, pCANTAB-3 scFvD1.3 rescued with M13KO7, 100 μM IPTG; sample 6, pCANTAB-3 scFvD1.3 rescued with M13KAT (no IPTG); sample 7, pCANTAB-3 scFvD1.3 rescued with M13KO7 gIIIΔ No2 (a different gene 3 deletion derivative) (no IPTG). Panel (A) samples contain the equivalent of 8 μl of phagemid culture supernatant per track, and 80 μl of the fd supernatant (tenfold lower phage yield than the phagemid). Panel (B) phagemid samples are those used in panel (A) at a fivefold higher sample loading (equivalent to 40 μl of culture supernatant per track) to enable visualization of the fusion band in samples rescued with parental M13KO7.

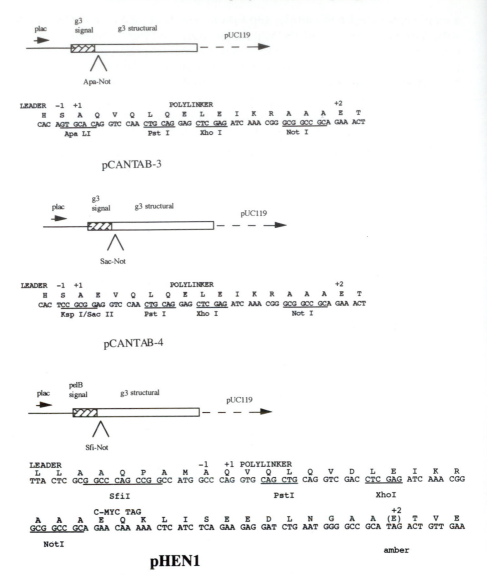

Figure 3. Cloning sites for, top to bottom, pCANTAB-3, pCANTAB-4 and pHEN1. Each vector is based on pUC119 into which gene 3 of bacteriophage fd has been inserted with cloning sites in its leader and N-terminal regions. In each case expression is under the control of the plac promoter, with gene 3 protein fusions being directed to the periplasm by the gene 3 leader for pCANTAB-3 and pCANTAB-4 and the pelB leader for pHEN1.

Ronald H. Jackson et al.

an amber codon can be introduced into clones in pCANTAB-3 and pCANTAB-4 by modifying the VKFORNOT primers used in *Protocol 7*.

2.3 Rescue phage

The most widely used phage for superinfection is M13K07 (20) which will preferentially package phagemid DNA. A standard protocol is given in *Protocol 2*. With phagemid pCANTAB-3 a yield of c. 10^{11} phage per ml is obtained. The incorporation of gene 3 fusion molecules relative to native g3p can be varied by the strength of induction of the *lac* promoter (*Protocol 2*, step **5**).

In many cases, for instance when selecting low-affinity antibodies, perhaps from a library of antibodies derived from an unimmunized mouse, or when increased sensitivity is required for an ELISA, the expression of all copies of g3p as fusion molecules is desirable. To facilitate this, a derivative of M13K07 (M13KAT) has been prepared which contains a deletion in gene 3 (M13KAT is grown on an *E. coli* strain supplying the gene 3 function and gives a lower titre than M13K07). As shown in *Figure 2*, an increased number of intact g3p fusion molecules are obtained when phagemids are rescued with M13KAT as shown by Western blotting of fusion proteins.

Protocol 2. Rescue of phagemids pHEN1, pCANTAB-3 or pCANTAB-4 by superinfection with helper phage[a]

1. Grow cells containing the phagemid in 2TYAG[b] at 30°C overnight or to mid-log phage ($A_{600} = 0.5-1.0$) if inoculation is on the same day as rescue.

2. Add 60 µl of cell culture to 6 ml 2TYAG in a 50 ml sterile plastic centrifuge tube (e.g. Falcon 2070) and grow at 35°C with fast shaking for 1 h.

3.[a]Centrifuge at 3500 *g* for 5 min (bench-top centrifuge) and resuspend the bacterial pellet in 10 ml 2TYA[d] to give approximately 10^8 cells/ml (using $A_{600} = 1$ at 5×10^8 cells/ml).

4. Add M13K07 (or other helper phage) to a final concentration of 10^8-10^9 p.f.u./ml.

5. Grow cells for 45 min to 1 h with moderate shaking (300 r.p.m.) at 35°C. To each tube add:[e]

   • kanamycin (25 mg/ml)   20 µl

   • isopropylthiogalactoside (IPTG)[f] 10 mM   20-100 µl

6. Grow overnight with fast shaking. Pellet cells and precipitate phage from the supernatant using PEG (see *Protocol 1*). For ELISA it will often be

283

Protocol 2. *Continued*

possible to obtain good signals using the supernatant directly, diluted 1:1 with PBS containing 4% skimmed milk powder (see *Protocol 10*).

---

[a] This protocol is for 10 ml cultures, the volumes can be adjusted proportionately for different scale preparations. It is essential that the cultures are vigorously aerated. If the volumes are increased, larger vessels are needed.

[b] 2TYAG is 16 g tryptone, 10 g yeast extract, 5 g NaCl per litre H_2O, containing 100 µg/ml ampicillin and 2% (w/v) glucose.

[c] Instead of this step, catabolite repression by glucose can be partially overcome by the addition of IPTG to 20–50 µg/ml. However, end-point ELISA signals are roughly tenfold lower than when glucose is removed, indicating that induction of the gene 3 protein fusion is incomplete.

[d] 2TYA is as 2TYAG but omitting glucose.

[e] For rescue of phagemids using phage expressing partner chains in the dual combinatorial approach (see Section 3.2), 30 µl of 5 mg/ml tetracycline is added in place of kanamycin.

[f] Addition of IPTG is optional, since removal of glucose derepresses synthesis of the gene 3 protein fusion, but the addition does increase expression levels two- to threefold.

---

3. Preparation and cloning of antibody DNA

Our initial studies on the cloning and selection of repertoires of antibody genes in bacteriophage fd have used single-chain Fv fragments. Protocols for preparing such repertoires from mice are described in detail below. However, Fab fragments, which may be preferred in many cases, have also been displayed on the surface of bacteriophage fd (9). Also, oligonucleotide primers have been designed for amplifying human V genes (21) allowing the development of strategies for the isolation of human antibody fragments specific for antigens from libraries prepared from unimmunized humans (22).

3.1 Single-chain Fv DNA

An overall scheme for preparing libraries of antibody scFv fragments in pCANTAB-3 or fd-tet-DOG1 is shown in *Figure 4*. For simplicity, the procedure described below is for the generation of a random combinatorial library. The method can be readily adapted for the generation of hierarchical libraries as in Section 6.1. The procedure relies on separate amplification of VH and VL genes which are then linked by a PCR assembly procedure. This minimizes the use of restriction enzymes.

3.1.1 Preparation of mRNA

Preparation of cytoplasmic mRNA from spleen (for example from an immunized mouse) or from monoclonal cells is performed using standard procedures for preparing mRNA for cDNA synthesis (18). Alternatively, a rapid boiling method may be used to extract mRNA from monoclonal cells (23). Splenocytes are prepared from spleen by procedures described for preparation for cell fusion (24).

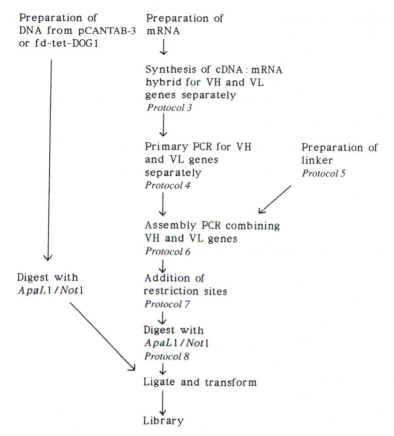

Preparation of
DNA from pCANTAB-3
or fd-tet-DOG1

Preparation of
mRNA

Synthesis of cDNA : mRNA
hybrid for VH and VL
genes separately
Protocol 3

Primary PCR for VH
and VL genes
separately
Protocol 4

Preparation of
linker
Protocol 5

Assembly PCR combining
VH and VL genes
Protocol 6

Digest with
*ApaL*1 / *Not*1

Addition of
restriction sites
Protocol 7

Digest with
*ApaL*1 / *Not*1
Protocol 8

Ligate and transform

Library

Figure 4. Flow diagram for the cloning of single-chain Fv fragments into pCANTAB-3 or fd.tet.DOG1.

3.1.2 Reverse transcription

Reverse transcription of the mRNA encoding VH and VL genes in separate reactions then generates cDNA:mRNA hybrids by using primers complementary to the 3' end of the CH or the VL domain (*Table 1*).

Protocol 3. Preparation of cDNA:mRNA hybrid

1. Set up the following reverse transcription mix containing six primers:
   - H$_2$O (DEPC treated)[a]   20 µl
   - 5 mM dNTP[b]   10 µl
   - 10 × first strand buffer[c]   10 µl
   - 0.1 M dithiothreitol   10 µl

285

Protocol 3. *Continued*

- FOR primers[d] (10 pmol/μl; *Table 1*) 2 μl (each)
- RNasin[e] (40 U/μl) 4 μl

2. Dilute 10 μg RNA to 40 μl final volume with DEPC-treated water. Heat at 65°C for 3 min and then place on ice for 1 min (to remove secondary structure).

3. Add to the RNA the reverse transcription mix and 4 μl 'Super RT'[f] and incubate at 42°C for 1 h.

4. Boil the reaction mix for 3 min, cool on ice for 1 min, and then spin in a microcentrifuge at 11 600 g for 5 min to pellet debris. Transfer the supernatant to a new tube. Store at −20°C.

[a] DEPC-treated water is prepared by adding diethylpyrocarbonate to H_2O (0.1% v/v), and incubating at 37°C for 2 h. DEPC-treated water should be autoclaved before use to inactivate the diethylpyrocarbonate.
[b] 5 mM dNTP is an equimolar mixture of dATP, dCTP, dGTP, and dTTP with a total concentration of 5 mM nucleotide (i.e. 1.25 mM each dNTP).
[c] 10 × first strand buffer is 1.4 M KCl, 0.5 M Tris–HCl pH 8.1 at 42°C, 80 mM $MgCl_2$.
[d] The primers anneal to the 3' end. One reaction is set up containing each of the light chain primers MJK1FONX, MJK2FONX, MJK4FONX and MJK5FONX, and each of the heavy chain primers MIGG1,2 and MIGG3 (for preparation of IgG DNA) shown in *Table 1*. If IgM DNA is to be prepared MIGM is used as heavy chain primer.
[e] RNasin is a ribonuclease inhibitor obtained from Promega Corporation.
[f] Super RT is available from Anglian Biotech Ltd.

3.1.3 Primary PCR

The cDNA:mRNA hybrid can then be used in the 'primary' PCR to amplify the library of VH or VL domains.

Protocol 4. Primary PCR of antibody genes

1. For each PCR and negative control the following reactions are set up (i.e. one reaction for each of the four VK primers MJK1FONX, MJK2FONX, MJK4FONX and MJK5FONX and four identical VH PCRs, together with controls (no DNA) for each; a total of 13 reactions).

   - H_2O 34.5 μl
   - 10 × Vent buffer[a] 5 μl
   - 100 × Vent BSA[a] 0.5 μl
   - 5 mM dNTP[b] 1.5 μl
   - FOR primer[c] (10 pmol/μl) 2.5 μl
   - BACK primer[d] (10 pmol/μl) 2.5 μl

2. Irradiate this mix with ultraviolet light at 254 nm for approximately 5 min.

3. Add 2.5 μl cDNA:mRNA hybrid (from *Protocol 3*), and overlay with 2 drops of paraffin oil (Sigma).

4. Place on a cycling heating block preset at 94°C.

5. Add 1 μl Vent DNA polymerase under the paraffin.

6. Amplify using 25 cycles of 94°C 1 min, 60°C 1 min, 72°C 2 min followed by incubation at 60°C for 5 min.

7. Purify on a 2% (low-melting-point agarose/Tris–acetate–EDTA gel.[f] Carefully excise the VH and VK bands (using a fresh sterile scalpel or razor blade for each) and transfer each band to a separate sterile micro-centrifuge tube. The use of suitable molecular weight markers will aid selection of the correct bands. Both the VH and VK bands are approximately 350 bases long. Use Geneclean[e] to purify each DNA band. Recover the DNA in 20 μl H_2O per original PCR.

[a] Vent DNA polymerase, Vent buffer, and Vent BSA are supplied by New England Biolabs. Vent BSA (100×) is at 10 mg/ml. Vent DNA polymerase has a proof-reading function.
[b] As in footnote *b, Protocol 3*.
[c] FOR primers shown in *Table 1*: VH1FOR-2 for heavy chains, MJK1FONX, MJK2FONX, MJK4FONX, or MJK5FONX for light chains.
[d] BACK primers shown in *Table 1*: VH1BACK for heavy chains, VK2BACK for light chains.
[e] Geneclean is from Bio101, used according to the manufacturer's instructions. Use a dedicated kit, aliquotted into single-use quantities.
[f] To avoid contamination, it is essential to depurinate the electrophoresis apparatus, combs, etc. with 0.25 M HCl overnight before use.

The primers described in *Table 1*, have been shown to generate a diverse library when used to prepare a repertoire of antibodies from a mouse immunized with 2-phenyl-5-oxazolone (2). As more information becomes available the primers will be refined to ensure that libraries of antibody fragments are fully representative.

3.1.4 Assembly of scFv fragments

The separately amplified VH or VL domains are then linked in an assembly step (*Protocol 6*) to incorporate the linker $(Gly_4Ser)_3$ generated in *Protocol 5*. *Figure 5* illustrates the methodology and the products of the primary PCR and the assembly.

Protocol 5. Preparation of linker[a]

1. Set up linker reaction mix:[b]

- H_2O 36.3 μl
- 10 × Vent buffer[c]   5 μl
- 100 × Vent BSA[c]   0.5 μl

Protocol 5. *Continued*

• 5 mM dNTP^d	2 µl
• LINKFOR primer (10 pmol/µl)^e	2.5 µl
• LINKBACK primer (10 pmol/µl)^e	2.5 µl
• fd-scFvD1.3 DNA (10 ng/µl)^a	1 µl
• Vent DNA polymerase^c	0.2 µl

2. Cover with paraffin and place on a cycling heating block at 94°C. Amplify the linker DNA by PCR for 25 cycles of 94°C 1 min, 65°C 1 min, 72°C 2 min.

3. Incubate at 60°C for a further 5 min in the heating block.

4. Purify on a 2% lower-melting-point agarose/Tris–acetate–EDTA gel (use loading dye without bromophenol blue, since the linker DNA sequence required is 93 bp long). Excise the correct (93 bp) gel fragment. Place in the upper chamber of a SPIN-X column^f and centrifuge in a microcentrifuge for 5 min at 11 600 *g*.

5. Ethanol-precipitate the DNA and dry the pellet carefully before re-suspending in 5 µl H$_2$O.

^{a} The linker was originally prepared by oligonucleotide synthesis to generate the sequence:
5′GGCACCACGGTCACCGTCTCCTCAGGTGGAGGCGGTTCAGGCGGAGGTGGCT
CTGGCGGTGGCGGATCGGACATCGAGCTCACTCAGTCTCCA3′
Having once made this construct and inserted it into a vector or a clone, it is convenient to prepare additional material from this by the PCR amplification of the linker-containing vector. In this protocol, fd-scFvD1.3 DNA (1) was used. However, the linker oligonucleotide sequence can be cloned into any convenient vector for subsequent amplification.
^{b} Each linker reaction provides sufficient linker for approximately three assembly reactions. It is often convenient to set up 10 reactions at one time.
^{c} Vent BSA, Vent buffer, and Vent DNA polymerase are supplied by New England Biolabs.
^{d} As in *Protocol 3*, footnote *b*.
^{e} The sequences of LINKFOR and LINKBACK are shown in *Table 1*.
^{f} SPIN-X columns are available from Costar. As an alternative to using SPIN-X columns the gel band can be purified using a Mermaid kit (Geneclean; Bio101 Inc.).

Protocol 6. Assembly of single-chain Fv antibody fragments

1. Estimate the quantities of VH and VL DNA prepared by the primary PCR reactions (*Protocol 4*) and the quantity of linker DNA prepared in *Protocol 5* using agarose gel electrophoresis. Adjust the volumes of VH, VL, and linker DNA added in step **2** below to give roughly equal masses of DNA fragments added to the assembly reaction (approximately 50 ng).

2. Set up four reactions for assembling the products of the primary VK PCR reactions with the products of the VH PCR reactions (one for each

Ronald H. Jackson et al.

of the four primary VK PCR reactions). Each assembly reaction is performed in two stages. For each first-stage reaction set up the following mixture:

- VH DNA[e] (from *Protocol 4*)   x μl (see **1** above)
- VL DNA[e] (from *Protocol 4*)   y μl (see **1** above)
- Linker DNA (from *Protocol 5*)   z μl (see **1** above)
- 10 × Vent buffer[a]   2.5 μl
- 100 × Vent BSA[a] (10 mg/ml)   0.25 μl
- 5 mM dNTP[b]   2 μl
- sterile H_2O up to 25 μl
- Vent DNA polymerase[a] (1 U/μl)   1 μl

3. Place on a cycling heating block and incubate for 20 cycles of 94°C for 1.5 min and 65°C for 3 min.

4. To each reaction add 25 μl of the following mixture for the second stage:

- 10 × Vent buffer   2.5 μl
- 100 × BSA   0.25 μl
- 5 mM dNTP   2 μl
- VH1BACK (*Table 1*; 10 pmol/μl)   5 μl
- VKFOR[c] (10 pmol/μl)   5 μl
- sterile H_2O up to 25 μl
- Vent DNA polymerase (1 U/μl)   1 μl

5. Amplify the DNA using 30 cycles of 94°C for 1 min, 50°C for 1 min, and 72°C for 2 min with a final extension step at 72°C for 10 min.

6. Electrophorese the product of each reaction on a 1.4% low-melting-point agarose/Tris–acetate–EDTA gel. Excise the band corresponding to the assembled product in each case (c. 720 bp). Pool the excised DNA bands and use Geneclean[d] to purify the DNA from the band. Recover the Geneclean product in 40 μl of distilled H_2O.

[a] See *Protocol 5*, footnote c.
[b] See *Protocol 3*, footnote b.
[c] VKFOR is the corresponding light chain forward primer that was used in the primary PCR, i.e. MJK1FONX, MJK2FONX, MJK4FONX, or MJK5FONX.
[d] Geneclean is from Bio101 Inc. and is used according to manufacturer's instructions.
[e] For the generation of hierarchical libraries, the VH and VL DNA are derived from different sources. Either the VH or VL DNA is obtained by PCR amplification of DNA from a single antibody fragment clone or group of clones, by the procedure in *Protocol 4*, where this domain is required to be kept constant. It is important that this DNA is purified as in *Protocol 4*, step 7 to avoid contamination with the original partner chain. The DNA encoding the complementary domain is derived as in *Protocol 4* for the primary PCR.

Table 1. Oligonucleotides for PCR reactions

cDNA synthesis and primary PCR oligos (restriction sites underlined):

VH1FOR-2	TGA GGA GAC <u>GGT GAC C</u>GT GGT CCC TTG GCC CC
VH1BACK	AGG TSM AR<u>C TGC AGS</u> AGT CWG G
MJK1FONX	CCG TTT GAT TTC CAG CTT GGT GCC
MJK2FONX	CCG TTT TAT TTC CAG CTT GGT CCC
MJK4FONX	CCG TTT TAT TTC CAA CTT TGT CCC
MJK5FONX	CCG TTT CAG CTC CAG CTT GGT CCC
VK2BACK	GAC ATT <u>GAG CTC</u> ACC CAG TCT CCA
MIGG1,2	CTG GAC AGG GAT CCA GAG TTC CA
MIGG3	CTG GAC AGG GCT CCA TAG TTC CA
MIGM	CC CTG GAT GAC TTC AGT GTT GTT CTG

PCR oligos to make linker:

LINKFOR	TGG AGA CTG GGT <u>GAG CTC</u> AAT GTC
LINKBACK	GGG ACC AC<u>G GTC ACC</u> GTC TCC TCA

Oligos for addition of restriction sites:

VH1BACKAPA	CAT GAC CAC A<u>GT GCA C</u>AG GTS MAR CTG CAG SAG TCW GG
JK1NOT10	GAG TCA TTC <u>TGC GGC CGC C</u>CG TTT GAT TTC CAG CTT GGT GCC
JK2NOT10	GAG TCA TTC <u>TGC GGC CGC C</u>CG TTT TAT TTC CAG CTT GGT CCC
JK4NOT10	GAG TCA TTC <u>TGC GGC CGC C</u>CG TTT TAT TTC CAA CTT TGT CCC
JK5NOT10	GAG TCA TTC <u>TGC GGC CGC C</u>CG TTT CAG CTC CAG CTT GGT CCC

Ambiguity codes: M = A or C
R = A or G
S = G or C
W = A or T

3.1.5 Amplification and digestion

A further amplification is performed using primers that incorporate restriction sites; in the example given, *ApaL*I and *Not*I sites (*Protocol 7*). The amplified DNA is then digested with *ApaL*I and *Not*I (*Protocol 8*) to allow cloning into pCANTAB-3 or fd-tet-DOG1 cut with the same enzymes.

Protocol 7. Incorporation of restriction enzyme sites

1. Set up four reactions for the incorporation of restriction enzyme sites, one for each of the VKFORNOT primers together with controls (no DNA).

   - Assembled product from *Protocol 6*   5 µl
   - 10 × Taq polymerase buffer[a]   5 µl
   - 50 mM MgCl$_2$ 1 µl
   - 5 mM dNTP[b]   2 µl

- VH1BACKAPA (*Table 1*; 10 pmol/μl)   2.5 μl
- VKFORNOT[c] (10 pmol/μl)   2.5 μl
- sterile H$_2$O   up to 50 μl
- Taq DNA polymerase[d] (2.5 U/μl)   0.5 μl

2. Cover with paraffin and place on a block preset at 94°C.

3. Amplify DNA by PCR with 30 cycles of 94°C 1 min, 55°C 1 min, 72°C 2 min followed by an incubation for 10 min at 72°C.

4. Analyse 5 μl of the products by electrophoresis on a 1.4% agarose/Tris–acetate–EDTA gel. If a strong band is seen at approximately 720 bp, the products are pooled for digestion (*Protocol 8*).

[a] 10 × Taq polymerase buffer is 100 mM Tris, pH 8.3, 500 mM KCl, 0.1% gelatin.
[b] As in *Protocol 3*, footnote *b*.
[c] VKFORNOT is one of the primers JK1NOT10, JK2NOT10, JK4NOT10, or JK5NOT10 (*Table 1*) which incorporate *Not*I restriction sites. VH1BACKAPA incorporates an *Apa*LI restriction site.
[d] Taq DNA polymerase is from Perkin–Elmer Cetus.

Protocol 8. Restriction enzyme digestion of assembled products

1. Add 200 μl of phenol (TE buffer saturated) to each of the DNA products from the PCR reactions in *Protocol 7*. Mix well. Incubate at room temperature for 10 min. Mix well. Centrifuge for 5 min in a microfuge at 13 000 r.p.m. Place the upper aqueous layer in a fresh tube. Add 200 μl of TE buffer to phenol. Mix well. Centrifuge for 5 min at 11 600 g. Carefully remove the upper aqueous layer and combine it with the first aqueous extract.

2. Ethanol-precipitate the DNA contained in the aqueous extract. Wash the pellet twice with 70% EtOH. Dry the DNA pellet but do not overdry. Resuspend in 70 μl H$_2$O.

3. Digest the DNA product overnight with *Not*I at 37°C using the reaction mixture below:

- DNA   70 μl
- NEB *Not*I buffer × 10   10 μl
- NEB BSA × 10   10 μl
- *Not*I (10 U/μl)   10 μl

4. Ethanol-precipitate the *Not*I digested DNA. Resuspend the pellet in 80 μl H$_2$O.

5. Digest with *Apa*LI using the reaction mixture below:

- DNA digested with *Not*I   80 μl

Protocol 8. *Continued*

- NEB buffer 4   10 μl
- *ApaL*I   10 μl

Incubate at 37°C for 4 h, adding fresh enzyme at hourly intervals as *ApaL*I has a short half-life at 37°C.

6. Purify on 1.5% low-melting-point agarose/Tris–acetate–EDTA gel. Excise the band due to *Not*I/*ApaL*I-digested DNA; purify the DNA using Geneclean and recover the DNA in 20 μl of H$_2$O.

3.1.6 Ligation and transformation

Ligation of DNA (e.g. using a DNA ligation kit, Amersham International) and electroporation of *E. coli* cells (25) are performed using standard procedures. Alternatively, transformation of competent cells (26) can be used.

3.2 Cloning of Fab fragments

Fab fragments may have advantages over scFv fragments. They are more stable at 37°C than the corresponding scFv fragments (27) and often show higher affinities for antigen (28). Fab fragments and Fv fragments have also been displayed on the surface of bacteriophage (9; T. Simon, unpublished). Fab fragments were expressed in two ways:

(a) cloning both chains into fd-tet-DOG1 but expressing only one chain as a fusion with gene 3 protein, the other being expressed as a soluble chain

(b) cloning one chain into fd-tet-DOG1 and expressing the complementary chain from phagemid pHEN1 as a soluble fragment on superinfection with the clone in fd-tet-DOG1

In the latter dual combinatorial approach (9), the two libraries, e.g. of 10^7 VHCH1 chains and 10^7 light chains, are crossed by rescuing the phagemid library with the phage library, generating 10^{14} different VHCH1/VLCL combinations. Clones of VHCH1 fragments and of light chains isolated by this dual combinatorial approach can, in theory, then be recrossed to generate further specificities. Rescue of phagemids using phage-expressing fusions to make dual combinatorial libraries requires a minor modification of the standard protocol (*Protocol 2*, footnote *e*).

Also, since chains have to be stably associated to be selected, the selection process will favour stably associated Fab fragments (or Fv fragments if VH and VL domains are displayed).

3.3 Cloning of proteins other than antibodies

Genes encoding other proteins can be cloned into the vectors described in this article simply by amplifying the appropriate gene with primers for PCR which

1) PRIMARY PCR amplify heavy and light chains

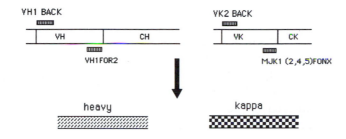

2) ASSEMBLY PCR – assemble heavy and light chains via linker

linker = (gly.gly.gly.gly.ser)$_3$

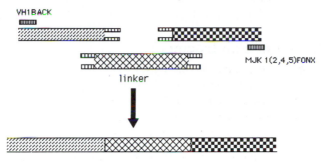

3) ADDING RESTRICTION SITES BY PCR

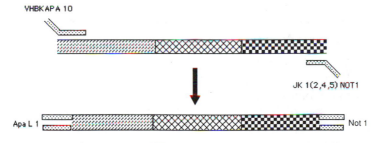

DIGEST WITH Apa L1 AND Not 1 AND CLONE INTO VECTOR

Figure 5. Preparation of library of DNA inserts encoding single chain Fv fragments. (1) cDNA:mRNA hybrids are used as template with the primers VH1BACK and VH1FOR2 to amplify the heavy chain VH region and with the primers VK2BACK and MJK1(2,4,5)FONX to amplify the kappa light chain VK region. (2) The linker fragment is used to assemble the VH chain fragment with the VK chain fragment using the complementarity of the ends of the linker fragment to the 3' end of the VH fragment and the 5' end of the VL fragment. This assembled product is amplified in a second stage using the primers VH1BACK and MJK1(2,4,5)FONX. (3) Restriction sites are added by PCR amplification using the primers VHBKAPA10 and JK1(2,4,5)NOT10. The amplified product is digested with ApaL 1 and Not1 and cloned into pCANTAB3 or fd-tet-DOG1 digested with the same enzymes.

are extended at the 5' end of the sequences complementary to the gene with sequences incorporating the appropriate restriction site. For instance, alkaline phosphatase was cloned into fd-tet-DOG1 following amplification with primers specific for the *E. coli phoA* gene (10).

4. Selection of protein variants displayed on the surface of bacteriophage

Protein variants expressed on the surface of bacteriophage have been selected on the basis of their affinity for antigen using chromatography, panning, or adsorption to cells. Elution from affinity matrices has been achieved by specific elution using the antigen (or a related compound) or non-specific elution using, for example, 100 mM triethylamine. Washing procedures remove non-specifically bound phage. The phage binds to and is eluted from the matrix according to the affinity or the nature of the binding interaction. Specifically eluted phage are then used to infect male *E. coli* cells expressing the F pilus, allowing recovery of phage containing DNA encoding proteins with the desired binding characteristics.

4.1 Affinity chromatography

In our laboratories, we have found that elution from affinity matrices with 100 mM triethylamine can be used to isolate phage displaying antibodies directed against a number of different antigens. This procedure may be a useful starting point for the isolation of phage by affinity chromatography.

Protocol 9. Selection of antibodies displayed on phage by affinity chromatography

1. Prepare phage[a] using polyethylene glycol precipitation, and resuspend in 10 ml of PBS containing 2% skimmed milk powder (MPBS).
2. Prepare a 1 ml column of antigen or antigen derivative coupled to Sepharose.[b] Wash with 20 ml of PBS followed by 10 ml of MPBS.
3. Apply phage to the column. Reload the initial flowthrough if the overall yield is important.
4. Wash the column with 10 ml MPBS.
5. Wash the column with 50 ml PBS.
6. Wash the column with 30 ml 50 mM Tris–HCl, 500 mM NaCl, pH 8.5.[c]
7. Wash the column wit 30 ml 50 mM Tris–HCl, 500 mM NaCl, pH 9.5.[c]
8. Elute phage bound to antigen with 10 ml 100 mM triethylamine.[d] Collect in 2 ml 1 M Tris, pH 7.4.

9. Make several tenfold dilutions of the eluate and add to 1 ml of log-phase *E. coli* TG1 cells. Incubate at 37°C for 45 min and plate on 2YT agar plates containing the appropriate antibiotic.

10. For further rounds of selection[e] take approximately 10^6 colonies derived from the eluate of the initial column and scrape the colonies obtained after overnight growth into 10 ml of 2YT medium. Dilute 100 μl of this suspension into 10–100 ml of 2YT medium. Grow overnight and repeat this protocol from step **1**.

11. Screen for antigen-binding phage using an appropriate assay such as ELISA (*Protocol 10*).

[a] Approximately 10^{10}–10^{12} antibody-expressing phage should be applied to each 1 ml column.
[b] CNBr-Sepharose 4B is available from Pharmacia Biosystems Ltd. No alternative chromatography matrices have been tested but others may be suitable.
[c] Keep these washes. If no antigen-binding phage is found in the eluate, assay phage from these washes. Some antigen–antibody interactions may be disrupted under these conditions. Alternatively, a single wash with 200 ml of PBS may be used.
[d] Treatment at pH 12 with 100 mM triethylamine for 15 min reduces the infectivity of phage by more than tenfold. Rapid neutralization is therefore essential.
[e] It is usually necessary to do at least two rounds of selection if isolating phage from a library.

The method has been used to obtain a millionfold enrichment of a phage antibody directed against lysozyme from a background of fdCAT1 vector phage (1). If elution at high pH is unsuitable and other non-specific eluants are used, such as acid pH or 25% ethylene glycol, the effect of eluant on infectivity should be tested before use, e.g. by incubating the vector phage in eluant and comparing the number of colonies obtained on infection of *E. coli* with the number obtained using untreated phage.

Additional selectivity in isolation of phage antibodies may be obtainable by elution with the antigen or a related molecule, if a suitable one is available. Specific elution with a derivative of the antigen 2-phenyl-5-oxazolone (4-ε-aminocaproic acid methylene-2-phenyl-5-oxazolone) was used to isolate phage expressing specificities directed against oxazolone from a repertoire of phage containing antibody genes derived from a mouse immunized with 2-phenyl-5-oxazolone coupled to chicken serum albumin (*Table 2*; ref. 2). The power of the selection technique is demonstrated by the large number of high-affinity antibodies selected from the repertoire.

Selection can be made not only on the basis of specificity but also on the basis of affinity. Separation is readily attainable by affinity chromatography between phage expressing an antibody with a dissociation constant of 10^{-8} M and one with a dissociation constant of 10^{-5} M (2). The isolation of the latter antibody from an immune repertoire demonstrates that antibodies with affinities characteristic of the primary immune response can be isolated using phage technology.

Table 2. Affinity selection of 2-phenyl-5-oxazolone binding phage

	Clones binding to phOx		
	Pre-column	After first round	After second round
Random combinatorial library	0/568 (0%)	48/376 (13%)	175/188 (93%)
Hierarchical libraries: 1	6/190 (3%)	348/380 (92%)	–
2	0/190 (0%)	23/380 (7%)	–

This table shows selection of phage with hapten-binding activities from random combinatorial and hierarchical libraries in the vector fd-tet-DOG1 derived from mice immunized with 2-phenyl-5-oxazolone coupled to chicken serum albumin. The random combinatorial library was prepared as in Section 3.1, *Protocols 3–8*; the hierarchical libraries were prepared keeping either the VH chain (library 1) or the VL chain (library 2) constant as in *Protocol 6*, footnote *e*. Phage were prepared as in *Protocol 1*. For affinity selection, a 1 ml column of phOx-BSA-Sepharose was washed with 300 ml phosphate-buffered saline (PBS), and 20 ml PBS containing 2% skimmed milk powder (MPBS). 10^{12} TU phage were loaded in 10 ml MPBS, washed with 10 ml MPBS and finally 200 ml PBS. The bound phage were eluted with 5 ml 1 mM 4-ε-aminocaproic acid methylene-2-phenyl-5-oxazolone (phOx-CAP). About 10^6 TU eluted phage were amplified by infecting 1 ml log-phase *E. coli* TG1 and plating on 2YT agar plates containing 15 μg/ml tetracycline. For a further round of selection, colonies were scraped into 10 ml 2xTY medium and then phage were prepared as in *Protocol 1*. Affinity chromatography was exactly as above. Other details are in ref. 2.

4.2 Panning

Affinity chromatography using elution with a gradient of antigen offers the prospect of fine resolution on the basis of affinity for antigen. However, columns have a large binding capacity for antigen and therefore antibodies with a wide range of affinities are able to bind. Preliminary results from our laboratories suggest that panning procedures can give superior selection of high-affinity over low-affinity antibodies when binding followed by non-specific elution is used. The phage antibody was bound to antigen coated on to a large surface area of plastic (for example Falcon 3001 Petri dishes 35 × 10 mm or Nunc Maxisorp tubes). It is then washed to remove non-specifically bound material and antigen-binding phage eluted using specific or denaturing eluants and infected into *E. coli*. The phage eluted is enriched for those encoding antibodies specific for the antigen. Coating of tubes or plates with low amounts of antigen selectively enriches high-affinity antibodies. In most cases more than one round of enrichment is performed, to obtain a high proportion of antigen-binding phage and minimize the number of assays which need to be performed. The simplicity of panning is advantageous for multiple rounds of selection and also for the rapid testing of the effectiveness of different elution conditions.

Antibodies directed against cell-surface antigens can be isolated by selective adsorption of phage on the surface of cells. Similarly, it may be possible to incorporate negative selection with cells to remove undesired cross-reactivities with cell-surface markers.

5. Analytical procedures for proteins expressed on phage

There are diverse ways of assaying and analysing the binding functions of proteins displayed on phage but the methods below have wide applicability.

5.1 ELISA and other immunoassays

Phage antibody clones can be assayed directly for the ability to bind specific antigens by immunoassay techniques such as ELISA. Detection of phage antibodies with antiserum raised in sheep against bacteriophage fd can be used for most antigens, giving very sensitive ELISA assays (*Protocol 10*). *Protocol 10* should be readily adaptable to assay any phage antibody by substituting an appropriate antigen.

Protocol 10. ELISA using phage antibodies

1. Coat an ELISA plate (e.g. Falcon 3912) with 10 µg/ml antigen[a] in 50 mM NaHCO$_3$, pH 9.6 overnight at room temperature.

2. Prepare phage supernatants as in *Protocol 1*, steps **1** and **2**. They can be concentrated if required by PEG precipitation, as described in *Protocol 1*. Make to 2% (w/v) skimmed milk powder, from a 4% solution in 10 mM phosphate, 0.15 M NaCl (PBS).

3. Rinse the wells three times with PBS and block with 300 µl per well of 2% (w/v) skimmed milk powder in PBS for 2 h at 37°C. This can be done while precipitating phage.

4. Rinse the wells three times with PBS and transfer 200 µl of supernatants or concentrated phage into wells. Incubate for 2 h at room temperature.

5. Wash the wells for 2 min three times with each of PBS/0.2% Tween-20 and PBS (to remove detergent).

6. Add 200 µl of anti-fd serum raised in sheep (1/1000 dilution) to each well. Incubate at 1 h at room temperature.

7. Wash as in step **5**.

8. Add 200 µl peroxidase-conjugated rabbit anti-goat immunoglobulin (Sigma; 1/5000 dilution) to each well. Incubate for 1 h at room temperature.

9. Wash as in step **5**.

10. Add 200 µl ABTS[b] containing 1 µl of H$_2$O$_2$ per 10 ml to each well and

Protocol 10. *Continued*

develop until the absorbance at 405 nm is suitable (A_{405} = 0.2–1.0). Read absorbance in a plate reader.

a The normal antigen concentration for coating is 10 µg/ml. With some antigens higher concentrations give improved results (e.g. for ELISA with lysozyme we use 1 mg/ml).
b ABTS is 2,2'-azinobis-(3-ethylbenzthiazoline) sulphonic acid (Sigma) in 54 ml 50 mM citric acid and 46 ml 50 mM trisodium citrate, at a concentration of 0.55 mg/ml.

Binding to antigen of phage antibodies has also been detected by radioimmunoassay using antigen labelled with [125]I. Phage antibodies that have bound labelled antigen are detected using sheep anti-fd, which is then precipitated with donkey anti-sheep immunoglobulin bound to magnetic particles (Amerlex M, Amersham International).

5.2 Detection of clones by oligonucleotide probing and PCR

Colonies expressing phage antibodies can be analysed by standard colony blot procedures (17), for instance to distinguish clones containing the gene of interest from those containing vector DNA when subcloning genes into phage vectors. Alternatively, PCR procedures can be used to verify the presence and size of inserts in clones using specific primers and a sample of DNA derived by picking a colony into 20 µl of distilled water and boiling for 5 min. The diversity of clones generated can be assessed by *Bst*NI digestion of the PCR amplified insert (2).

5.3 Analysis of fusion proteins by Western blotting

Fusion proteins with g3p have been analysed by Western blotting using a rabbit antiserum directed against g3p (29) as in *Figure 2*. Intact fusion proteins are indicated by a band corresponding to an apparent molecular mass combining the apparent molecular mass of g3p (55 000–71 000) and the mass of the displayed protein. A band approximating to the mass of g3p is observed on blots due to the presence of wild-type g3p or the degradation of fusion molecules with g3p.

6. Further manipulation of isolated clones

6.1 Hierarchical libraries

It is relatively easy to recombine the heavy chain from one antibody with a range of light chains from other antibodies, or vice versa, and screen for antibodies with altered affinities or specificities. For instance the VH domain from one antibody directed against oxazolone was recombined with a library of VL domains isolated from a mouse immunized with the same antigen by combining the gene encoding the VH domain with a library of VL domains in the assembly process described in *Protocol 6*, footnote *e* (2). This allowed the isolation of a number of light chain partners that were not isolated in the screening of the original library. This hierarchical approach may be important for the improvement of affinity and specificity of antibodies and antibody fragments whether initially isolated from repertoires using phage technology or rescued from cell lines expressing monoclonal antibodies.

6.2 Affinity maturation of antibodies

Improved variants of antibodies and other proteins generated by *in vivo* or *in vitro* mutagenesis techniques can be selected. Since the genes encoding the variants are contained in the same package as the antibodies displayed on the surface, large libraries can be prepared by random mutagenesis techniques and antibodies with improved properties isolated. Protocols for selection will vary according to the desired improvement.

6.3 Protein engineering

There is the potential for the improvement of proteins other than antibodies using phage technology. For example, catalytically active alkaline phosphatase expressed on the surface of bacteriophage has been specifically bound to and eluted from an arsenate-Sepharose affinity column (10). If improvement of an enzyme or other protein is desired, large numbers of random mutants can be screened for molecules with improved properties, provided that there is a suitable selection system.

6.4 Expression as soluble proteins

For detailed study of proteins it will usually be necessary to express them in soluble form. Clones of genes in the vector pHEN1, which contains an amber codon at amino acid 2 in gene 3, can be expressed directly as soluble protein following transformation into non-suppressing strains of *E. coli* such as HB2151 without recloning. This allows purification of large quantities of antigen for reasons such as the determination of binding affinities.

7. Conclusion

Display of proteins on the surface of bacteriophage gains its power as a method by linking proteins, presented for binding to ligand, with the DNA encoding them. The phage system mimics the natural processes by which the immune system produces molecules which bind tightly to ligands, allowing large numbers of variants to be surveyed and leading to the more rapid isolation of improved protein molecules. The monoclonal antibody technique of Kohler and Milstein (30) has enabled isolation of highly specific antibodies of importance in research, diagnosis, and therapy. We expect phage antibody technology to extend the variety of specific antibodies obtainable and to simplify their isolation greatly.

References

1. McCafferty, J., Griffiths, A. D., Winter, G., and Chiswell, D. J. (1990). *Nature,* **348,** 552–4.
2. Clackson, T., Hoogenboom, H. R., Griffiths, A. D., and Winter, G. (1991). *Nature,* **352,** 624–8.
3. Crissman, J. W. and Smith, G. P. (1984). *Virology,* **132,** 445–55.
4. Glaser-Wuttke, G., Keppner, J., and Rasched, I. (1989). *Biochim. Biophys. Acta,* **985,** 239–47.
5. Parmley, S. F. and Smith, G. P. (1988). *Gene,* **73,** 305–18.
6. Scott, J. K. and Smith, G. P. (1990). *Science,* **249,** 386–90.
7. Devlin, J. J., Panganiban, L. C., and Devlin, P. E. (1990). *Science,* **249,** 404–6.
8. Cwirla, S. E., Peters, E. A., Barrett, R. W., and Dower, W. J. (1990). *Proc. Natl Acad. Sci. USA,* **87,** 6378–82.
9. Hoogenboom, H. R., Griffiths, A. D., Johnson, K. S., Chiswell, D. J., Hudson, P., and Winter, G. P. (1991). *Nucl. Acids Res.,* **19,** 4133–37.
10. McCafferty, J., Jackson, R. H., and Chiswell, D. J. (1991). *Prot. Engng.,* **4,** 955–61.
11. Bass, S., Greene, R., and Wells, J. A. (1990). *Proteins,* **8,** 309–14.
12. Kang, A. S., Barbas, C. F., Janda, K. D., Benkovic, S. J., and Lerner, R. A. (1991). *Proc. Natl Acad. Sci. USA,* **88,** 4363–6.
13. Zacher, A. N., Stock, C. A., Golden, J. W., and Smith, G. P. (1980). *Gene,* **9,** 127–40.
14. Orlandi, R., Gussow, D. H., Jones, P. T., and Winter, G. (1989). *Proc. Natl Acad. Sci. USA,* **86,** 3833–7.
15. Bird, R. E., *et al.* (1988). *Science,* **242,** 423–6.
16. Huston, J. S., *et al.* (1988). *Proc. Natl Acad. Sci. USA,* **85,** 5879–83.
17. Chaudhary, V. K., Batra, J. K., Gallo, M. G., Willingham, M. C., FitzGerald, D. J., and Pastan, I. (1990). *Proc. Natl Acad. Sci. USA,* **87,** 1066–70.
18. Sambrook, J., Fritsch, E. F., and Maniatis, T. (ed.) (1989). *Molecular cloning, a laboratory manual* (2nd edn). Cold Spring Harbor Laboratory Press, Cold Spring Harbor, New York.
19. Munro, S. and Pelham, H. (1986). *Cell,* **46,** 291–300.
20. Vieira, J. and Messing, J. (1987). In *Methods in enzymology* (ed. R. Wu and L. Grossman), Vol. 153, pp. 3–11. Academic Press, San Diego.
21. Marks, J. D., Tristem, M., Karpas, A., and Winter, G. (1991). *Eur. J. Immunol.,* **21,** 985–91.
22. Marks, J. D., Hoogenboom, H. R., Bonnert, T. P., McCafferty, J., Griffiths, A. D., and Winter, G. (1991). *J. Mol. Biol.,* **222,** 581–97.
23. Clackson, T., Gussow, D., and Jones, P. T. (1991). In *PCR: a practical approach,* (ed. M. J. McPherson, G. R. Taylor, and P. Quirke). IRL Press, Oxford.
24. Harlow, E. and Lane, D. (1988). *Antibodies, a laboratory manual.* Cold Spring Harbor Laboratory Press, Cold Spring Harbor, New York.
25. Dower, W. J., Miller, J. F., and Ragsdale, C. W. (1988). *Nucl. Acids Res.,* **16,** 6127–45.

26. Hanahan, D. (1985). In *DNA cloning: a practical approach,* (ed. D. M. Glover), Vol. I, p. 109. IRL Press, Oxford.
27. Glockshuber, R., Malia, M., Pfitzinger, I., and Pluckthun, A. (1990). *Biochemistry,* **29,** 1362–7.
28. Bird, R. E. and Walker, B. W. (1991). *TIBTECH,* **9,** 132–7.
29. Stengele, I., Bross, P., Garces, X., Giray, J., and Rasched, I. (1990). *J. Mol. Biol.,* **212,** 143–9.
30. Kohler, G. and Milstein, C. (1975). *Nature,* **256,** 52–3.

13

Expression of proteins in prokaryotic systems—principles and case studies

GEOFFREY T. YARRANTON and ANDREW MOUNTAIN

1. Introduction

The rapid developments in recombinant DNA techniques have resulted in the identification and isolation of many novel genes, some of known function and some of unknown function. Almost invariably there is a need to express the gene in a heterologous cell system in order to produce:

- material for structure–function studies
- diagnostic reagents, e.g. monoclonal or polyclonal antibodies
- material for *in vivo* activity testing

Several alternative systems for the expression of foreign genes have been developed including mammalian cells, insect cells, fungal cells, bacterial cells, and transgenic animals or plants. The choice of expression system for a given gene depends upon the likely properties of the encoded protein, e.g. protein modifications needed for biological activity, as well as the objective of the study, e.g. structure–function analysis or production of diagnostic reagents. Other important considerations for the investigator are the facilities available, times and cost involved in generating the amounts of recombinant protein required.

The most widely used and convenient system for the production of foreign proteins remains that based on the simple prokaryote, *Escherichia coli*. The advantages of this system are:

- ease of gene manipulation
- availability of reagents, including gene expression vectors
- ease of producing quantities of protein (up to a gram in simple shake-flask culture
- speed
- adaptability of the system, thereby allowing solution of expression problems

Disadvantages of *E. coli* for foreign gene expression are:

- inability to carry out post-translational modifications typical of eukaryotic cells
- the propensity of foreign proteins to form insoluble inclusion bodies when expressed intracellularly (1)

Recent advances have shown that it is possible to engineer the *E. coli* host cell genetically to carry out at least one post-translational modification, namely myristoylation of a foreign protein (2). This suggests that other advances in this area are likely. The problem of protein insolubility can also be addressed either by refolding the denatured protein *in vitro*, varying the conditions of expression (e.g. temperature), or secreting the protein into the periplasm or culture medium.

Expression of any foreign gene in *E. coli* begins with the insertion of a cDNA copy of the gene into an expression vector. Many forms of expression vector are available and usually comprise:

- a plasmid origin of DNA replication
- antibiotic selectable marker
- strong promoter and transcriptional terminator separated by a multi-cloning site (expression cassette)
- DNA sequence encoding a ribosome-binding site

There is very little difference between the various strong promoters now available (ptac, λpL, T_7) except that the method of transcriptional regulation varies. ptac and T_7 expression based systems are controlled by the chemical inducer IPTG, whereas the λ promoters are controlled by a temperature switch. Gene expression can also be controlled through the use of inducible copy-number vectors. The dual-origin vector (3) relies upon a temperature switch to increase plasmid copy number from five/cell to 100–200 copies/cell, and this induces foreign gene expression. Examples showing the use of these vectors will be given below.

The translation of foreign mRNA remains a significant problem associated with gene expression in *E. coli*. The basic rules for achieving translation of a mRNA are simple:

- complementarity between the Shine–Dalgarno (SD) sequence and the 3' end of the 16S ribosomal RNA
- a 6–10 bp spacing between the SD and the initiation codon (AUG)
- an AT-rich base composition between the SD and AUG

Regions of internal complementarity within the mRNA, particularly those involving the SD and AUG should be avoided, since these significantly reduce the frequency of translational initiation. In instances where translation is problematical, a variety of approaches are available to overcome the block (4).

Accumulation of foreign protein intracellularly may also be affected by proteolysis. This is particularly relevant to small peptides or to protein fragments. Two approaches have been used to overcome this problem, namely the use of host strains with reduced protease levels and the production of fusion proteins, brought about by the in-phase translational fusion of the gene sequence to the C-terminus of a highly expressed homologous protein. The fusion protein product is often purified using a property of the homologous fusion protein partner and cleaved to release the peptide or protein fragment of interest. Fusion vectors have been designed to produce unique chemical or protease cleavage sites between fusion partner and product for this purpose. The fusion protein approach has an additional advantage, in that it also overcomes the problem of translational initiation, since the N-terminus of the fusion protein is a highly expressed homologous protein.

Recent advances in the development of secretion systems for *E. coli* have increased the likelihood of recovering active soluble recombinant protein product. In general, secretion is achieved using an N-terminal hydrophobic signal sequence from a bacterial protein (e.g. OmpA, PelB), which is engineered into the recombinant gene sequence in place of its existing signal. The signal sequence is sufficient in some cases to direct the protein through the inner membrane and into the periplasm. *E. coli*, however, possesses an outer membrane that poses an additional barrier to culture medium accumulation of secreted protein. In some cases, proteins seem to pass through the outer membrane relatively easily, in other cases periplasmic accumulation is observed. The rules determining protein secretion in *E. coli* are not yet determined and hence an empirical approach needs to be taken when attempting this.

2. Expression of calf prochymosin using the dual-origin vector system

2.1 Vector design

Constitutive or copy-number inducible expression vectors carry similar elements required for achieving high-level gene expression, these are listed in *Table 1*. The dual-origin vector pMG168 (*Figure 1*) carries a cDNA gene copy of bovine prochymosin (5), inserted downstream of the ptrp promoter and upstream of the bacteriophage T_7 transcription terminator. This cDNA gene copy has been redesigned to remove the 5′ untranslated region of the mRNA and the coding sequence for the signal peptide of preprochymosin. For efficient expression of foreign proteins it is important to remove 5′ untranslated regions and have precisely engineered 5′ gene sequences. This can be achieved using PCR technology or oligonucleotide assembly. In either case, redesigned regions must be sequenced prior to evaluating expression. The expression cassette of pMG168 was derived from a pAT153-based expression

Table 1. Requirements for efficient expression

Strong promoter	ptac, ptrp, λpL, pT$_7$ are all readily available promoters, for use in expression vectors
Promoter regulation	ptrp–tryptophan/IAA, ptac–IPTG, λpL–temperature, pT$_7$–IPTG
Transcription terminator	T$_7$ term, rrn T$_1$ T$_2$—located downstream of the inserted cDNA, this element terminates transcription
Ribosome-binding site (Shine–Dalgarno sequence)	e.g. AAGG, located upstream of the AUG initiation codon, this sequence has complementarity with the 3' end of the 16S ribosomal RNA
SD–AUG spacing and base composition	Determined empirically, this spacing is crucial to high-level expression; the optimum distance is usually 6–10 bp, the composition AT-rich

vector pCT70 (*Figure 2*), in which the translation of prochymosin mRNA had been optimized. The methodology for translational optimization is described below.

2.2 Translational optimization

Translational efficiency can only be determined empirically. The general rules have been discussed in Section 1. The initial prochymosin expression plasmid

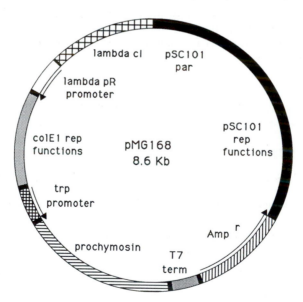

Figure 1. Dual-origin expression vector pMG168. This vector comprises two origins of replication, one from the plasmid pSC101 (five copies/cell) and one from ColE1. The latter origin is expressed conditionally from the λpL promoter, being inactive at 30°C and active at 37°C. Fully induced copy numbers range from 100 to 500/cell.

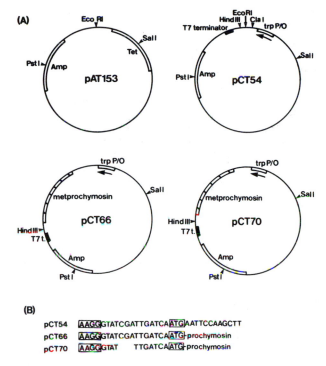

Figure 2. Plasmid maps for constitutive copy-number vectors. (A) pAT153-based expression vectors comprising ptrp promoter, multi-cloning site, and T_7 terminator. pCT66 and pCT70 carry a gene encoding calf prochymosin. (B) DNA sequence encoding the SD–ATG region in each plasmid.

(pCT66) shows undetectable expression of calf prochymosin when cell extracts are evaluated, as described in *Protocol 1*, using *E. coli* host strain HB101.

Protocol 1. SDS-polyacrylamide gel electrophoresis of *E. coli* extracts

Proteins in extracts prepared from *E. coli* are analysed by SDS-PAGE (15). A vertical protein gel apparatus is used and 1 mm thick gels are poured between glass plates. The percentage acrylamide depends on the size of the expressed protein. A 5% stacking gel is employed.

1. Prepare the following solutions:
   - acrylamide/bis-acrylamide solution: 30% (w/v) acrylamide; 0.8% (w/v) bis-acrylamide
   - 4 × lower Tris buffer: 1.5 M Tris–HCl, pH 8.8; 0.4% (w/v) sodium dodecyl sulphate

Protocol 1. *Continued*

- 4 × upper Tris buffer: 0.5 M Tris–HCl, pH 6.8; 0.4% (w/v) sodium dodecyl sulphate
- 10% ammonium persulphate
- sample buffer: 10% (w/v) glycerol; 5% (w/v) β-mercaptoethanol; 3% (w/v) sodium dodecyl sulphate; 0.2% (w/v) bromophenol blue dye; 1 × upper Tris buffer
- 4 × tank buffer: 30 g Tris base; 144 g glycine; 2.5 litres of distilled water
- electrophoresis buffer: 250 ml 4 × tank buffer; 740 ml distilled water; 10 ml 10% (w/v) SDS

2. For a 10% acrylamide gel, mix the solutions in the following amounts: 10 ml 4 × lower Tris buffer, 13.3 ml acrylamide/bis solution, 16.12 ml distilled water, 0.03 ml TEMED, and 0.2 ml ammonium persulphate.

3. Pour the gel to within 4 cm of the top of the gel plate. Gently cover the surface of the gel with butanol (water-saturated) by adding a few drops at the side of the gel from a Pasteur pipette. This protects the gel from drying out.

4. When the gel has set, prepare the 5% acrylamide stacking gel by mixing the following: 5 ml 4 × upper Tris buffer, 2 ml acrylamide/bis solution, 12.7 ml distilled water, 0.03 ml TEMED, and 0.2 ml ammonium persulphate.

5. Pour off the butanol from the lower gel, pour on the stacking gel, add the gel comb, and wait for the gel to set.

6. Assemble the gel apparatus, fill upper and low chambers with electrophoresis buffer. Remove any air-bubbles at the bottom of the gel.

7. Prepare protein samples by boiling cell pellets in two volumes of sample buffer for 2 min. Centrifuge samples for 1 min in a microfuge to pellet cell debris.

8. Load 5–10 µl of sample and electrophorese at 15 V/cm for approximately 3 h or until the dye front reaches the bottom of the gel.

9. Remove the gel from between the glass plates and stain for 1 h. Place the container on a shaking platform at low speed.

10. Destain the gel in 7% (w/v) acetic acid solution, add two pieces of sponge to absorb eluted stain, this improves the destaining process. Agitate gently on a shaking platform.

The SD–AUG composition and distance in the expression cassette is: AAGGGTATCGATTGATCAATG and within this region there are several unique restriction sites. To optimize the translation of the prochymosin mRNA, pCT66 plasmid DNA was digested with *Cla*I, giving the linear form of the plasmid, followed by limited DNase digestion using the single-strand-specific DNase S1 as described in *Protocol 2*.

Protocol 2. Nuclease S1 treatment of linear plasmid DNA

1. Prepare the following solutions:
   - 4 × nuclease S1 buffer: 100 mM sodium acetate, pH 4.5; 4 mM zinc acetate; 1 M sodium chloride
   - buffer-saturated phenol: 0.1 M Tris–HCl, pH 7.5 saturated phenol; 0.1% 8-hydroxyquinoline

2. Incubate 1 μg of linear plasmid DNA (restriction enzyme digested) in a final reaction mixture volume of 20 μl S1 buffer (unit strength) containing 100 units of S1 nuclease at 37°C.

3. Remove 2 μl samples at 2 min intervals and add 2 μl 200 mM EDTA to stop the reaction.

4. Neutralize the reaction mixture by the addition of 4 μl 0.5 M Tris–HCl, pH 7.4.

5. Extract twice with equal volumes of buffer-saturated phenol. Take the upper aqueous layer that separates when the mixture is centrifuged in an Eppendorf centrifuge for 1 min.

6. Extract the aqueous layer with an equal volume of chloroform. Take the upper aqueous layer, after microfuging for 1 min.

7. Isopropanol precipitate the DNA (see *Protocol 4A*, steps **8–9**).

Nuclease S1 digestion removes more nucleotides than those in the single-stranded tails, hence a random set of deletions can be obtained, in which the distance and base composition between the SD and AUG is varied. Re-ligated plasmid DNA can be re-digested with *Cla*I prior to transformation (*Protocol 3*), this removes the parental form of the plasmid. Recombinant plasmids lacking the *Cla*I site can be screened for prochymosin expression as described previously (*Protocol 1*). Using this approach, pCT70 (*Figure 2*) was isolated, in which the SD–AUG distance is 11 nucleotides: AAGGGTATT-TGATCAATG, and expression of prochymosin represents 5% of total cell protein.

The improvement in gene expression level from undetectable to 5% of total cell protein, achieved by varying the SD–AUG distance, is not uncommon. Wherever possible the 5' mRNA sequence extending to the first 10–20 codons should be analysed by a simple computer program for regions of complementarity, particularly any involving either the SD or AUG sequence.

If stable complementarity is found, then alterations in sequence, making use of degeneracy of the genetic code, should be employed.

2.3 Dual-origin vector: expression

The optimized expression cassette comprising the calf prochymosin gene from pCT70 can be subcloned into a dual-origin expression vector to generate pMG168 (*Figure 1*). The advantages of the dual-origin vector system are

- plasmid stability (even in the absence of antibiotic selection)
- increased expression level due to plasmid copy-number amplification

Host strains of *E. coli* are transformed with dual-origin vectors at 30°C as described below.

Protocol 3. Bacterial transformation

A. *Constitutive copy number vectors*

1. Thaw out a tube of frozen competent cells at room temperature.
2. Add plasmid DNA (10–100 ng) or ligation mixture to cells on ice and leave for 15 min.
3. Heat shock cells at 37°C for 2 min. Add 0.5 ml of L-broth (see *Protocol 4*, step 1) and incubate for 30 min at 37°C.
4. Plate out cells on to selective L-agar containing the relevant antibiotic selection.
5. Incubate for 16 h at 37°C.

B. *Dual-origin vectors*

1. Thaw out a tube of frozen competent cells at room temperature.
2. Add plasmid DNA (10–100 ng) or ligation mixture to cells on ice and leave for 15 min.
3. Heat shock cells for 2 min at 30°C. Add 0.5 ml of L-broth and incubate for 30 min at 30°C.
4. Plate out cells on selective agar plates containing the relevant antibiotic.
5. Incubate for 20 h at 30°C.

It is essential not to transform at 37°C since this induces plasmid copy number and will result in cell death. Plasmid DNA can be prepared from temperature-induced cells (*Protocol 4*) using the alkaline-lysis method.

310

Protocol 4. Plasmid DNA preparation: constitutive copy-number and dual-origin vectors

A. *Constitutive copy-number vectors*

1. Prepare the following solutions:
   - L-broth: 5 g NaCl; 5 g tryptone (Difco); 2.5 g yeast extract (Difco); distilled water to 500 ml final volume; sterilize by autoclaving
   - solution I: 50 mM glucose; 25 mM Tris–HCl, pH 8.0; 10 mM EDTA
   - solution II: 0.2 M NaOH; 1% SDS
   - solution III: 5 M potassium acetate, pH 4.8

2. Inoculate 500 ml of L-broth in a 2 litre flask with the bacterial host strain. This should be supplemented with the relevant antibiotic, e.g. ampicillin (100 μg/ml) to the appropriate concentration. Incubate the culture overnight at 37 °C with vigorous shaking.

3. Harvest bacterial cells by centrifugation at 15 000 g for 10 min in a Sorvall GSA rotor.

4. Resuspend the pellet of cells in 6 ml of solution I.

5. Add 12 ml of solution II, mix gently.

6. Add 9 ml of solution III, mix well (not vortex) and leave for 10 min on ice.

7. Centrifuge for 10 min at 10 000 g in a Sorvall SS-34 rotor.

8. Remove supernatant and add 0.6 volume of isopropanol, mix well by inverting the tube.

9. Centrifuge for 10 min at 10 000 g in a Sorvall SS-34 rotor to sediment the precipitated DNA.

10. Resuspend the dried DNA pellet in 6 ml of 10 mM Tris–HCl, pH 7.5, 1 mM EDTA buffer. Add 7 g CsCl, mix well to dissolve and then add 100 μl 10 mg/ml ethidium bromide. Centrifuge DNA samples in appropriate tubes at 15 °C, in a Beckman 70Ti rotor for 16 h, at 200 000 g.

11. After centrifugation view the DNA bands with the aid of a UV light (305 nm). Collect the lower band by puncturing the side of the tube with a needle and syringe. Remember to make a small hole in the top of the tube before puncturing the side.

12. Remove the ethidium bromide by three equivolume extractions with CsCl-saturated isopropanol.

13. Precipitate the plasmid DNA from the CsCl solution with 0.6 volume of isopropanol (not saturated). Wash twice with 70% ethanol solution, dry, and resuspend in 10 mM Tris–HCl 1 mM EDTA, pH 7.5. This procedure generates ~1 mg plasmid DNA per 500 ml of culture.

Protocol 4. *Continued*

B. *Dual-origin vectors*

1. Prepare solutions as described above.

2. Inoculate 20 ml of L-broth in a 250 ml flask with the bacterial strain of interest. Antibiotic selection is optional since these plasmids are very stable. Grow the culture overnight at 30°C. Note the culture temperature, temperatures above 34°C will induce copy-number amplification and cells will die.

3. Next morning inoculate 250 ml of L-broth with 10 ml of overnight culture. Grow at 30°C until the absorbance $A_{600} = 0.4$. Switch cells to 37°C and leave overnight with shaking.

4. Harvest bacterial cells and proceed as described in part A (step **3** onwards).

Prochymosin expression directed from pMG168 transformed into *E. coli* strain E103S was analysed as described in *Protocols 5 and 1*.

Protocol 5. Dual-origin vector: induced protein production

1. Inoculate 20 ml of L-broth in a 250 ml flask with the bacterial strain of interest and incubate at 30°C overnight.

2. Dilute the overnight culture 20-fold into L-broth and incubate (100 ml) in a 500 ml flask at 30°C.

3. Follow culture growth by increase in A_{600}; when this reaches 0.4, switch the temperature rapidly to 42°C by placing the flask in a waterbath at 60°C. Follow the temperature rise using a thermometer inserted into the culture fluid. When the temperature reaches 42°C, switch the flask to a shaking incubator at 37°C.

4. To sample the culture at given times, 1 ml samples are removed, centrifuged for 1 min in a microfuge and the cell pellet suspended in 200 μl of sample buffer. This mixture is boiled for 2 min and then can be stored frozen prior to SDS-PAGE.

A Coomassie-blue stained SDS-PA gel of cell extracts made at various time points after induction is shown in *Figure 3*. The appearance of a novel protein band of the correct mol. wt for prochymosin is indicated. This can be confirmed by Western blotting using polyclonal or monoclonal antibodies, specific for prochymosin, as described in *Protocol 6*.

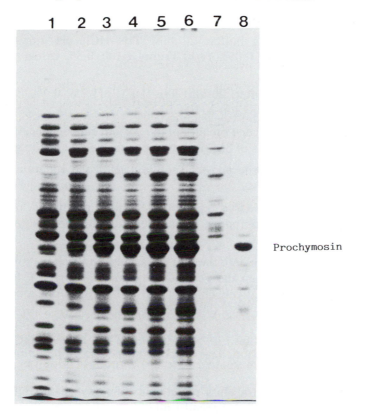

Figure 3. Polyacrylamide-SDS gel of extracts prepared from cells carrying pMG168. Cells were grown at 30°C and induced for vector copy-number and protein production by a temperature switch, as described in *Protocol 5*. Each lane comprises total protein extracts made at the following times after induction: lane 1, 0 min; lane 2, 30 min; lane 3, 60 min; lane 4, 90 min; lane 5, 120 min; lane 6, 180 min. At 180 min post-induction a sample was removed, the insoluble and soluble fractions prepared and analysed. Lane 7 is a sample of the soluble fraction, lane 8 the insoluble fraction. The prochymosin band is indicated.

Protocol 6. Western blot procedure

The verification that expression of a given protein has been achieved is usually made by Western blotting. Cell extracts are subjected to SDS-PAGE as in *Protocol 1*, and then the resolved protein bands are transferred to a nitrocellulose or other membrane. A particular protein band is then identified by reaction with monoclonal or polyclonal antisera specific for that protein (16).

1. Prepare the following solutions:
   - running buffer: 3.03 g Tris base; 14.4 g glycine; 200 ml methanol; distilled water to 1 litre

Protocol 6. *Continued*

- low-salt blocking buffer: 20 mM Tris–HCl, pH 7.6; 0.9% (w/v) sodium chloride; 0.05% (v/v) Triton X-100; 0.5% casein (BDH Hammarsten)

- low-salt wash buffer: 20 mM Tris–HCl, pH 7.6; 0.9% (w/v) sodium chloride; 0.05% Triton X-100

2. Measure the length and breadth of the gel and cut one piece of nitrocellulose and two pieces of 3MM filter paper to the size of the gel.

3. Soak the Scotch-Brite pads or sponge pads of the apparatus in running buffer.

4. Assemble a 'sandwich' comprising 3MM filter paper, gel, nitrocellulose filter, and 3MM filter paper. Ensure that all air-bubbles between gel and filter are removed.

5. Insert the 'sandwich' into the blot apparatus, ensuring that the nitrocellulose filter is between the gel and the anode (+ charge). SDS–protein complexes are negatively charged and will move to the anode.

6. Electroblot for either 60 V, 0.22 A for 3 h, or 30 V, 0.1 A overnight.

7. Remove the filter and wash in two changes of low-salt blocking buffer, during a 30 min period. This blocks non-specific protein-binding sites on the filter.

8. Place the filter on a glass plate in a humidity chamber (freezer box containing wet tissues). Spread 1 ml of low-salt buffer containing a 1/100 dilution of antiserum on to the side of the filter on to which the proteins were transferred. Incubate for 1 h at room temperature.

9. Wash the filter in four changes of low-salt wash buffer during a 1 h period, with gentle shaking.

10. Detect antibodies bound to protein on the filter using either anti-Ig conjugated to enzyme, e.g. horseradish peroxidase, or ^{125}I-labelled protein A.

The propensity of foreign proteins expressed in *E. coli* to form insoluble inclusion bodies is a major problem for the recovery of activity. Extracts prepared from induced cells can be separated into soluble and insoluble fractions, as described in *Protocol 7*. For prochymosin, insoluble and soluble fractions have been analysed by SDS-PAGE (*Figure 3*). All the prochymosin produced using this system is found in the insoluble fraction. Resolubilization of the inclusion bodies followed by renaturation results in the recovery of active chymosin (6). Using the dual-origin vector system, more than 10% of total cell protein can be accumulated as prochymosin.

Protocol 7. Preparation of soluble and insoluble cell fractions

1. Prepare the following solutions:
   - lysis buffer: 100 mM Tris–HCl, pH 6.5; 5 mM EDTA; 1 mM PMSF; 260 μg/ml lysozyme
   - pancreatic DNase I: 1 mg/ml in distilled water
   - wash buffer: 100 mM Tris–HCl, pH 6.5; 5 mM EDTA
   - sodium deoxycholate: 10% solution in distilled water
   - magnesium chloride: 1 M $MgCl_2$ in distilled water

2. Withdraw 1 ml of induced cell culture and centrifuge in an Eppendorf microfuge for 1 min. Discard the supernatant and resuspend the cell pellet in 0.45 ml of lysis buffer.

3. Incubate cell suspension at 0°C for 5 min.

4. Add sodium deoxycholate to a final concentration of 0.1% and incubate for a further 5 min at 0°C.

5. Add pancreatic DNase I and magnesium chloride to final concentrations of 25 μg/ml and 10 mM, respectively. Incubate for 5 min at 0°C.

3. Direct intracellular expression of immunoglobulin Fv fragments

The antigen-binding activity of immunoglobulins can be found in the heavy and light chain variable regions (VH and VL), when these are expressed as short (110 amino acid) fragments that are allowed to associate. Together, the VH and VL fragments constitute an Fv fragment. The expression of antibody fragments to generate functional Fv fragments in *E. coli* poses two problems:

- how to accumulate protein fragments of ony 110 amino acids in size, which are subject to proteolysis
- how to reconstitute a functional, heterodimeric recombinant molecule

This section describes the intracellular production of Fv molecules and Section 4 describes the secretion of antibody fragments.

3.1 Primary expression analysis

The VH and VL genes encoding antibody-binding activity recognizing egg white lysozyme (7) were ligated into the dual-origin vector, such that expression was driven by the ptrp promoter and relied upon a temperature shift to induce increased copy-number and gene expression.

   Independent expression of VH and VL rather than co-expression was

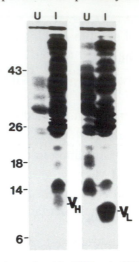

Figure 4. Autoradiograph of polyacrylamide-SDS gel of VH and VL gene expression. Cells were grown and induced as described in *Protocol 5*. Molecular weight markers are shown on the left-hand side of the gel. Extracts prepared from cells that were uninduced (U) and induced (I) are shown.

attempted, with the aim of subsequently reassociating the independent chains to obtain antigen-binding activity.

Dual-origin expression vectors were transformed (*Protocol 3*) into *E. coli* IB373 a K12 strain derived from E103S. This strain has been used at Celltech for the expression of recombinant proteins and appears to have low proteolytic activity. Using the method of [^{35}S]methionine pulse-labelling *in vivo* (*Protocol 8*), the expression of VH and VL was demonstrated (*Figure 4*). The expression of VL appeared better than VH but on Coomassie-stained SDS-PA gels neither protein could be detected.

Protocol 8. Pulse-labelling cells for *de novo* protein synthesis

1. Prepare the following solutions
   - M9 minimal medium, per litre: 6 g disodium hydrogen phosphate; 3 g potassium dihydrogen phosphate; 0.5 g sodium chloride; 1 g ammonium chloride; 10 ml 0.01 M calcium chloride;[a] 1 ml 0.01 M magnesium sulphate;[a] 0.4% sterile glucose[a]
   - labelling medium: M9 minimal medium containing 1% methionine assay medium purchased from Difco
2. Inoculate an overnight culture (10 ml labelling medium) from a single colony grown on selective agar plates.
3. Next morning dilute the culture 1 in 20 into fresh labelling medium and shake vigorously until $A_{600} = 0.3$.

316

4. Remove 1 ml of culture into an Eppendorf tube, add 14 μCi [^{35}S]methionine, and incubate for 2 min.

5. Pellet cells by centrifuging for 30 sec and resuspend the cell pellet in two volumes of sample buffer (*Protocol 1*). Boil for 2 min.

6. Analyse samples by SDS-PAGE as described in *Protocol 1*.

7. Following electrophoresis, remove the gel and dry on to 3MM paper using a vacuum gel dryer.

8. Expose dried gel overnight or longer to X-ray film.

a Sterilized separately.

The conclusions from this expression study were:

(a) VH and VL mRNAs are translated with different efficiencies even though they have identical SD–AUG spacings and sequence

(b) VL expression is efficient but accumulation of the translated product is low, probably as a consequence of proteolysis

To solve these problems two approaches were investigated:

(a) modification of the SD sequence to enhance the translation of VII mRNA

(b) screening of different host strains for improved protein accumulation

3.2 Improving the translation of VH mRNA

To improve the translation of VH mRNA a new expression plasmid was constructed in which the four-nucleotide (AAGG) SD sequence was replaced by a nine-nucleotide sequence (5′-AAGGAGGTA), thus extending the homology between 16S ribosomal RNA and the mRNA. For rapid analysis of translational efficiency, the constructs were evaluated in a commercially available *in vitro* transcription–translation system based on that developed by Pratt (8). In this system, plasmid DNA is added to an *E. coli* cell extract and the incorporation of [^{35}S]methionine into protein product is measured by SDS-PAGE. A comparison of the expression of VH and VL in expression vectors with a 4 bp or 9 bp SD sequence in the *in vitro* system was made by SDS-PAGE. Improved expression of VH was observed while little improvements in VL expression was detected. *In vivo* evaluation of these expression plasmids was carried out to test the predictive value of these *in vitro* results.

3.3 Improving the *in vivo* accumulation

Dual-origin vectors expressing VH and VL were transformed into a variety of host cell strains made competent for transformation as described in *Protocol 3*. Protein accumulation was evaluated by SDS-PAGE (*Protocols 1* and *5*).

The proteolytic turnover of foreign proteins and protein fragments is controlled in *E. coli* by the products of the *lon* gene and *htpr* gene. The *lon* gene product is an ATP-dependent protease (9) whereas the *htpr* gene product is a sigma subunit of RNA polymerase, primarily involved in directing the transcription of heat-shock genes (10). *E. coli* B strains are naturally *lon⁻* and have been used for foreign protein production for this reason. No accumulation of VH or VL could be detected in an *E. coli* B strain. Accumulation of VL detected as a stained band on Coomassie-stained SDS-PA gel was observed for an *htpr⁻* strain and this was further improved through the use of a *lon⁻htpr⁻* double mutant. The accumulation of VH was detected in the double mutant also, but only when the vector with improved translation of VH was used (*Figure 5*). This demonstrates that the *in vivo* results parallel the *in vitro* transcription–translation results.

Analysis of the solubility of VH and VL using the method described in *Protocol 7* demonstrated that both are in the insoluble fraction. However, these can be solubilized and refolded to give active, functional antigen-binding activity (11). The yield of pure recombinant Fv after refolding and association is 0.3 mg, starting with 1 litre of culture expressing VH and 0.15 litre of culture expressing VL. This represents an overall yield of 0.8% of the starting material.

In conclusion, the accumulation of these short protein fragments can be achieved through the use of the dual-origin vector system in combination with *lon⁻htpr⁻* mutations in the host strain. The *E. coli* strains CAG345 and

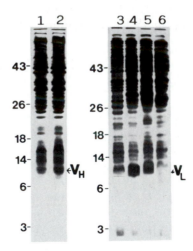

Figure 5. Polyacrylamide-SDS gel of VH and VL protein accumulation in protease-deficient strains of *E. coli*. Molecular weight markers are shown on the left of each gel. Extracts were prepared from cells 5 h post-induction. Lane 1, VH expression construct with 4-bp homology to 16S rRNA; lane 2, VH expression construct with 9-bp homology to 16S rRNA; lane 3, CAG629 (*lon⁻htpR⁻*) expressing VL uninduced; lane 4 CAG629 expressing VL induced; lane 5, CAG345 expressing VL induced; and lane 6, uninduced.

CAG629, *htpr⁻* and *htpr, lon⁻*, respectively, can be obtained from Dr C. Gross, University of Wisconsin. In some cases improved translation can be achieved using an SD region with extended complementarity to the 16S rRNA.

4. Expression of antibody fragments using a secretion system

4.1 Vector design

The factors that govern the design of plasmid-based secretion vectors are very similar to those outlined above for intracellular accumulation of product. Secretion of heterologous proteins, however, is often even more deleterious for *E. coli* than their intracellular accumulation. For this reason it is even more important to use a tightly controlled promoter in plasmid-based secretion systems in order to achieve stable plasmid maintenance prior to inducing expression. The plasmid vector pACtac has been used successfully in the authors' laboratory for the production of a variety of antibody fragments. *Figure 6* shows a plasmid pCTR008, which is a pACtac derivative for the production of a chimeric cDNA Fab' fragment of antibody A5B7.

pACtac is a derivative of pACYC184 (12), employing both its replication functions (which maintain approximately 15 copies/cell) and its chloramphenicol-resistance selectable marker. Transcription of the target gene from the tac promoter is tightly controlled by the product of the *lacI^Q* gene, which is also carried on pACtac. *lacI^Q* has a mutation in the promoter of *lacI* which results in elevation of its expression (13). Transcription from the tac promoter is induced by the addition of IPTG. Although several other promoter/regulator gene combinations have been used successfully in *E. coli* secretion systems, the tac promoter/*lacI^Q* system (or lac promoter/*lacI^Q*) has the advantage of readily allowing partial induction by using submaximal concentrations of IPTG. Such partial inducibility is desirable because it offers the possibility of optimizing the balance betwen secretion of product and cell lysis. The latter often results from high-level expression of heterologous secreted proteins, to an extent that varies between different proteins. In pACtac the efficient (dual) transcription termination signal from the *E. coli* rrnB operon is positioned to prevent readthrough of transcription initiated at the tac promoter into the *lacI^Q* gene or plasmid replication functions.

Secretion of the Fab' heavy and light chains expressed by pCTR008 is directed by the *ompA* signal sequence. The constructs were engineered to give precise fusions between the C-terminal end of this signal sequence and the mature N-termini of the heavy and light chains. For each fusion, the signal sequence coding region is preceded by the *ompA* translation initiation sequence (extending 18 nucleotides 5' to the ATG start codon) with a single nucleotide change, this being a T in place of the A six nucleotides 5' to the

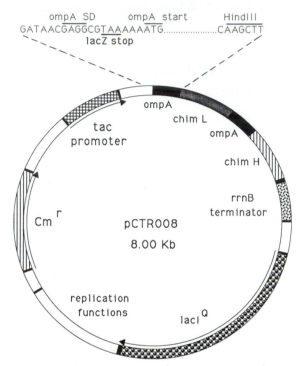

Figure 6. Secretion vector for the co-expression of Fab fragments. Expression is directed by ptac and this is controlled by the plasmid-borne *lacI*^{Q} gene.

ATG (see *Figure 6*). This single change was introduced to give translational coupling of the *ompA*–light chain fusion to an efficiently expressed upstream *lacZ* peptide in pCTR008. The co-translational coupling between cistrons is an effective way of achieving efficient translation (14).

4.2 Induction and detection of product

Detailed protocols for growth of cultures, induction of expression, and preparation of cell fractions for the detection of product on SDS-PAGE are given in *Protocols 9* and *10*.

Protocol 9. Secretion system: induced protein production

HBM (high biomass medium) is used routinely for expression experiments in the authors' laboratory. Although more difficult to prepare than conventional complex media, HBM gives greater yields because it allows induction at a higher population density. If desired, pilot experiments to investigate secretability of proteins can be performed using L-broth, for which inductions should be at an A_{600} of 0.5–0.8.

1. Prepare HBM as follows:

- 7 g $(NH_4)_2SO_4$
- 6.24 g $NaH_2PO_4.2H_2O$
- 40 g yeast extract
- 20 g glycerol
- 10 ml 'trace elements'
- 0.05 ml antifoam Mazu DF 843
- pH adjusted to 7.0 with 4 M NaOH
- volume made up to 1 litre with deionized water

'Trace elements' contains the following per litre:

- NaOH   15 g
- EDTA   60 g
- $MgSO_4.7H_2O$ 20 g
- $CaCl_2.6H_2O$ 5 g
- $ZnSO_4.4H_2O$ 2 g
- $MnSO_4.4H_2O$ 2 g
- $CuSO_4.5H_2O$ 0.5 g
- $CoCl_2.6H_2O$ 0.095 g
- $FeSO_4.7H_2O$ 10 g
- H_3BO_4 0.031 g
- Na_2MoO_4 0.002 g
- KCl   74.5 g

2. Retransform cells with DNA of the expression construct and select transformants on L-agar plates containing 30 μg/ml chloramphenicol and incubated at 30°C for 24 h.

3. Inoculate a single colony into 4 ml HBM plus 30 μg/ml chloramphenicol in a 25 ml universal, and grow overnight with shaking at 30°C.

4. Inoculate 3 ml of the overnight culture into 100 ml HBM containing 30 μg/ml chloramphenicol in a 250 ml baffled flask, and grow at 30°C with shaking (approximately 250 r.p.m.).

5. Follow culture growth by increase in A_{600}. Induce between A_{600} 5.0 and 8.0 by adding IPTG to 1 mM for full induction, and 20–50 μM for partial induction.

6. Continue incubation with shaking at 30°C for between 4 and 24 h.

Protocol 10. *E. coli* cell fractionation

1. Measure the final A_{600} of cultures (i.e. 4–24 h after induction). Collect cells from a volume of culture equivalent to 4 ml of $A_{600} = 2$ by centrifugation at 7000 g for 15 min at 5°C. Decant supernatant.

2. To prepare a whole-cell fraction (comprising contents of cell and periplasm) resuspend the pellet in 500 μl of 30 mM Tris–HCl pH 8, 20% sucrose. Take 10 μl for loading on SDS-PAGE.

3. To prepare a periplasmic fraction, add 2 μl of 0.25 M EDTA to 450 μl of whole-cell fraction (to permeabilize the outer membrane by removal of metal ions) and incubate at 30°C for 15 min. Transfer to a cold Eppendorf tube and spin for 5 min in a cold room. Discard the supernatant and resuspend in 500 μl ice-cold deionized water (to disrupt the outer membrane by osmotic shock and release the contents of the periplasm) and whirlimix for 30 sec. Incubate for 10 min on ice and spin for 5 min in the cold room. Remove the supernatant, which is the periplasmic fraction. Take 20 μl for loading on SDS-PAGE.

4. To prepare a washed cell fraction, resuspend the pellet from step 3 in 500 μl of cold deionized water and spin for 5 min in a cold room to collect cells. Discard the supernatant, resuspend the cells again in 500 μl of deionized water, collect the cells again, and resuspend in 500 μl of deionized water. Take 10 μl of this washed cell fraction for loading on SDS-PAGE.

5. The supernatant from step 1 is the culture supernatant fraction. Take 20–40 μl for loading on a protein gel. If the yield of protein is low, concentrate the supernatant up to 20-fold using Amicon concentration cells before running 20–40 μl samples on SDS-PAGE.

6. To prepare these samples for electrophoresis on protein gels, the volume of each is adjusted to 30 μl with deionized water, 8 μl of 5× sample buffer are added, and the mixture is boiled for 5 min. The samples can be stored frozen prior to SDS-PAGE.

In the case of pCTR008, Western blotting with anti-human Fd or anti-human κ antibodies on cell fractions prepared from uninduced cultures shows no detectable heavy or light chain. These results suggest that the pACtac system exhibits very tight control of transcription in the absence of IPTG. Comparison of the A_{600}s of induced and uninduced cultures gives a fairly reliable indication of successful induction. In the case of pCTR008, the A_{600} continues to increase for several hours following induction, then declines through cell lysis to an A_{600} of approximately 10 (for HBM medium) after about 20 h. The A_{600} of the uninduced culture is approximately 30 at the equivalent time. Expression experiments with other proteins suggest that this pattern usually indicates good yields of protein in the culture supernatant. For some proteins full induction of such secretion constructs leads to a rapid cessation of growth—presumably due to deleterious effects of the heterologous protein on the integrity of the cells' inner membrane—and this usually indicates poor or zero yield of product in the culture supernatant. In such cases it is often possible to achieve superior cell viability and product yields by partial induction.

Some of the factors that determine the secretability of heterologous proteins from *E. coli* have become clearer over the past 2 years. It now appears likely that secretion through the inner membrane requires the protein to be in a relatively unfolded and soluble state. The signal sequence and *E. coli* 'chaperon' proteins (chaperonins) co-operate, maintaining the protein in a translocation-competent state. The probable reason for the failure of many heterologous proteins to be secreted is that they cannot be maintained in this state long enough for entry into the secretion pathway. Many other proteins are translocated through the inner membrane but fail to be released from it, and show no cleavage, or inappropriate cleavage, of the single peptide.

The yield of several heterologous proteins (including antibody fragments) is greater when expression/secretion is performed at a temperature of 30°C or lower, rather than at 37°C, presumably because at the lower temperature the protein is more likely to either adopt a translocation-competent state before secretion or a soluble state after secretion. In the authors' laboratory, secretion experiments are routinely performed at 30°C.

The factors that determine whether heterologous secreted proteins are accurately processed, whether they are released from the inner membrane into the periplasm, whether they are soluble within the periplasm, and then whether they penetrate through the outer membrane are not yet clear. Antibody Fv, single-chain Fv, and Fab fragments all appear to be fairly readily secreted into the periplasm, to show accurate signal processing (as determined by N-terminal sequencing), and to be capable of release into the culture supernatant. Penetration of the outer membrane appears, for such fragments, to be concentration dependent. At early time points after full induction most of the fragment appears to be in the periplasm. As the concentration of fragment in the periplasm increases, it probably permeabilizes the outer membrane and begins to leak into the culture supernatant. At high fragment concentrations cell lysis begins to occur. For all antibody fragments expressed, a proportion of product is not released from the outside of the inner membrane. The magnitude of this proportion varies between different antibodies and different fragments: for A5B7 chimeric Fab' it comprises about 80% of total product, while for a single-chain Fv of the antibody B72.3 it is around 10%. *Figure 7* shows a typical induction profile for *E. coli* strain W3110 secreting A5B7 chimeric Fab'.

The yield of antibody fragment appearing in the culture supernatant also varies between different antibodies and different fragments, but yields of 50–100 mg/l in shake flasks (with HBM medium) have been achieved for Fvs,

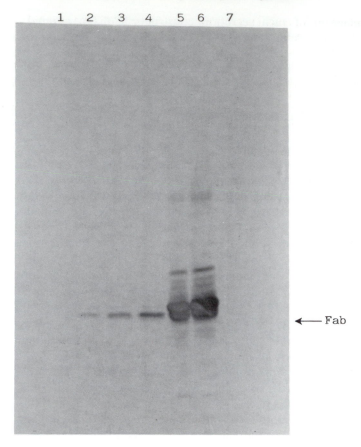

Figure 7. Western blot analysis of antibody secretion. Supernatant and cell extracts of *E. coli* W3110 pAC.A5B7 were analysed by SDS-PAGE and then by Western blotting, using an anti-murine Fab polyclonal antibody. The Fab heavy and light chains co-migrate in this gel system. Samples were moved at the following times after IPTG induction: lane 1, 0 min; lane 2, 1 h; lane 3, 4 h; lane 4, 6 h; lane 5, 8 h; lane 6, cell extract at 5 h; lane 7, cell extract at 8 h.

single-chain Fvs, and Fabs, including the chimeric Fab' of A5B7 from pCTR008.

References

1. Schein, G. H. (1989). *Bio/Tech.*, **7**, 1141.
2. Duronio, R. J. *et al.* (1990). *Proc. Natl Acad. Sci. USA*, **87**, 1506.
3. Wright, E. M., Humphreys, G. O., and Yarranton, G. T. (1986). *Gene*, **49**, 311.
4. Yarranton, G. T. (1991). In *Transgenesis—applications of gene transfer*, (ed. J. A. H. Murray), Cambridge Series on Biotechnology. Open University Press, Milton Keynes.

5. Emtage, J. S., *et al.* (1983). *Proc. Natl. Acad. Sci. USA,* **80,** 3671.
6. Marston, F. A. O., Angal, S., Lowe, P. A., Chan, M., and Hill, C. R. (1988). *Biochem. Soc. Trans.,* **16,** 112.
7. Roberts, S., Cheetham, J. C., and Rees, A. R. (1987). *Nature,* **328,** 731.
8. Pratt, J. M. (1984). In *Transcription and translation: practical approach,* Chapter 7, p. 179. IRL Press, Oxford.
9. Chung, C. H. and Goldberg, A. L. (1981). *Proc. Natl Acad. Sci. USA,* **78,** 4931.
10. Baker, T. A., Grossman, A. D., and Gross, C. A. (1984). *Proc. Natl Acad. Sci. USA,* **81,** 6779.
11. Field, H., Yarranton, G. T., and Rees, A. R. (1988). In *Vaccines 1988,* pp. 29–34. Cold Spring Harbor Publications, New York.
12. Chang, A. C. Y. and Cohen, S. N. (1978). *J. Bacteriol.,* **134,** 1141.
13. Mueller-Hill, B., Crapo, L., and Gilbert, W. (1968). *Proc. Natl Acad. Sci. USA,* **59,** 1259.
14. Schoner, B. E., Hsiung, H. M., Belagaje, R. M., Mayne, N. G., and Schoner, R. G. (1984). *Proc. Natl Acad. Sci. USA,* **81,** 5403.
15. Laemmli, U. K. (1970). *Nature,* **227,** 680.

14

Expression of proteins in yeast

MASAKAZU KIKUCHI and MORIO IKEHARA

1. Introduction

With the advent of recombinant DNA technology, it is now possible to express and produce many proteins in *Escherichia coli* (see Chapter 13). The *E. coli* expression system, however, has not always been found suitable for protein production, such as in the case of the hepatitis B virus surface antigen (1, 2). Further, post-translational processing and modification differ in *E. coli* and mammalian cells. Although mammalian cells can secrete biologically active proteins directly into the culture medium, genetic manipulation and cultivation on a large scale are not easy.

The success achieved in the transformation of *Saccharomyces cerevisiae* by plasmids in 1978 (3) opened up the possibility of using this micro-organism as an alternative host for protein production. *Saccharomyces cerevisiae* is a unicellular eukaryote with characteristics of both prokaryotes and eukaryotes. In addition, *S. cerevisiae* is not pathogenic, as evidenced by its longtime use in the brewing and baking industries. The capability of producing proteins free from toxic contaminants is one of the advantages of the yeast system. Yeasts also have a secretory system in common with mammalian cells and are able to secrete proteins into the culture medium in active form. A great deal of biological and genetic data on the characteristics of yeast have been accumulated, and the mutation, manipulation, and cultivation of yeasts are rather easy to carry out compared to mammalian cells. Basic techniques in manipulating yeasts are the same as those for *E. coli*. Many publications and reviews have covered the manipulation of genes using *E. coli*. For a comprehensive coverage of the subject, the reader is referred to *Molecular cloning: a laboratory manual* (4) and to Chapters 11–13 of this volume.

Success in protein engineering really depends on the selection of a suitable host strain and expression system for the gene of interest. The procedure for purification and characterization of the expressed protein can also have an influence on the results of protein engineering. This chapter deals primarily with specific techniques necessary for expressing genes in yeast, based on the results of our own work. These techniques will be presented by reference to a particular 'case study', human lysozyme, and will include the construction of an expression/secretion plasmid for human lysozyme, transformation, cultivation of yeast transformants, and purification and analysis of the recombinant human lysozyme.

2. Construction of expression/secretion vectors

There are many factors that are known to affect expression levels of heterologous genes in yeast, such as promoters, terminators, plasmid copy number, and plasmid stability. This is similar to the situation in *E. coli*. In addition, expression levels generally depend on the conditions of the yeast culture, which include composition of culture medium, temperature and pH, and sometimes on the toxicity or secretory characteristics of the products.

2.1 Selection of expression systems

The yeast expression system is now much more commonly used for both the intracellular expression and secretion of proteins. In many cases, expressed proteins in yeast cells are accumulated as inclusion bodies which are insoluble and inactive, as in *E. coli*. The advantage of yeast intracellular expression over that of *E. coli* is exemplified in the case of the hepatitis B virus vaccine, the first recombinant DNA product to be produced in yeast and subsequently commercialized.

Yeasts have a well-defined secretory system similar to that found in mammalian cells. The avantage of a secretion system lies in the fact that it allows the production of proteins in soluble and active forms and simplifies their purification. Furthermore, yeasts have the necessary post-translational machinery necessary for modification of some proteins. Some mammalian proteins are believed to require glycosylation for their activity and stability. This type of modification occurs in the secretory pathway in the endoplasmic reticulum (ER) and the Golgi apparatus. Secretory proteins are known to be synthesized as preproteins carrying a so-called signal sequence at their N-termini. Signal sequences help proteins translocate through the membrane of the ER. To secrete the protein in question into the medium, a signal sequence must be present at the N-terminus. Several signal sequences are available for secretion of heterologous protein in yeast, such as yeast signal sequences, natural signal sequences of heterologous proteins, designed and synthetic sequences, and fused signal sequences.

2.2 Selection of vectors

Yeast cloning vectors are generally constructed as *E. coli*–yeast shuttle vectors, which comprise replicating plasmids (YRp-type), episomal plasmids (YEp-type), centrometric plasmids (YCp-type), and integrating plasmids (YIp-type). The *E. coli*–yeast shuttle vectors, except integrating plasmids, have sequences that allow replication in both *E. coli* and *S. cerevisiae*. YRp-

Masakazu Kikuchi and Morio Ikehara

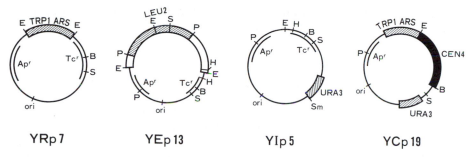

Figure 1. Typical yeast vectors. The pBR322 sequences are represented by solid lines, 2 μm circular DNA sequences by open double lines, yeast chromosomal sequences by hatched lines, and yeast chromosomal centromere sequence by a filled double line. B, *Bam*H1; E, *Eco*R1; H, *Hind*III; P, *Pst*1; S, *Sal*1; Sm, *Sma*1.

type vectors replicate with high copy-number but are not suitable for protein production because of their instability. YEp-type vectors contain the replication origin of yeast 2 μm plasmid DNA, which is stabilized by two proteins encoded by the *REP1* and *REP2* genes of 2 μm DNA (5). The YEp-1 type vectors, which lack the *REP1* and *REP2* genes, are stably maintained only in the strain cir$^+$ harbouring endogenous 2 μm DNA. The YEp-2 type vectors (e.g. pCV20, pCV21) contain the *REP1* and *REP2* genes and are maintained even in the strain cir^0 which does not harbour 2 μm DNA. These types of vectors are most stable and harbour the highest copy-number in the strain cir^0. YCp-type vectors are stably maintained in yeast cells, but the copy number is reduced to 1–2 copies/cell. YIp-type vectors contain the replication origin of *E. coli* but not that of *S. cerevisiae*. For the expression of protein, vectors must be inserted into a specific site of the chromosome by recombination. Transformation frequency of this type of vector is generally low, as shown in *Table 2*, but linearization of the vectors increases the transformation frequency. Typical vectors are summarized in *Figure 1* and *Table 1*. Gene cloning and gene amplification are usually carried out in *E. coli*. A number of

Table 1. Typical vectors

| Plasmid | Size (kbp) | Selectable markers | | Available cloning sites |
		Yeast	*E. coli*	
YRp7	5.7	*TRP1*	Ap^r Tc^r	*Bam*H1, *Sal*1
YEp13	10.7	*LEU2*	Ap^r Tc^r	*Bam*H1, *Hind*III
YCp19	10.6	*TRP1, URA3*	Ap^r	*Bam*H1, *Eco*R1, *Sal*1
YIp5	5.4	*URA3*	Ap^r Tc^r	*Bam*H1, *Eco*R1, *Hind*III, *Sal*1

329

yeast plasmids are summarized by Parent *et al.* (6) and the reader can select a suitable vector by consulting the paper.

2.3 Selection of promoters and terminators

The selection of a promoter is the most important decision to be made in obtaining the efficient expression of proteins in yeast. The promoter region of yeasts contains the upstream activator sequence (UAS), the TATA element (TATA), and initiator element, I. The size of a yeast promoter is relatively large compared with that of a prokaryote promoter (7) and the essential sequence for yeast promoter is found in the region about 500 bp upstream from the RNA start site. Generally, constitutive promoters of genes encoding glycolytic enzymes have been used to obtain high levels of expression of heterologous proteins. Promoters available are PGK (phosphoglyceric acid kinase), GLD, GPD, or TDH3 (glyceraldehyde-3-phosphate dehydrogenase), ADH (alcohol dehydrogenase), ENO1 (enolase), and TP1 (triose phosphate isomerase). To regulate the expression of proteins, the PHO5 (acid phosphatase) promoter, and GAL1 and GAL10 (galactose metabolism) promoters are used.

A terminator is often important, which influences the expression level of a protein. The most commonly used terminators are PGK, ADH, ENO1, and TPI.

2.4 Construction of an expression/secretion plasmid for human lysozyme

Human lysozyme expressed intracellularly in *E. coli* and *S. cerevisiae* lacks activity (8, 9). Thus, the strategy we pursued was directed at the construction of a secretion plasmid that could be used for human lysozyme with *S. crevisiae* (10). To construct the human lysozyme secretion vector, the YEp-type vector, a GLD or TDH3 promoter, DNA encoding the hen's egg white lysozyme signal sequence, and the human lysozyme gene were selected (10). The lysozyme gene and signal sequences were chemically synthesized using an automatic DNA synthesizer (Model 380B Applied Biosystems), selecting codons preferentially used in the highly expressed genes (10). *Figure 2* illustrates the strategy used to construct pGEL125.

Recently, DNAs encoding mutant proteins have been readily produced by site-directed mutagenesis, cassette mutagenesis, saturation mutagenesis, etc. With these mutated DNAs, expression plasmids may be constructed according to the procedures described.

Protocol 1. Ligation of *XhoI*–*Taq*I fragments of the synthetic DNA (I) and synthetic gene (II)

1. Ligate the two fragments at 15°C for 16 h by mixing the following:

- the synthetic DNA (I) (0.5 μg)   10 μl

Masakazu Kikuchi and Morio Ikehara

- the synthetic DNA (II) (1 μg)   5 μl
- 10 × T4 buffer[a]   3 μl
- 10 mM ATP   3 μl
- H_2O 7 μl
- T4 ligase (300 U/μl)   2 μl

2. Keep the reaction mixture at 65°C for 10 min to denature T4 ligase.
3. After cooling, mix the following:
   - 1 M NaCl   3 μl
   - XhoI (10 U/μl)   7 μl
   - the reaction mixture   28 μl[b]

   and keep the mixture at 37°C for 3 h.
4. Electrophorese the whole mixture on a 2% agarose gel in TAE[c] at 100 V for 2 h.
5. Recover the product by electrophoretic elution.
6. Precipitate the product DNA by adding cold EtOH and centrifuge.
7. Add 20 μl of TE[d] buffer to the precipitate of XhoI–XhoI fragment (III).

[a] 10 × T4 buffer: 0.66 M Tris–HCl, pH 7.5; 0.05 M $MgCl_2$; 0.05 M dithiothreitol.
[b] A part of the reaction mixture (2 μl) is examined by electrophoresis to ascertain whether or not ligation has taken place.
[c] TAE: 0.04 M Tris–acetate; 0.002 M EDTA.
[d] TE buffer: 10 mM Tris–HCl, pH 8.0; 1 mM EDTA, pH 8.0.

Protocol 2. Construction of pGEL125

1. Digest pGLD906-1 (11) at 37°C for 1 h by mixing the following to obtain an XhoI fragment:
   - pGLD906-1 (1 μg/μl)   2 μl
   - high buffer[a]   1 μl
   - H_2O 6 μl
   - XhoI (10 U/μl)   1 μl

2. Ligate 10 ng of the synthetic gene (III) and 10 ng of the XhoI fragment of pGLD906-1 in the same manner as described in *Protocol 1*.
3. Transform E. coli DH1 with the reaction mixture obtained from step **2** by the method of Hanahan (12).
4. Prepare plasmid pGEL125 by the alkaline extraction procedure.

[a] High buffer: 50 mM Tris–HCl, pH 7.5; 10 mM $MgCl_2$; 1 mM dithiothreitol; 100 mM NaCl.

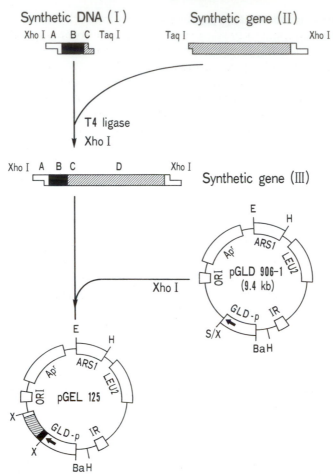

Figure 2. Construction of pGEL125. A, the 5'upstream region of the translation start site of *Saccharomyces cerevisiae* phosphoglycerate kinase gene; B, the signal sequence coding region for hen egg white lysozyme; C, the region encoding two NH_2-terminal amino acids (Lys–Val) of human lysozyme; D, human lysozyme gene lacking two amino acids (Lys–Val) coding region. E, *Eco*RI; H, *Hind*III; Ba, *Bam*HI; X, *Xho*I.

3. Expression and secretion

Levels of expression in yeast are generally not so high, although high levels of expression have been obtained for hepatitis B core antigen (40% of total cell protein) (13) and human superoxide dismutase (30–70% of total cell protein) (14).

3.1 Selection of host strains

Yeast strains that carry auxotrophic markers such as *trp*1, *leu*2, *ura*3, and *his*4 are commonly used for transformation and selection. Sometimes the G418 resistance gene, dihydrofolate reductase gene, hygromycin B resistance gene, etc. are used as dominant selection markers. Levels of protein production are often improved by using host strains which carry the *pep*4–3 mutation (15). The *PEP*4 gene encodes yeast proteinase A which activates vacuolar proteinase. Thus, the *pep*4–3 mutation decreases the activity of several vacuolar proteinases, resulting in an improvement in the level of protein production. Further, a respiratory-deficient mutant (*rho*⁻ strain) has been shown useful for increased expression of proteins (16). The improvement of host strains by mutation is also a useful strategy for obtaining high levels of expression. Almost all strains are available from the Yeast Genetic Stock Center, Department of Biophysics and Medical Physics, University of California, Berkley, CA 94720, USA.

3.2 Transformation

Two methods are available for transforming yeast using spheroplasts and alkaline-metal-treated cells as recipients. Both methods have some advantages and disadvantages. The method employing spheroplasts was established by Hinnen *et al.* in 1978 (3). This method yields almost the same transformation frequency with each experiment, but the manipulation is rather complicated and time consuming, and often produces polyploids by cell fusion. Transformation also depends on the condition of the competent cells prepared from the recipient strain.

The second method was established by Ito *et al.* in 1983 (17) who found that growing yeast cells take in exogenous DNA after treatment with alkaline metals such as lithium, caesium, etc. This method (lithium method) does not produce polyploids and is simple to manipulate. In particular, the advantage of the method is that colonies appear on the surface of the agar after transformation. Thus, colonies appearing on the plate can be used directly for replica-plating and colony hybridization. Most of the colonies appear in the agar with the spheroplast method. A disadvantage of the lithium method is that it has a low transformation frequency compared with the spheroplast method. *Table 2* summarizes transformation frequencies by both methods and provides standards for your experiment.

Protocol 3. Transformation of *S. cerevisiae* AH22R⁻ using the spheroplast method (18)

1. To transform yeast spheroplasts, prepare the following media and solutions:
   - YPDA: yeast extract, 10 g/l; polypeptone, 20 g/l; glucose, 20 g/l; adenine (0.4 g/l) (added when necessary)
   - selective medium: yeast nitrogen base without amino acid, 6.7 g/l; glucose, 20 g/l; sorbitol, 220 g/l; histidine, 100 mg/l; adenine, 30 mg/l; uracil, 30 mg/l; add agar (20 g/l) if preparing plates

Protocol 3. *Continued*

- overlay medium: yeast nitrogen base without amino acid, 6.7 g/l; glucose, 20 g/l; sorbitol, 220 g/l; agar, 30 g/l
- solution A: 0.1 M sodium citrate, pH 5.8; 0.01 M EDTA; 1.2 M sorbitol
- solution B: 1.2 M sorbitol; 10 mM $CaCl_2$
- solution C: 20% polyethyleneglycol 4000; 10 mM Tris–HCl, pH 7.0; 10 mM $CaCl_2$
- solution D: 1.2 M sorbitol; YPDA

2. Grow the yeast host strain AH22R⁻ overnight in 4 ml of YPDA at 30°C with shaking.

3. Harvest the cells by centrifugation at 600 *g* for 5 min and wash with 4 ml of solution A.

4. Resuspend the cells in 3.6 ml of solution A, add 0.4 ml of solution A containing Zymolyase® (1 mg/ml) and incubate at 30°C for 1.5 h.[a]

5. Centrifuge at 600 *g* for 5 min, and wash the cells twice with 4 ml of solution B.

6. Resuspend the cells in 0.1 ml of solution B and shake gently for 2 min.

7. Add plasmid DNA of pGEL125 (4–10 µl solution containing 1–2 µg plasmid DNA) and allow to stand at 25°C for 15 min.

8. Add 2 ml of solution C, allow to stand at 25°C for 15 min and centrifuge at 600 *g* for 5 min.

9. Suspend in 0.3–0.5 ml of solution D and allow to stand at 30°C for 20 min.

10. Mix 0.1 ml portion and 8 ml of overlay medium.[b]

11. Pour the mixture on to selective medium plate.

12. Incubate plates at 30°C for 5–10 days.

---

[a] Spheroplast formation should be examined under the microscope after treatment with Zymolyase® as follows: dilute the reaction mixture tenfold (add 0.5 ml of water to 0.05 ml of the mixture) and examine microscopically if the cells swell and burst.

[b] Keep at 46°C in a waterbath.

3.3 Culture conditions

Yeast generally grows at 20–35°C and at pH 3.0–8.0. The most suitable culture conditions are found at about 25–28°C and at pH 4.5–6.5. For cultivating yeast, YPDA medium is generally used as nutrient medium, and Burkholder medium (*Table 3*) (20) and Bacto yeast nitrogen base medium (produced by Difco) as minimal media. Other types of culture media have

Table 2. Transformation frequency

Method	Recipient[a] strain	Plasmid DNA	Transformants per 1 μg of DNA	Reference
Spheroplast	AH22	pYeleu10[b]	1–90	3
	MC16	pJDB248[b]	10^4–10^5	18
	D13-1A	YRp7	500–5000	19
Lithium	D13-1A	YRp7	400–500	17
	D13-1A	YEp6	5	17
	AH22	pJDB248	100–200	17

[a] Markers of *S. cerevisiae* strains used are as follows: AH22 (a *leu*2–3, *leu*2–112, *his*4–519, *can*1); MC16 (α *leu*2–3, *his*4–712[FS], *ade*2–1, *lys*2–1, SUF2); D13-1A (a *his*3–532, *trp*1, *gal*1).
[b] pYeleu10 and pJDB248 belong to YIp- and YEp-type plasmids.

Table 3. Burkholder medium composition

Glucose	20 g
Asparagine	2 g
$(NH_4)_2SO_4$	2 g
KH_2PO_4	1.5 g
$CaCl_2.2H_2O$	0.33 g
$MgSO_4.7H_2O$	0.5 g
KI	100 μg
H_3BO_3	60 μg
$MnSO_4.7H_2O$	30 μg
$ZnSO_4.7H_2O$	300 μg
$CuSO_4.5H_2O$	40 μg
$FeCl_3.6H_2O$	250 μg
$Na_2MoO_4.2H_2O$	25 μg
Thiamine–HCl	200 μg
Pyridoxine	200 μg
Nicotinic acid	200 μg
Ca-pantothenate	200 μg
Biotin	2 μg
Inositol	10 mg
Distilled water	1 litre
	pH 5.0

When preparing Burkholder medium, stock solutions prepared beforehand are usually added. Each stock solution of inorganic salts, trace elements, and vitamins is made concentrated fourfold, fourfold, and 1000-fold higher than the concentration of the medium and preserved.

been devised for specific purposes. Minimal medium is used for selection and maintenance of transformants. Amino acids and other nutrients are often added to the minimal medium, depending on the auxotrophy of the host strain. Each of the amino acids and nucleic acid bases is dissolved, sterilized, and preserved.

The difficulty is that a medium suitable for one strain is not always suitable for another, and it is often difficult to determine the requirements for a given strain. Thus, in such cases, modification of medium composition is required for obtaining a high level of expression.

Protocol 4. Cultivation of *S. cerevisiae* AH22R⁻/pGEL125

1. Inoculate 5 ml of the modified Burkholder medium[a] in a test-tube with a loopful of *S. cerevisiae* AH22R⁻/pGEL125.
2. Grow the cells at 30°C for 72 h with shaking.
3. Transfer 1 ml of the culture to 4 ml of the modified Burkholder medium in a test-tube and continue shaking at 30°C for 1 day.
4. Transfer 2 ml of the culture to 18 ml of modified Burkholder medium in a 200 ml flask.
5. Grow the cells at 30°C for 48–96 h with shaking.

[a] One litre of medium contains 0.4 g of KH_2PO_4, 10 g of glucose, 5 g of asparagine, and 80 g of sucrose.

---

4. Purification of protein product

At the end of the growth period, the cells are separated from the growth medium by centrifugation. Protein that has accumulated in the cells is usually released by disrupting the cells with sonication, using a French press, lytic enzymes, etc. A high concentration of guanidine–HCl or urea is used to solubilize inclusion bodies in which case, refolding of the protein is required.

Standard techniques for purifying recombinant proteins are employed, such as ion-exchange column chromatography, high performance liquid chromatography (HPLC), gel filtration, affinity chromatography, isoelectric focusing, etc. Purified proteins are then analysed for biological activity, molecular weight, amino acid composition, N-terminal and C-terminal amino acids, and if feasible, three-dimensional structure.

Protocol 5. Purification of secreted human lysozyme (21)

1. Harvest cells from 1 litre of culture of *S. cerevisiae* AH22R⁻/pGEL125 by centrifugation.

2. Pass the supernatant through a cation-exchange column (CM-Toyopearl 650C, 1.6 cm ϕ × 36 cm) equilibrated with 50 mM sodium phosphate buffer (pH 6.5).

3. Wash the column with 500 ml of the same buffer.

4. Elute the human lysozyme with the same buffer containing 0.5 M NaCl.

5. Collect 5 ml fractions and assay for lytic activity using *Micrococcus luteus* as a substrate.

6. Combine the fractions containing the human lysozyme.

7. Concentrate and deionize the lysozyme solution by ultrafiltration with DIAFLO® ultrafiltration membrane YM5 (Amicon Corp.) and wash the protein with 50 mM sodium phosphate buffer (pH 6.5).

8. Adsorb the human lysozyme fraction on Asahipak 502C for HPLC. Elute the human lysozyme using a linear gradient of 0–0.18 M sodium sulphate in 0.05 M sodium phosphate buffer (pH 6.5).

9. Collect and combine the lysozyme fractions, concentrate, and deionize.

5. Conclusion

The combination of an expression system with mutagenesis techniques has been of great value in the study of protein engineering. It is commonly accepted that protein engineering consumes more time than the usual recombinant DNA technology. Thus, the need for saving time should be taken into account when starting a new project. The selection of the proper expression system for obtaining biologically active proteins is probably the most important factor that will determine whether the study will proceed correctly and smoothly. This selection process depends largely on the characteristics of the particular protein and may differ from protein to protein. Moreover, it is desirable to obtain information on the three-dimensional structure of the protein as rapidly as possible, to guide further engineering experiments.

An appropriate expression system, rapid and simple purification procedures, and rapid means of analysing the three-dimensional structure undoubtedly make engineering of proteins a promising field for the future. For human lysozyme, a fully active engineered protein can be obtained using the yeast secretion system (10). Native and mutant human lysozymes thus obtained can be purified by a simple two-step procedure (21–23). The protein can be crystallized within 1 week to several months, and the three-dimensional structures of mutants analysed within 1–2 weeks each. These rapid procedures have led to considerable progress in understanding the structure and function of signal sequence (24, 25), *in vivo* folding mechanisms (21, 22), and the structure/activity profile of human lysozyme (23). These results clearly indicate that the selection of an efficient expression system, purification of proteins, and structural analysis represent the major hurdles in the engineering of proteins.

References

1. Valenzuela, P., Medina, A., Rutter, W. J., Ammerer, G., and Hall, B. D. (1982). *Nature,* **298,** 347.
2. Miyanohara, A., Toh-e, A., Nozaki, C., Hamada, F., Ohtomo, N., and Matsubara, K. (1983). *Proc. Natl Acad. Sci. USA,* **80,** 1.
3. Hinnen, A., Hicks, J. B., and Fink, G. R. (1978). *Proc. Natl Acad. Sci. USA,* **75,** 1929.
4. Sambrook, J., Fritsch, E. F., and Maniatis, T. (1989). *Molecular cloning, a laboratory manual* (2nd edn). Cold Spring Harbor Press, Cold Spring Harbor, New York.
5. Broach, J. R. (1982). *The molecular biology of yeast Saccharomyces,* (ed. J. Strathern, E. Jones, and J. R. Broach), p. 445. Cold Spring Harbor Laboratory, Cold Spring Harbor, New York.
6. Parent, S. A., Fenimore, C. M., and Bostian, K. A. (1985). *Yeast,* **1,** 83.
7. Struhl, K. (1987). *Cell,* **49,** 295.
8. Muraki, M., *et al.* (1986). *Agric. Biol. Chem.,* **50,** 713.
9. Hayakawa, T., *et al.* (1987). *Gene,* **56,** 53.
10. Yoshimura, K., *et al.* (1987). *Biochem. Biophys. Res. Commun.,* **145,** 712.
11. Itoh, Y., Hayakawa, T., and Fujisawa, Y. (1986). *Biochem. Biophys. Res. Commun.,* **138,** 268.
12. Hanahan, D. (1983). *J. Mol. Biol.,* **166,** 557.
13. Kniskern, P. J., *et al.* (1986). *Gene,* **46,** 135.
14. Hallewell, R. A., *et al.* (1987). *Bio/tech.,* **5,** 363.
15. Bitter, G. A., Chen, K. A., Banks, A. R., and Lai, P. H. (1984). *Proc. Natl Acad. Sci. USA,* **81,** 5330.
16. Kaisho, Y., Yoshimura, K., and Nakahama, K. (1989). *Yeast,* **5,** 91.
17. Ito, H., Fukuda, Y., Murata, K., and Kimura, A. (1983). *J. Bacteriol.,* **153,** 163.
18. Beggs, J. D. (1978). *Nature,* **275,** 104.
19. Struhl, K., Stinchcomb, D. T., Scherer, S., and Davis, R. W. (1979). *Proc. Natl Acad. Sci. USA,* **76,** 1035.
20. Burkholder, P. R. (1943). *Am. J. Biol.,* **30,** 206.
21. Taniyama, Y., Yamamoto, Y., Kuroki, R., and Kikuchi, M. (1990). *J. Biol. Chem.,* **265,** 7570.
22. Taniyama, Y., Yamamoto, Y., Nakao, M., Kikuchi, M., and Ikehara, M. (1988). *Biochem. Biophys. Res. Commun.,* **152,** 962.
23. Kuroki, R., Taniyama, Y., Seko, C., Nakamura, H., Kikuchi, M., and Ikehara, M. (1989). *Proc. Natl Acad. Sci. USA,* **86,** 6903.
24. Yamamoto, Y., Taniyama, Y., Kikuchi, M., and Ikehara, M. (1987). *Biochem. Biophys. Res. Commun.,* **149,** 431.
25. Yamamoto, Y., Taniyama, Y., and Kikuchi, M. (1989). *Biochemistry,* **28,** 2728.

15

Genetic engineering of the processing and secretory pathway of an inflammatory cytokine in mammalian cells: a case study

MICHAEL KRIEGLER

1. Introduction

In addition to the induction of tumour regression, tumour necrosis factor (TNF) has been implicated as the causative agent in a number of pathologies, including cachexia, septic shock, rheumatoid arthritis, autoimmunity, and induction of HIV expression. Our early studies of TNF led us to propose that this complex physiology might be manifest by different forms of TNF: the 17 kDa secretory form and a precursor 26 kDa transmembrane form we first identified and characterized (1). It was shown that TNF is first synthesized as a transmembrane molecule that is cleaved by a specialized 'TNF convertase' to the 17 kDa form prior to its release from the cell. In an attempt to separate the biological activities of the 17 kDa and 26 kDa forms, we genetically engineered the molecule and thus generated both an uncleavable 26 kDa mutant form and a solely secretable 17 kDa mutant form of the wild-type molecule and studied their biological activities. In these studies we found that an uncleavable mutant of TNF kills tumour cells by cell-to-cell contact, and that TNF need not be internalized by its target cell to kill (2). Thus, the 26 kDa integral transmembrane form of TNF may function *in vivo* to kill tumour cells and other targets locally, in contrast to the systemic bioactivity of the secretory component.

In this report, the principles and methods applied in the TNF experiments, many of which can be broadly generalized, will be described. The reasoning, experimental tactics, and relevant methods employed in achieving the expression of wild-type TNF in mammalian cells and the genetic engineering of the TNF molecule in a manner such that the producer cell handles the protein in a predictable, fundamentally different manner than the wild-type gene product will be discussed. Some experimental protocols are not contained within this case study due to lack of space. The reader can refer to ref. 3 for additional details.

2. Factors to consider when expressing foreign genes in mammalian cells, and the case of TNF

2.1 Vector selection and recombinant cistron construction

Choice of the eukaryotic expression vector is the key in engineering gene expression in mammalian cells. The expression vector must be suitable for the cell type to be engineered to express the gene, matching the correct *cis*-acting elements with the appropriate *trans*-acting factors present in the host cell. The variety of vectors available and appropriate cell types have been described elsewhere (3). In the case of TNF expression we chose to express the gene in a retroviral vector in a murine 3T3 cell. The intention was to generate recombinant retroviruses encoding TNF for use in gene therapy studies and the parentage of the host cell type appropriate for packaging recombinant retroviruses was a 3T3 cell. We employed a variety of retroviral vectors, some of our own design, some designed by others (*Figures 1* and *2*). These retroviral vectors contain promoters in their long terminal repeats (LTRs) and these promoters function particularly well in murine cells.

The placement of a gene with respect to the promoter in an eukaryotic expression vector to obtain efficient gene expression is far less critical than is its placement in the majority of prokaryotic expression vectors. Basically, the gene must be positioned downstream of the promoter, there should be no initiation codons in any reading frame between the promoter and the correct initiation codon of the gene to be expressed, and the promoter and the structural gene should not be separated by nucleic acid stretches of high G/C content. The initiation codon should not be more than a few hundred nucleotides from the promoter, and the termination codon of the structural gene to be expressed should be followed by a functional polyadenylation signal. In the case of TNF, the gene was placed several hundred base pairs downstream from the transcription start site (within which is contained the retroviral packaging signal information), between the two long terminal repeats (LTRs) of the expression vector, with the polyadenylation signal being provided by the downstream LTR.

2.2 DNA transfer methods and the case of TNF

There exist a variety of methods for gene transfer into mammalian cells. Some are more appropriate for short-term or transient transfection, others are more appropriate for stable or long-term transfection. Generally speaking, transient transfection results in higher levels of expression but the protein of interest is only expressed for 1–3 days. Stably transfected cell lines are

A.

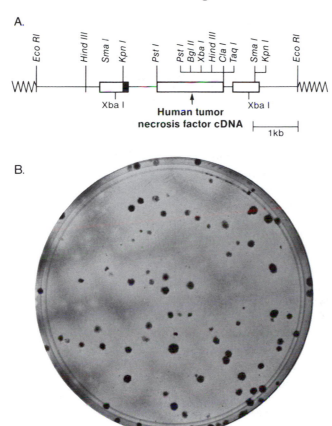

B.

Figure 1. Structure and bioactivity of TNF retroviral genome. (A) Restriction map of pFVXMTNF retroviral genome; (B) plaque assay of pFVXMTNF-transfected ψam cells.

effectively immortal. Although stably transfected cell lines initially express the transferred gene at lower levels than those achieved by transient transfection, they can be coaxed to express the transferred gene at much higher levels through a process of gene co-amplification. This process will be discussed in a subsequent section and is also described in detail elsewhere (3).

Stable transfection can be achieved through a number of different methods. The method of choice is calcium-phosphate-mediated gene transfer. However, electroporation and retrovirus-mediated gene transfer are also useful. Again, the decision as to which method to use depends in large part on the cell type to be transfected. If the cell type has not been transfected previously, you will have to determine which approach is most appropriate through trial and error.

Transient transfection can be achieved through two distinct methods,

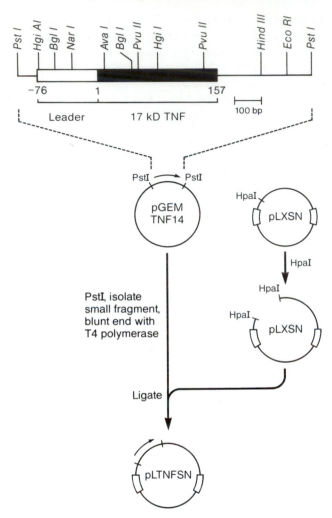

Figure 2. Assembly of the TNF-expressing, G418 resistance-conferring, retroviral vector. The method for insertion of the TNF cDNA into the LXSN retroviral vector is shown.

DEAE-dextran-mediated gene transfer and calcium-phosphate-mediated gene transfer. The DEAE dextran method is the method of choice for a majority of cells. It is both simple and efficient. This is due to the relative ease of preparation of the DNA/DEAE dextran mixture for transfection, as well as the high efficiency of transient gene transfer and expression that can be accomplished with this technique.

With this procedure, a DEAE dextran mixture is prepared and the DNA sample of interest is added, mixed, and then transferred to the cells in culture. The ease of sample preparation facilitates the preparation of a large number of samples, as many as 1000 or more. Thus, this method is the preferred method for analysing the numerous samples of plasmid DNA required in the expression cloning of either secreted molecules or cell-surface molecules. The chemical events that lead to the cellular uptake of DNA are not clearly understood. Nevertheless, the transfection efficiency of this method can be improved dramatically by including chloroquine in the treatment of the trans-fected cells. It is felt that chloroquine serves to neutralize the pH of the lysozymes of the transfected cells, thus inhibiting the degradation of the DNA internalized by the cell as the DNA makes its way to the nucleus.

2.2.1 The DEAE dextran transfection method

The DEAE dextran method (*Protocol 1*) has been reported to yield transfec-tion efficiencies of as high as 80%. DNA introduced into cells with this method appears to undergo mutations at a higher rate than that observed with calcium-phosphate-mediated transfection (4, 5, 6). This method is not useful for isolating stable transfectants.

Protocol 1. Transfection of COS cells by the DEAE dextran method

Materials

- DEAE dextran (Sigma; mol. wt 5×10^5): stock solution 0.1% (1 mg/ml) in TBS, filter-sterilize, store at 4°C
- Chloroquine (Sigma; mol. wt 319.89): stock solution 100 mM in H_2O, filter-sterilize, aliquot in 0.5 ml amounts, store frozen at -20°C, protect from light
- phosphate-buffered saline (PBS) with Ca^{2+} and Mg^{2+}
- Tris–EDTA (TE), pH 8.0: 0.1 M Tris base, 0.001 M EDTA
- 10 × 'A': 8.00 g NaCl; 0.38 g KCl; 0.20 g Na_2HPO_4; 3.00 g Tris base; adjust pH to 7.5 with HCl, bring up to 100.0 ml with glass-distilled H_2O, filter-sterilize, store at room temperature
- 100 × 'B': 1.5 g $CaCl_2.2H_2O$; 1.0 g $MgCl_2.6H_2O$; bring up to 100.0 ml with glass-distilled H_2O, filter-sterilize, store at room temperature
- Tris-buffered saline (TBS)—prepare fresh: 10 ml 10 × 'A'; 1 ml 100 × 'B'; 89 ml H_2O; filter-sterilize, store at 4°C

Method

The following transfection protocol is adapted for COS cells, but it can be used for a variety of cell types. We use Falcon 3046 six-well cluster dishes with a surface area of 9.6 cm^2 and maximum volume capacity of 3 ml.

1. Day 1: seed cells at a concentration of 2×10^4 cells/cm^2 in a total volume of 2 ml/well (1.92×10^5 cells/well of a six-well cluster dish). Cells should

Protocol 1. *Continued*

be about 75% confluent when used to seed the dishes. Do not use old or very dense cell passages.

2. Day 2: using sterile technique, reagents, tubes, and tips, resuspend 0.5 μg DNA in 10–25 μl TE. Add 50 μl DEAE dextran (in TBS). Final DEAE dextran concentration should be about 0.04%.

3. Prior to proceeding with transfection, observe cell monolayers microscopically. Cells should appear about 60–70% confluent and well disturbed. Bring all reagents to room temperature.

4. Aspirate off growth media and wash monolayer once with 3 ml of PBS followed by one wash with 3 ml of TBS.

5. Aspirate off TBS solution and add 100–125 μl of the appropriate DNA/DEAE-dextran/TBS mixture to the wells (final concentration of DNA is 4–5 μg/ml).

6. Incubate dishes at room temperature inside a laminar-flow hood. Rock the dishes every 5 min for 1 h, making sure that the DNA solution covers the cells. This is critical and must be done to avoid dehydration of the cells.

7. After the 1 h incubation period, aspirate off the DNA solution and wash once with 3 ml of TBS followed by 3 ml of PBS.

8. Remove the PBS solution by aspiration and replace with 2 ml of complete growth medium containing 100 μM chloroquine. Incubate the dishes in an incubator set at 37°C + 5% CO_2 for 4 h.

9. Remove the medium containing chloroquine and replace with 2–3 ml of complete growth medium (no chloroquine).

10. Incubate the transfected cells for 1–3 days, after which the cells will be ready for experimental analysis. The exact incubation period will depend on the intent of the experiment. We find that optimal expression of transfected cytokine genes and cell-surface protein occurs at 3 days posttransfection.

2.2.2 Calcium phosphate transfection method

The most commonly used method to transfer DNA into recipient cells is the co-precipitation of the DNA of interest with calcium phosphate, after which the precipitate is added to the cells. With this technique, DNA entering the cell is taken up into phagocytic vesicles (7). Nevertheless, sufficient quantities of DNA enter the nucleus of the treated cells to allow relatively high frequencies of genetic transformation. This technique was employed originally by Graham and van der Eb to increase the infectivity of adenoviral DNA, but it was popularized by Wigler and colleagues (8) and Maitland and McDougall (9).

This procedure has been shown to be appropriate for the transfer into a variety of cell lines of single-copy genes present in the total genomic DNA derived from a donor cell or tissue sample. It has been the method of choice for identifying a number of cellular oncogenes as well as cellular genes for which a biochemical selection strategy exists. Using a variety of cell types, transfection efficiencies of up to 10^{-3} have been obtained. This is the method of choice for the generation of stable transfectants.

Over time, a number of variations of the basic technique have been developed. If the experiment involves the transfer of plasmid DNA, high molecular weight genomic DNA isolated from a defined cell or tissue source may be included. The addition of such DNA, called carrier DNA, often serves to increase the efficiency of transformation by the plasmid DNA. Upon arrival of the plasmid DNA/carrier DNA calcium phosphate co-precipitate to the nucleus of the treated cell, the plasmid DNA appears to integrate into the carrier DNA, often in a tandem array, and this assembly of plasmid and carrier DNA, called a transgenome, subsequently integrates into the chromosome of the host cell.

Another procedural option available, is the addition of a chemical shock step to the transfection protocol. Either dimethylsulphoxide or glycerol is appropriate. The optimal concentrations and lengths of treatment vary from cell line to cell line. The use of these agents can affect cell viability dramatically. We do not include a chemical shock in our protocol and prefer to allow transfected cells to sit overnight in the presence of the co-precipitate.

Chen and Okayama (10) optimized this transfer technique carefully. They, like us, do not employ a chemical shock. They reported that incubation of the cells and the co-precipitate is optimal at 35°C in 2–4% CO_2 for 15–24 h; that circular DNA is far more active than linear DNA, and that an optimal, finer precipitate is obtained when the DNA concentration is between 20 and 30 μg/ml in the precipitation mix.

If the protocol presented here does not yield the desired result, it may be wise to alter the incubator temperaure, CO_2 concentration, and DNA concentration to those just cited. However, those temperature and CO_2 concentrations are not optimal for cell growth and should be maintained only temporarily.

Protocol 2. Transfection by calcium phosphate method

Materials
- 2 × HBS (HEPES-buffered saline): salt solution—3.7 g KCl; 10.0 g D(+)glucose; 1.0 g Na_2HPO_4; bring up to 50.0 ml with glass-distilled H_2O. Then, in a separate beaker, add: 1.0 ml salt solution; 1.0 g HEPES (N-2-hydroxyethylpiperazine-N'-2-ethane sulphonic acid); 1.0 g NaCl; bring up to 85 ml with glass-distilled H_2O, pH to 7.05–7.10 with 1 M NaOH, bring up to 100 ml with glass-distilled H_2O

Protocol 2. *Continued*

- 2.5 M CaCl$_2$: 36.8 g CaCl$_2$.2H$_2$O; bring up to 100 ml with glass-distilled H$_2$O
- distilled H$_2$O: pH of water must be in the range 7.05–7.10
- TE (Tris–EDTA), pH 7.05: 0.01 M Tris base; 0.001 M EDTA
- filter-sterilize all reagents through a 0.2 μm filter prior to use; store in sterile 50.0 ml conical polypropylene tubes

Method
Generally, a large calcium phosphate cocktail mixture is prepared to transfect many plates simultaneously. Described here is the protocol for 1 ml (or 1 × 100 mm dish equivalent) of solution. Scale-up the amounts as necessary, and allow for an appropriate amount of sample-transfer errors. Adherence to sterile technique is critical. Use sterile reagents, tips, and tubes.

1. Day 1: seed 1.3 × 10^6 cells per 100 mm dish. Cells should be about 75% confluent when used to seed the dishes. Do not use 'old' or very dense cell passages.

2. Day 2: add 1–20 μg DNA (1 mg/ml in sterile TE, pH 7.05) to 0.45 ml sterile H$_2$O.
   Note: first 'sterilize' DNA by ethanol precipitation with NaCl (0.1 M final aqueous concentration) and 2× volume 100% ethanol.

3. Add 0.5 ml 2× HBS. Mix well.

4. Add 50 μl of 2.5 M CaCl$_2$ and vortex immediately.

5. Allow the DNA mixture to sit undisturbed for 15–30 min at room temperature. If a precipitate has not formed, use a sterile 1 ml cotton-plugged pipette to bubble air into the mixture; the precipitate will usually form. If it does not, it is likely that one of the reagents was improperly made or an incorrect volume was added.

6. Add 1 ml of the DNA transfection cocktail directly to the medium in the 100 mm dish (plated with cells on day 1).

7. Incubate the dishes containing the DNA precipitate for 16 h at 37°C. Remove the medium containing the precipitate and add fresh complete growth medium.

8. Allow the cells to incubate for 24 h. Post-incubation, the cultures may be split into selective media. We generally split our cultures 1:5; however, if you want to isolate individual colonies for further analysis, split your cultures 1:10 and 1:100 as well. (For experimental details on drug selection, see the section on selection and amplification.)

3. Dominant selectable markers for co-transfection and gene co-amplification in the case of TNF

More often than not the gene to be introduced into the host cell does not manifest a readily selectable phenotype. As a result it is nearly impossible to identify cells that have taken up and are expressing the foreign DNA in the absence of an accessory selection strategy. The solution to this problem is the incorporation of a dominant selectable marker gene, either into the eukaryotic expression vector or the co-transfection of the non-selectable cistron with a cistron encoding a dominant selectable marker gene whose expression is driven from another promoter. The first scheme works because both the non-selectable gene and the dominant selectable gene are present on the same recombinant DNA molecule and thus the transfer of one gene ensures the transfer of the other. In this situation, generally speaking, both genes are expressed in the transfected cell. Therefore, selection for the dominant selectable marker gene virtually ensures the isolation of cell lines expressing the non-selectable gene as well. The second scheme works because, under the conditions employed in the DNA transfection procedure, multiple recombinant DNA molecules are taken up by the transfected cell and stably incorporated into the host genome. Thus, the recombinant molecule encoding the non-selectable gene and the recombinant molecule encoding the dominant selectable marker are both integrated into the host chromosome. Again, selection for the dominant selectable marker gene virtually ensures the isolation of cell lines expressing the non-selectable gene as well.

A subset of the genes employed as dominant selectable markers can, under the appropriate conditions, function as amplifiable markers as well. Co-transfection with such amplifiable markers followed by a clearly defined, somewhat time consuming selection procedure can serve to increase the expression of a transferred gene by several orders of magnitude. This method is commonly employed by the pharmaceutical industry in the large-scale manufacture of proteins in mammalian cells. A wide variety of dominant selectable, amplifiable markers are available. A complete list of markers and their appliction is provided elsewhere (3).

In the case of TNF two different types of vector/cDNA constructions were utilized. In our first experiments we inserted the TNF cDNA into a retroviral vector, described previously, that contained all of the *cis*-acting elements necessary for the efficient expression and rescue of the gene (see above; *Figure 1*). Next we inserted the TNF cDNA into the same retroviral vector that also contained the gene encoding neo[r] which serves as a dominant selectable marker that confers resistance to the drug G418 (genticin; *Figure 2*). The first type of retroviral vector, described above, was co-transfected into the recipient cell with a second plasmid, itself containing a cistron encoding G418 resistance whose expression was driven by a promoter derived

347

from the human β-actin gene. To assay for the expression of the TNF gene we developed a unique assay for TNF gene expression, a so-called plaque assay. In this assay, post-transfection tissue culture cells are placed in G418 to select for drug-resistant colonies, a procedure that takes about 10 days. Post-selection, the medium is removed from the monolayer and the remaining colonies are overlain with TNF-sensitive cells, which are subsequently over-laid with soft agar-containing medium. The cells are allowed to incubate overnight, after which the monolayers are examined by eye. TNF-producing colonies secrete TNF which subsequently kills the TNF-sensitive cells on top of and surrounding the producer colonies (*Figure 1*). Those colonies not producing TNF do not manifest this plaque-forming phenotype. The TNF-producing colonies can be readily isolated by picking the cells off the dish, through the soft agar, with a Pipetman. The picked cells are subsequently expanded in 24-well dishes and, after they proliferate, are ready for further analysis.

4. Analysis of transfected, expressing cell lines, and the case of TNF

Once cell lines producing the protein of interest have been identified, the precise level of expression and cellular distribution of the molecules can be determined. The methods of choice for these types of analyses are metabolic labelling followed by immunoprecipitation, Western analysis, and subcellular fractionation followed by these two methodologies. Metabolic labelling is useful for the analysis of rates of production and turnover of a given protein, as well as for an analysis of the processing of that molecule. Pulse-chase analysis, a variation of metabolic labelling, is particularly useful in determining whether there is a precursor/product relationship between similar molecules in the same cell. After a brief labelling period with a radioactive amino acid, the labelled cells are exposed to a cold-chase solution of the same, unlabelled amino acid. Harvesting the cells at various time-points after the labelling pulse, and subsequently displaying the immunoprecipitated proteins on a polyacrylamide gel, enables you to follow the processing and eventual degradation of the molecule of interest. Western analysis is the method of preference for accurate mass determination of a given protein when insufficient amounts of the protein are available for quantitation by staining of a polyacrylamide gel by Coomassie brilliant blue or silver staining. The key advantage of this method is its extreme sensitivity. With a high-quality antiserum the detection sensitivity of a Western blot lies in the picogram range.

All of these approaches have been employed in the analysis of TNF-producing cell lines. Examination of the nucleotide sequence of the TNF cDNA revealed the presence of but two methionine residues, at positions

1 and 6 of the TNF leader, what we now know to be the transmembrane domain. In contrast, there are two cysteines in the presequence and two cysteines in the region encoding the mature or processed TNF. Lastly, the only tyrosines in the molecule are found in the C-terminal portion of the molecule, that encoding the portion of the molecule which is ultimately released from the cell (*Figure 3*). These bits of information are important to keep in mind because they enabled us to evaluate the results of both *in vivo*

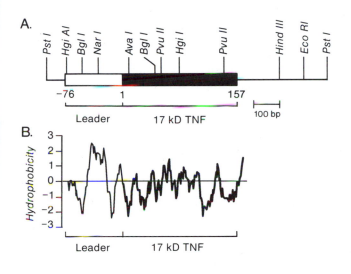

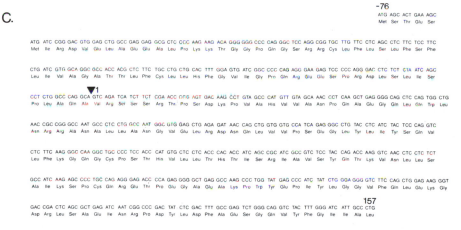

Figure 3. Sequences and charge properties of human tumour necrosis factor. (A) Restriction endonuclease map of human tumour necrosis factor cDNA; (B) hydrophobicity/hydrophilicity profile of human tumour necrosis factor by the method of Kyte and Doolittle (11) using a width of 10 residues and a jump of 1 residue; (C) nucleic acid and amino acid sequences of human tumour necrosis factor.

expression experiments and *in vitro* translation expression experiments to determine:

(a) the serological identity of the previously unknown TNF precursor
(b) the precursor–product relationship between the 26 kDa transmembrane form of TNF and the 17 kDa secretory component
(c) the orientation of the transmembrane form of TNF in the membrane
(d) the percentage of TNF present in the plasma membrane of TNF-producing cells
(e) the approximate size of the cytosolic domain (the amino-terminal portion) of the transmembrane form of the TNF molecule.

In metabolic labelling/immunoprecipitation experiments with polyclonal anti-TNF antibody we identified two immunoreactive proteins, one of 26 kDa and one of 17 kDa. The 17 kDa molecule is of the size expected for the TNF secretory component and the 26 kDa molecule is of the size expected for a translation product derived from the entire structural gene. The identity of this immunoreactive molecule was determined by a number of criteria. First, both the 17 kDa and 26 kDa molecules could be labelled with [^{35}S]cysteine but only the 26 kDa molecule could be labelled with [^{35}S]methionine, information consistent with the predicted amino acid sequence of the TNF molecules. In addition, competition/immunoprecipitation experiments in which metabolically labelled cell lysate immunoprecipitations were challenged with cold, recombinant 17 kDa TNF revealed that the cold 17 kDa molecule could compete with both the 17 kDa and 26 kDa molecules, indicating that the two molecules share epitopes (*Figure 4*). Pulse-chase experiments conducted as described above revealed that the 26 kDa molecule served as the precursor for the 17 kDa molecule and that much, but not all, of the 26 kDa molecule was cleaved to the 17 kDa molecule in the cytoplasm of the producer cell prior to release (*Figure 5*).

Subcellular fractionation studies of TNF-producing cells, followed by Western analysis of the fractions, indicated that the 26 kDa form of TNF was membrane-associated. Radioiodination of intact cells producing TNF, followed by immunoprecipitation of the lysates, revealed that some of the 26 kDa form of TNF was present on the cell surface and thus could function as a suitable target for the radioiodination reagent.

4.1 Immunoprecipitation analysis

Immunoprecipitation analysis is a very powerful technique. When combined with radiolabelling of cells, immunoprecipitation analysis allows the synthesis of a protein to be monitored and, in pulse-chase experiments, the kinetics of processing and turnover of the protein to be monitored as well. Radiolabelling followed by immunoprecipitation is not the method of choice for quantitating the total amount of a given protein in a cell. For protein quanti-

Michael Kriegler

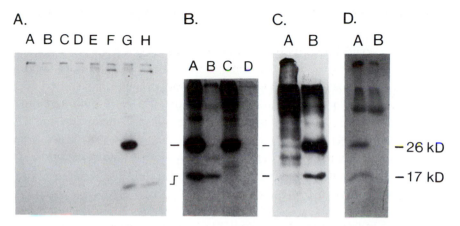

Figure 4. Immunoprecipitation and competition analysis of 17 kDa and 26 kDa TNF. Cells were labelled with [^{35}S]cysteine or [^{35}S]methionine, subjected to immunoprecipitation, electrophoresis in a 12% polyacrylamide gel, and autofluorography. Autoradiograms of these gels are shown. (A) Immunoprecipitation analysis of [^{35}S]cysteine-labelled ψam and TNF 6.8 cell lines with pre-immune and anti-17 kDa TNF antisera. Lane A, ψam cell lysate precipitated with pre-immune serum; lane B, ψam cell supernatant precipitated with pre-immune serum; lane C, ψam cell lysate precipitated with anti-TNF antiserum; lane D, ψam cell supernatant precipitated with anti-17 kDa TNF antiserum; lane E, TNF 6.8 cell lysate precipitated with pre-immune serum; lane F, TNF 6.8 cell supernatant precipitated with pre-immune serum; lane G, TNF 6.8 cell lysate precipitated with anti-17 kDa TNF antiserum; lane H, TNF 6.8 cell supernatant precipitated with anti-17 kDa TNF antiserum. (B) Immunoprecipitation with anti-17 kDa TNF antiserum of 17 kDa and 26 kDa TNF labelled with [^{35}S]cysteine and [^{35}S]methionine from TNF 6.8 cells. Lane A, [^{35}S]cysteine-labelled cell lysate; lane B; [^{35}S]cysteine-labelled cell supernatant; lane C, [^{35}S]methionine-labelled cell lysate; lane D, [^{35}S]methionine-labelled cell supernatant. (C) Competition binding analysis of [^{35}S]cysteine-labelled TNF 6.8 cell lysates. Lane A, [^{35}S]cysteine-labelled cell lysate and unlabelled recombinant TNF immunoprecipitated with anti-17 kDa TNF antiserum; lane B, [^{35}S]cysteine-labelled cell lysate precipitated with anti-17 kDa TNF antiserum. (D) Competition binding analysis of [^{35}S]cysteine-labelled cell lysates of induced human monocytes. Lane A, [^{35}S]cysteine-labelled induced human monocyte cell lysate immunoprecipitated with anti-17 kDa TNF antiserum; lane B, [^{35}S]cysteine-labelled induced human monocyte cell lysate and unlabelled (50 μg) recombinant TNF precipitated with anti-17 kDa TNF antiserum.

tation Western analysis should be used. An efficient, broadly applicable technique for immunoprecipitation analysis is presented here.

Protocol 3. Labelling and immune precipitation of recombinant proteins

Materials
- Protein A–Sepharose CL-4B beads (Sigma)
- [^{35}S]methionine or [^{35}S]cysteine (New England Nuclear or Amersham)

351

Protocol 3. *Continued*

- dialysed fetal calf serum (FCS) (Sigma)
- methionine-minus or cysteine-minus media (MEM Select-Amine Kit, Gibco)
- lysis buffer: 20 mM Tris–HCl, pH 8.0; 200 mM LiCl (Sigma); 0.5% NP-40 (Sigma); 1 mM EDTA; filter-sterilize through a 0.2 μm filter and store at 4°C
- buffer B: 20 mM Tris–HCl, pH 8.0; 100 mM NaCl; 0.5% NP-40
- 2× gel-loading buffer: 4% SDS, 15% glycerol, 62.5 mM Tris–HCl, pH 6.8 0.005% bromophenol blue, 200 mM dithiothreitol (add dithiothreitol just prior to use from a 1 M stock)
- Nutator (Baxter)
- PBS with Ca^{2+} and Mg^{2+} (Sigma)

Method

1. Prepare Protein A–Sepharose CL-4B beads (hereafter called 'Protein A beads'):
   (a) Fill the bottle containing the beads with enough H_2O to quantitatively transfer the 1.0 g of the beads to a 15 ml test tube.
   (b) Allow the beads to swell for 30 min on ice, with occasional agitation to resuspend the beads.
   (c) When the beads have settled, the beads and H_2O should be 50:50 (v/v). Adjust the H_2O level if necessary.
   (d) Freeze aliquots of the bead slurry at −20°C.
2. Label adherent cells:
   (a) Wash a 60 mm dish of cells (subconfluent) twice with PBS.
   (b) Add 1.0 ml of methionine-minus or cysteine-minus medium containing 5% dialysed FCS. Incubate at 37°C for 30 min.
   (c) Add 100–200 μCi of the appropriate radioactive amino acid to the medium. Incubate at 37°C for 3 h, rocking the dish every 30 min.
   (d) Save the media (if desired) for immunoprecipitation. Store at −70°C if not processing immediately.
   (e) Add 1.0 ml of lysis buffer to the dish. Incubate at 4°C for 5 min. (The plasma membrane is lysed with this buffer. This liberates the cytoplasmic proteins and frees many proteins from the nucleic but does not destroy the ultrastructure of the nuclei, which remains attached to the dish.)
   (f) Transfer the lysate to a microfuge tube. Spin at 16 000 *g* at 4°C for 5 min. Transfer the supernatant to a new microfuge tube. Store at −70°C if not processing immediately.

3. Label non-adherent cells:

    (a) Wash the cells twice with PBS.

    (b) Resuspend the cells in methionine-minus or cysteine-minus medium, 5% dialysed FCS, at a concentration range of $1.0–10.0 \times 10^6$ cells/ml. Transfer cells into a 12-well plate, 1 ml of cell suspension per well. Incubate at 37°C for 30 min.

    (c) Add 100–200 μCi of the appropriate label, and continue to incubate at 37°C for 3 h. Rock the plate every 30 min.

    (d) Transfer the cell suspension to a microfuge tube. Spin at 16 000 *g* for 1 min.

    (e) Save the supernatant for immunoprecipitation. Store at −70°C if not processing immediately.

    (f) Resuspend the cell pellet in 1.0 ml of lysis buffer. Incubate for 5 min at 4°C with occasional rocking.

    (g) Spin the tube at 16 000 *g* for 5 min at 4°C. Transfer the supernatant to a new microfuge tube. Store at −70°C if not processing immediately.

4. Immunoprecipitate:

    (a) To each sample tube, add the appropriate amount of antibody (determined by antibody titration experiments).

    (b) Incubate at 4°C for 60 min on the Nutator.

    (c) Spin the sample tubes at 16 000 *g* for 15 min at 4°C.

    (d) Transfer the supernatants to new microfuge tubes, and add to each tube 30 μl of a Protein A bead slurry. Use disposable Pipetteman tips whose tips have been cut off with a clean razor blade to create a larger orifice. The larger orifice ensures quantitative transfer and will keep the beads from being crushed during transfer.

    (e) Incubate for 60 min at 4°C on the Nutator.

    (f) Spin for 15 sec in a microfuge and discard the supernatant in a radioactive-waste container.

    (g) Rinse the beads three times with 1.0 ml of lysis buffer and twice with 1.0 ml of buffer B. All spins should be for no longer than 15 sec.

    (h) Remove the supernatant from the final wash and leave the beads undisturbed in a volume of approximately 30 μl. Add 30 μl of 2× sample buffer. Boil for 3 min and load the samples on a polyacrylamide gel. Samples can be stored at −70°C.

5. Prepare SDS-polyacrylamide gels (see *Protocol 4*).

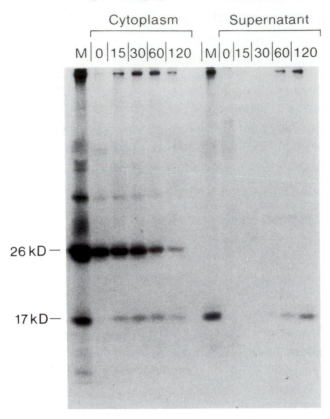

Figure 5. Pulse-chase analysis of 26 kDa and 17 kDa TNF. TNF-producing NIH 3T3 cells (line TNF 6.8) were labelled for 15 min with [^{35}S]cysteine and subsequently incubated with excess unlabelled cysteine for the times indicated above each lane. After chase, cell lysates and supernatants were subjected to immunoprecipitation with anti-17 kDa TNF antiserum. Immune precipitates were subjected to electrophoresis in a 12% polyacrylamide gel, after which the gel was subjected to autofluorography. An autoradiogram of that gel is shown.

Upon completion of your immunoprecipitation, or prior to Western analysis of a protein mixture or cell lysate, you will want to separate the protein products on a polyacrylamide gel. The protocol presented here is quick because relatively small gels are used. To be highly reproducible the gels should always be run under exactly the same conditions.

Protocol 4. Polyacrylamide gel electrophoresis of recombinant proteins

Materials
- vertical mini-gel apparatus: several designs are available; our laboratory uses a C.B.S. Scientific Vertical Mini-Gel Unit with 0.75 mm spacers, 10-

or 14-well combs, and notched glass plates, 11.3 cm (width) × 10.0 cm (height)

- gel-pouring stand (W.E.P. company)
- acrylamide/bis (N,N′-methylene-bis-acrylamide), 29:1 mixture (Bio-Rad): prepare a 50% w/v solution in deionized, warm H_2O; filter through a 0.45 μm filter and store in dark bottles at 4°C
- TEMED (N,N,N′,N′-tetramethylethylenediamine, electrophoresis-grade; Bio-Rad and other manufacturers)
- ammonium persulphate (APS) (Bio-Rad): add 1 g to 10 ml of deionized H_2O (10% v/v) and store at 4°C
- 10% SDS
- 1.5 M Tris–HCl, pH 8.8
- 0.5 M Tris–HCl, pH 6.8
- 10× Tris-glycine SDS electrophoresis buffer (10× TGS): 250 mM Tris, pH 8.3; 2.5 M glycine (Sigma); 1% SDS
- Rainbow protein molecular weight markers (Amersham, available as [14C]methylated or non-radioactive protein standards)—these markers run as discrete coloured bands, allowing the experimenter to follow the progress of electrophoresis while the gel is running and to assess the extent of protein transfer from an acrylamide gel to a filter during Western analysis
- 2× SDS gel-loading buffer: 4% SDS; 15% glycerol; 62.5 mM Tris–HCl, pH 6.8; 0.005% bromophenol blue; 200 mM dithiothreitol (add dithiothreitol just before use from a 1 M stock)
- microcapillary pipette tips (West Coast Scientific, Inc.)
- staining solution: 0.25% (v/v) Coomassie brilliant blue R250 (Sigma); 10% (v/v) glacial acetic acid; 45% (v/v) methanol
- destaining solution: 10% (v/v) glacial acetic acid; 45% (v/v) methanol
- Enlightning (New England Nuclear)

Method

1. Clean and dry the glass plates.
2. Assemble the glass plates by placing the side-spacers between the notched and unnotched plates. Clamp one side with the free clamp and the other side with the clamp mounted to the gel-pouring stand.
3. Add 5 ml of molten 1% agarose in H_2O, at 65°C, to the trough in the gel-pouring stand, and place the bottom of the glass-plate assembly in the trough to seal the bottom of the assembly.

Protocol 4. *Continued*

4. SDS-polyacrylamide mixtures for 12% and 15% gels:

	12%	15%
• 1.5 M Tris–HCl, pH 8.8	5.0 ml	5.0 ml
• 50% acrylamide	4.8 ml	6.0 ml
• 10% SDS	0.2 ml	0.2 ml
• H$_2$O	9.8 ml	8.6 ml
• 10% APS	0.2 ml	0.2 ml
	20.0 ml	20.0 ml

5. Add 20 μl of TEMED to the SDS-polyacrylamide mixture. Swirl rapidly and pour immediately into the gap between the plates. Leave sufficient space for the stacking gel (length of the teeth, plus 1 cm). With an 18-gauge needle and syringe, carefully overlay with H$_2$O or isobutanol.

6. After polymerization (30–45 min), pour off the overlay and rinse with H$_2$O to remove unpolymerized acrylamide. Drain as much H$_2$O as possible. Use a Kimwipe if necessary.

7. 5% stacker mixture:
   • 2.5 ml 0.5 M Tris–HCl, pH 6.8
   • 1.0 ml 50% acrylamide
   • 0.1 ml 10% SDS
   • 6.3 ml H$_2$O
   • 0.1 ml 10% APS

   10.0 ml total

8. Add 25 μl of TEMED to the stacker mixture. Swirl rapidly and pour into the gap between the plates over the resolving gel. Immediately insert the comb, avoiding trapped bubbles.

9. After polymerization (30–45 min), carefully pull out the comb and rinse the wells with H$_2$O from a bent 18-gauge needle fitted to a 5 ml disposable syringe.

10. Mount the gel assembly into the electrophoresis apparatus and add 1× TGS to the upper and lower chambers. Rinse the wells with electrophoresis buffer by gentle squirting with a Pasteur pipette.

11. Mix the samples with one volume of 2× SDS gel-loading buffer. Boil the samples for 3 min.

12. Load 12–15 μl in each well with a Pipetman using the disposable capillary-thin pipette tips. Load an equal volume of 1× SDS gel-loading buffer (diluting with sample buffer) in any unused well.

13. Connect to a power supply. Apply a voltage of 30 V until the dye enters

the resolving gel, then increase the voltage to 75 V. When the Rainbow markers have migrated the desired distance, turn off the power supply. Remove the gel from the electrophoresis apparatus.

14. Either proceed to Western analysis, or

    (a) Stain the gel with staining solution for 30 min; destain with destaining solution; dry, if desired, on a gel dryer.

    (b) Treat with Enlightning for 20 min; dry at 60°C on a gel dryer; expose to Kodak XAR-5 or Amersham Hyperfilm-ECL film at −70°C.

15. To reduce spurious background:

    (a) Reduce the amount of antibody used. Include only enough antibody to precipitate all of the protein of interest.

    (b) Pre-absorb the samples with Protein A beads. Add the beads to the sample. Incubate for 30 min. Pellet the beads and save the supernatant, then with the primary antibody and fresh beads, begin immunoprecipitation of the supernatant as described below.

    (c) Immunoprecipitate the sample with pre-immune serum from the same animal from which the primary antibody was produced (if available), then immunoprecipitate with the primary antibody.

    (d) Affinity-purify the antibody on an antigen-affinity column prior to use.

16. Poor incorporation of label: determine whether your medium is deficient in the appropriate amino acid.

4.2 Western-type techniques

In contrast to immunoprecipitation of radiolabelled proteins, Western analysis is quantitative and serves as a mass determination for a particular protein. This procedure is not useful in evaluating the rate of synthesis, turnover, or processing of a protein. This procedure is considerably quicker than immunoprecipitation analysis, especially in the analysis of a large number of samples, for two reasons. First, you do not need to immunoprecipitate each sample prior to gel loading; rather you just load your samples on to the gel. Secondly, Western analysis is ideally suited to non-isotopic, enhanced chemiluminescence detection techniques. In our hands these techniques are sufficiently sensitive to detect less than 7 pg of a 17 kDa molecule per sample with X-ray film exposure times of just a few seconds. An autoradiographic image of similar intensity of a parallel culture radiolabelled with [^{35}S]cysteine requires an exposure time of 24–48 h. Western analysis combined with non-isotopic detection is a very powerful technique and is presented here.

Having run a polyacrylamide gel with the protein samples to be analysed on it, those protein samples must be transferred to nitrocellulose for Western analysis. Unlike Southern and Northern transfer techniques, in which DNA and RNA molecules can be transferred by capillary action, in Western transfer the proteins are electrophoresed from the polyacrylamide gel on to the nitrocellulose filter. The procedure for this transfer is described here.

Protocol 5. Western transfer—protein transfer from acrylamide gels

Materials
- 10× stock Tris–glycine buffer, pH 8.5: 0.25 M Tris (Sigma); 1.92 M glycine (Sigma)
- transfer buffer: 0.025 M Tris, 0.192 M glycine, 20% methanol (Mallinckrodt; Baxter, New York)
- Trans-Blot apparatus (Bio-Rad)

Method
1. Soak one 20 cm × 14 cm sheet of Whatman 3MM paper with transfer buffer and lay on a dry Scotch Brite pad.
2. Place the protein gel on top of the Whatman paper, and remove any air-bubbles.
3. Briefly submerge a pre-wet 9.0 cm × 6.0 cm sheet of nitrocellulose in a small volume of transfer buffer and place it on top of the acrylamide gel. Clip the membrane along the top to indicate any lanes to be cut and probed with different antisera later.
4. Remove any air-bubbles from the gel/nitrocellulose membrane interface with gloved fingertips, then place a second 20 cm × 14 cm sheet of Whatman paper soaked in transfer buffer on top of the nitrocellulose membrane.
5. Put a dry Scotch Brite pad atop this Whatman paper. Close the unit and submerge it in the Trans-Blot tank containing *c.* 2.5 litres of transfer buffer.
6. The final arrangement of the gel/nitrocellulose sandwich in the Trans-Blot unit is listed in order:
   (a) cathode (−) BLACK
   (b) Scotch Brite pad
   (c) pre-soaked Whatman 3MM paper
   (d) protein gel
   (e) nitrocellulose membrane
   (f) pre-soaked Whatman 3MM paper
   (g) Scotch Brite pad
   (h) anode (+) RED
7. Run the Trans-Blot Cell for 1 h at a constant voltage of 40 V (current 215 mA).

After Western transfer, prior to exposure with the primary antibody, the nitrocellulose filter bearing the electrophoresed protein samples is treated for an extended period of time with blocking solution. This blocking solution, rich in protein, serves to occupy previously unoccupied protein-binding sites on the membrane. This creates a situation in which, when the blocked filter is exposed to the primary antibody, the primary antibody (and later the secondary antibody) only binds to the antigens that it is directed against on the filter and does not bind to the filter non-specifically. Blocking is the secret of the sensitivity of Western analysis. The extended blocking procedure described in *Protocol 6* works very well.

After the blocking process has been completed, the Western transfer filter is exposed to the antibody directed against the protein of interest (primary antibody). After incubation with the primary antibody, the Western transfer filter is incubated with the secondary antibody (*Protocol 6*). This secondary antibody is a polyclonal antibody directed against the first antibody and is covalently attached to a reporter molecule, in this case horseradish peroxidase. These secondary-antibody molecules bind to the primary-antibody molecules at multiple sites and, as a result, multiple reporter molecules become associated with a single primary-antibody molecule. These reporter molecules serve to catalyse a chemical reaction that oxidizes a special substrate, which, in the case of enhanced chemiluminescence (ECL), results in the generation of photons that serve to expose X-ray film whenever an antigen/primary antibody/secondary antibody/reporter molecule complex has formed (*Protocol 7*).

Protocol 6. Development of the Western blot with antibody

Materials

- 10× stock phosphate buffered saline, pH 6.8 (PBS): 0.2% KCl (Fisher Scientific); 8.0% NaCl (Fisher Scientific); 0.2% potassium phosphate monobasic (KH$_2$PO$_4$; Fisher Scientific); 1.14% sodium phosphate monobasic (NaH$_2$PO$_4$; Fisher Scientific)

- blocking solution (refrigerate if storing): 1.0 M glycine; 5% dry milk (Carnation); 5% fetal calf serum (Sigma); 1% ovalbumin (powder; Sigma)

- wash solution (refrigerate if storing): 1× stock PBS, pH 6.8; 1% fetal calf serum; 0.1% TWEEN 2; (Bio-Rad); 0.1% dry milk; 0.1% ovalbumin (crystalline; Sigma)

- T-PBS (for ECL wash of Western blots): 1× stock PBS, pH 6.8; 0.05% TWEEN 20

- goat anti-rabbit IgG-horseradish peroxidase (Bio-Rad)

Protocol 6. *Continued*

A. *Blocking method*

1. After transfer, block unoccupied sites on the nitrocellulose membrane with ~50 ml of blocking solution for 2 days at 4°C, with agitation; or for 30 min at room temperature, with agitation.

2. Wash the blocked filter with wash solution for 5 min. Change the wash solution and repeat washing twice more.

3. All subsequent antibody incubation and washing steps are done with wash solution unless noted. Perform all incubations and washes with the filter oriented protein-side up.

B. *Primary-antibody incubation*

1. Incubate the nitrocellulose membrane at room temperature in a tray with the primary antibody diluted in wash solution for 3 h, at room temperature, with agitation; or at 4°C overnight, with agitation. The dilution of primary antibody should be determined empirically. We routinely use a 1:1000 dilution of primary antibody: 20 μl primary antibody + 20 ml wash solution.

2. After the primary antibody incubation, wash the membrane for 5 min with wash solution. Repeat this wash step twice more.

3. Wash once for 15 min with T-PBS (ECL wash solution).

C. *Secondary-antibody incubation*

1. Incubate the nitrocellulose membrane at room temperature in a tray with the secondary antibody diluted in wash solution for 1 h with agitation. Use a 1:3000 dilution of goat anti-rabbit IgG-HRP complex: 10 μl goat anti-rabbit IgG-HRP + 30 ml T-PBS wash solution.

2. Wash the membrane for 15 minutes with T-PBS wash solution.

3. Repeat this wash step twice more.

The final step in the Western analysis is the addition of the horseradish peroxidase substrate to the processed, complexed filter. Using the ECL reagent post-incubation, film exposure times of a few seconds may be all that is necessary to detect a clear signal from the complexed reporter molecules.

Protocol 7. Analysis of Western blots by the ECL method

1. Mix an equal volume of freshly prepared reagent 1[a] and an equal volume of freshly prepared reagent 2[a] for a final volume equivalent of 0.125 ml/cm^2 membrane.

2. Place the ECL-detection reagent in a clean tray and agitate.

3. Add the filter and agitate gently for exactly 60 sec.

4. Place the wet filter, protein-side up, on the shiny side of Benchcote paper.

5. Wrap the damp filter with Saran Wrap.

6. Expose to hyperfilm-ECL (Amersham) in a darkroom. Under many circumstances, exposure times of 5–10 sec will be sufficient.

[a] ECL reagents 1 and 2 as described in the ECL kit supplied by Amersham International Ltd.

---

5. Analysis of membrane association and processing of proteins with *in vitro* translation reactions, and the case of TNF

Cloned genes can be transcribed and translated *in vitro* in the presence of dog pancreatic microsomes to facilitate the analysis of the association of the protein of interest with microsomal membranes. In many cases a bona fide membrane protein will insert itself co-translationally into the microsomal membrane or a bona fide secretory protein will be processed, co-translationally, by signal peptidase to its mature, secretory form. This procedure is quite straightforward and is adequately described in kits provided by various manufacturers. *In vitro* translation analysis of TNF was carried out as described here.

In vitro translation in the presence of dog pancreatic microsomes of *in vitro* transcripts of the wild-type TNF cDNA creates microsomes into which the 26 kDa TNF molecule is incorporated. When the *in vitro* translation reaction is labelled with $[^{35}S]$cysteine, incubated with proteinase K, and subjected to immunoprecipitation analysis, the 26 kDa band normally observed in labelled cell lysates and found after *in vitro* translation of the wild-type TNF cDNA transcript in the absence of proteinase K drops to 24 kDa, a net reduction of 2 kDa. These results reveal that a 24 kDa portion of the TNF molecule is protected within the microsome and that a 2 kDa portion is susceptible to proteolytic degradation. When the same experiment is performed with TNF labelled with $[^{35}S]$methionine *in vitro* in the presence of microsomes and incubated with proteinase K, the 24 kDa band disappears, indicating that the two methionines at amino acid positions 1 and 6 have been removed by the protease. This experiment indicated that the amino-terminus represented the cytosolic portion of the 25 kDa tansmembrane form of TNF and that the business end of the molecule, the carboxy-terminal domain, resided inside the microsome and, therefore, outside the cell (*Figure 6*).

Thus, the nature of the precursor for TNF, its subcellular location, and its orientation in the plasma membrane of a producer cell were determined by

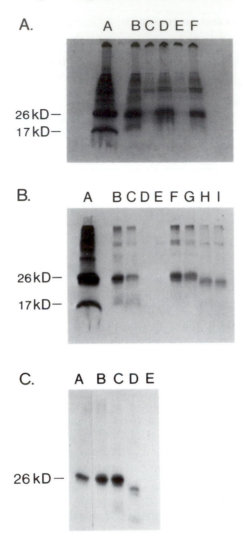

this simple procedure. We also determined that TNF existed in two molecular forms, a transmembrane form and a 'secretory' form (*Figure 7*). To determine if these two forms of the molecular manifest different bioactivities and, as a result, different biologies and physiologies, the activities of these molecules were separated by generating cell lines that produced pure populations of either one or the other. To accomplish this, many of the techniques mentioned above were used, in combination with site-directed mutagenesis of the structural gene. These experimental methods will not be further elaborated here since they are adequately described in the companion volume *Directed mutagenesis: a practical approach* (12).

Michael Kriegler

Figure 6. Analysis of *in vitro* synthesized 26 kDa TNF and its association with the micro-somal membrane. T7 transcripts encoding 26 kDa TNF were translated *in vitro*, immuno-precipitated, and subjected to electrophoresis in a 12% polyacrylamide gel. The resultant gels were subjected to autofluorography and autoradiography. Autoradiograms of those gels are shown. (A) Alkali treatment and ultracentrifugation analysis of [^{35}S]cysteine-labelled 26 kDa TNF translated *in vitro* in the presence of microsomes. All samples were immunoprecipitated with anti-17 kDa TNF antiserum. Lane A, TNF 6.8 cell lysate; lane B, *in vitro* translated 26 kDa TNF; lane C, neutral supernatants of pelleted microsomal prepara-tions; lane D, pellet of neutral microsomal preparations; lane E, supernatant of pelleted alkaline microsomal preparations; lane F, pellet of alkaline microsomal preparations. (B) Proteinase K digestion of [^{35}S]cysteine-labelled 26 kDa TNF synthesized *in vitro* in the absence and presence of microsomes. Lanes A, B, D, F, and H immunoprecipitated with anti-17 kDa TNF antiserum; lanes C, E, G, and I immunoprecipitated with anti-TNF 2 TNF leader antiserum. Lane A, TNF 6.8 cell lysate; lanes B and C, 26 kDa TNF synthesized in the absence of microsomes; lanes D and E, 26 kDa TNF synthesized in the absence of microsomes digested with proteinase K; lanes F and G, 26 kDa TNF synthesized in the presence of microsomes; lanes H and I, 26 kDa TNF synthesized in the presence of microsomes digested with proteinase K. (C) Determination of the polarity of integral transmembrane 26 kDa TNF in the microsomal bilayer. 26 kDa TNF was translated in the presence of microsomes and labelled with either [^{35}S]cysteine or [^{35}S]methionine. Half of each reaction was digested with proteinase K. All reactions were immunoprecipitated with anti-17 kDa TNF antibody. Lane A, 26 kDa TNF marker; lane B, [^{35}S]cysteine-labelled 26 kDa TNF undigested with proteinase K; lane C, [^{35}S]methionine-labelled 26 kDa TNF undigested with proteinase K; lane D, [^{35}S]methionine-labelled 26 kDa TNF digested with proteinase K; lane E, [^{35}S]methionine-labelled 26 kDa TNF digested with proteinase K.

6. Conclusion

Throughout this chapter a number of powerful procedures for transferring and expressing wild-type and engineered genes in mammalian cells have been described. The most important take-home message is that this entire pro-cedure is really quite simple and, having mastered the techniques mentioned here, one can rapidly analyse and control the synthesis, processing, and behaviour of virtually any gene transferred to a given mammalian cell. The key to success is, as is the case with most experimental enterprises, the possession of quality reagents. Perhaps the most valuable reagent in engineer-ing and expressing foreign genes in mammalian cells is, in addition to the gene itself, a high-titre, relatively specific antiserum to the molecule of interest. While this may take several months to generate, the end-product is well worth the wait and your patience will be rewarded in the eventual experimental outcome.

References

1. Kriegler, M., Perez, C., DeFay, K., Albert, I., and Liu, S. D. (1988). *Cell*, **53**, 45.
2. Perez, C., Albert, I., DeFay, K., Zachariades, N., Gooding, L., and Kriegler, M. (1990). *Cell*, **63**, 251.

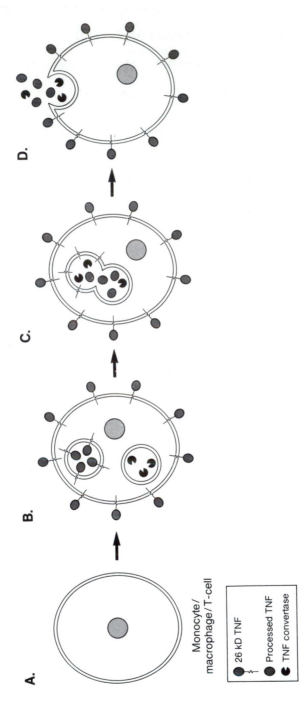

Figure 7. Schematic representation of TNF synthesis, processing, and release. The proposed time course of monocyte/macrophage/T-cell activation and release of newly synthesized TNF is shown. (A) Resting monocyte/macrophage/T-cell; (B) primed monocyte/macrophage/T-cell, synthesizing membrane TNF and TNF convertase; (C) activated monocyte/macrophage/T-cell, converting 26 kDa TNF to the 17 kDa secretory component; (D) TNF release.

Monocyte/
macrophage/T-cell

● ～ 26 kD TNF

● Processed TNF

◖ TNF convertase

3. Kriegler, M. (1991). *Gene transfer and expression: a laboratory manual.* W. H. Freeman, New York.
4. Calos, M. P., Lebkowski, J. S., and Botchan, M. R. (1983). *Proc. Natl Acad. Sci. USA,* **80,** 3015.
5. Razzaque, A., Mizusawa, H., and Seidman, M. (1983). *Proc. Natl Acad. Sci. USA,* **80,** 3015.
6. Ashman, C. R. and Davidson, R. L. (1985). *Somat. Cell. and Mol. Genet.,* **11,** 499.
7. Graham, F. L. and van der Eb, A. J. (1973). *Virology,* **52,** 456.
8. Wigler, M., Silverstein, S., Lee, L. S., Pellicer, A., Cheng, V. C., and Axel, R. (1977). *Cell,* **11,** 223.
9. Maitland, N. J. and McDougall, J. K. (1977). *Cell,* **11,** 233.
10. Chen, C. and Okayama, H. (1987). *Mol. Cell. Biol.,* **7,** 2745.
11. Kyte, J. and Doolittle, R. J. (1982). *J. Mol. Biol.,* **157,** 105.
12. McPherson, M. J. (ed.) (1991). *Directed mutagenesis: a practical approach.* IRL Press, Oxford.

16

Industrial applications of protein engineering and scale-up of protein production: α1-antitrypsin, a case study

RAINER BISCHOFF, MICHAEL COURTNEY,
and DENIS SPECK

1. Introduction

The advent of gene technology in the 1970s has led to the flourishing of the biotechnology industry in recent years. This expansion in commercial activity stems in large part from the ability to produce proteins in heterologous cell systems at industrial scale. Protein engineering is a more recent development which has important implications for the biotechnology industry. Clearly, the capability of designing novel proteins broadens the opportunity to identify new products with altered function or improved stability or potency. In the area of human therapeutics a number of so-called second-generation recombinant products are now in development, for example humanized monoclonal antibodies and modified tissue plasminogen activators (tPAs) with reduced plasma clearance rates. In some cases it remains to be seen whether the risk of immunogenicity associated with the use of a modified protein is offset by improved efficacy, but in general it would seem that protein engineering is destined to have a real impact on drug discovery.

In this chapter we shall focus on the practical aspects of growing a recombinant *Escherichia coli* strain in fermentation culture and purifying the recombinant protein product from the bacterial cell lysate. This methodology is of importance not only as a first step towards developing an industrial-scale process but also as a general approach for the preparation of engineered proteins for evaluation. In this context protocols for the production and purification of a molecular variant of the human plasma protease inhibitor, α1-antitrypsin (AAT) will be described.

The major physiological function of AAT is to limit the proteolytic action of neutrophil elastase during inflammation. A number of studies have shown that the amino acid residue at the P1 position of the active centre of AAT

(Met358) is crucial in defining the specificity of inhibition. For example, AAT-Pittsburgh, a natural variant with a Met358 → Arg substitution, is a strong inhibitor of a number of blood coagulation proteases, notably thrombin, kallikrein, and factor XIIa. Based on this activity profile, recombinant AAT-Pittsburgh is being evaluated as a possible therapeutic agent in septic shock.

Recently we described the properties of a novel engineered variant, AAT (Ala357, Arg358), which was designed to improve selectivity of inhibition of kallikrein, the key contact-phase protease which generates bradykinin and is thought to be responsible for the early hypotension observed in septic shock. Here we detail methods used to isolate from *E. coli* milligram quantities of this variant for biochemical and pharmacological evaluation.

2. Fermentation of a recombinant *E. coli* strain

Fermenters are stirred vessels that allow the control and regulation of physical and physiological parameters during the growth of micro-organisms. Such features are essential to obtain reproducible results at high cell densities (20–100 g/litre of cell dry weight). The product of interest can subsequently be isolated from the producing host in significant amounts for further studies.

In the following, we will describe a procedure which has been employed to produce the AAT variant (Ala357, Arg358) from recombinant *E. coli* cells at a 20 litre fermenter scale. The AAT gene is controlled by the leftward promoter of phage λ (P_L) and expression is driven by temperature shift from 30°C to 42°C (1). This inducible system permits control of gene expression, thus avoiding foreign product toxicity during the initial phase of growth.

2.1 Establishing the preculture

2.1.1 Storage of the producing strain

Many techniques have been described for the long-term storage of bacteria (2). In the case of recombinant strains which harbour a plasmid, there are two possibilities:

(a) storing the producing strain in a competent state (competent cells are those able to take up DNA at high frequency) and reintroducing the plasmid prior to cultivation (fresh transformants)

(b) storing the bacteria already containing the expression vector

For recombinant bacteria, a glycerol stock can be prepared as follows.

Protocol 1. Preparation of glycerol stock for recombinant bacteria

1. Grow cells in a rich or minimal medium to mid–late exponential phase, harvest by centrifugation and resuspend the pellet in the same volume and the same medium as before.

2. Add an equal volume of freezing mix (20% glycerol, 10 mM $MgSO_4$, 0.5% NaCl) and transfer 1 ml aliquots to 2 ml pre-sterilized plastic containers.
3. Leave the vials at $-20°C$ for at least 1 h (slow freezing) and transfer them subsequently to liquid nitrogen or a dry ice/ethanol bath. Store at $-70°C$.

Additional information with respect to cell storage can be found in ref. 3.

The preparation of E. coli competent cells and their subsequent transformation with an expression vector are given in Protocols 2 and 3.

Protocol 2. Preparation of E. coli competent cells

- Luria broth (LB) medium: 10 g:litre Bactotryptone (Difco); 5 g/litre yeast extract (Difco); 5 g/litre NaCl; 1.2 g/litre Tris–HCl, pH 7.5; 0.2 g/litre $MgCl_2$; autoclave together for 15 min at 121°C
1. Dilute overnight culture 1/50 into 0.5 litres of LB.
2. Aerate vigorously at 37°C until the $OD_{550} = 0.3$.
3. Chill on ice for 10 min.
4. Centrifuge for 10 min at 6000 g (4°C) in Sorvall GSA rotor.
5. Drain the pellet well (Pasteur pipette connected to water pump).
6. Resuspend into 200 ml ice-cold 0.1 M $CaCl_2$.
7. Stand the suspension on ice for 20 min and centrifuge as before, again drain the pellet well.
8. Resuspend into 5 ml ice-cold 0.1 M $CaCl_2$.
9. Stand the suspension on ice for 6 h or overnight.
10. Add 1 ml sterile glycerol (4°C) and mix well.
11. Freeze 0.2 ml aliquots in a dry ice/ethanol bath.
12. Store at $-70°C$.

Protocol 3. Transformation of competent E. coli cells

1. Take a 0.2 ml aliquot of competent cells and thaw in ice-water (10 min).
2. Add the DNA solution, not more than 10% cells per volume.
3. Mix at 0°C and leave on ice for 10 min.
4. Heat shock for 1 min at 37°C. Cool afterwards on ice for 5 min.
5. Add 1 ml LB and incubate at 37°C for 50 min to permit selective gene expression.
6. Plate 200 μl aliquots directly on selective medium. Expected transformation frequency $= 2 \times 10^7$ transformants per μg of pBR322.

2.1.2 Inoculation

A preculture is a cell suspension prepared to inoculate the fermenter (see *Figure 1*). Cells must be at a sufficient concentration and in an adapted physiological state to allow growth to continue in a larger volume with a minimal lag time.

In our example, cells are grown in 2 litre Erlenmeyer flasks containing 500 ml of LB medium supplemented with an antibiotic for plasmid selection. The cultures are maintained for 12 h at 30°C in a rotary shaker (New Brunswick Scientific) in order to reach a final OD_{600} = 1–2.

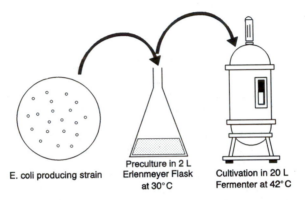

E. coli producing strain   Preculture in 2 L Erlenmeyer Flask at 30°C   Cultivation in 20 L Fermenter at 42°C

Figure 1. Schematic presentation of the fermenter inoculation procedure.

2.2 Batch fermentation

15 litres of culture medium containing glucose, a complex nitrogen source, and mineral salts are inoculated with the preculture in a 20 litre fermenter (LSL Biolafitte). pH is maintained at 7.0 with 4 M H_2SO_4 and 4 M NaOH throughout the fermentation period using a pH electrode piloting two delivery pumps. Dissolved oxygen (pO_2) is controlled to 20% saturation by varying the agitation speed. Above 600 r.p.m., pure O_2 is added to avoid foaming and cell shearing due to rapid agitation.

When the cell concentration reaches 4 g/litre cell dry weight, the temperature of the medium is shifted from 30°C to 42°C in order to induce gene expression. At a concentration of 15 g/litre cell dry weight, the culture is chilled to 4°C before harvesting to avoid proteolytic degradation. During the course of fermentation, samples are collected to measure the following parameters:

- optical density at 600 nm (OD_{600}), cell dry weight (DW), and AAT production (*Figure 2*)
- concentration of viable cells before and after induction (*Figure 3*)

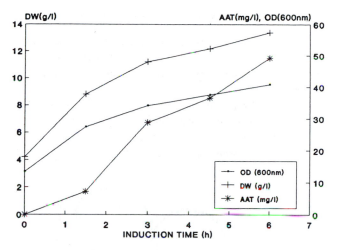

Figure 2. Accumulation of soluble AAT Ala357, Arg358 during the fermentation process after induction at 42°C. Samples corresponding to a 1 ml cell suspension at $OD_{600} = 100$ are taken at 1.5 h intervals. After centrifugation, the cell pellet is resuspended in 1 ml extraction buffer (Section 2.4) and sonicated three times for 30 sec each at 4°C. The crude extract is centrifuged in a Sigma centrifuge (2MK) and AAT Ala357, Arg358 in the supernatant is measured by radial immunodiffusion (RID) against a polyclonal antiserum raised against human plasma-derived AAT (Behring). Dry weight measurements are obtained by taking 5 ml samples of culture in pre-weighed haemolysis tubes. After 10 min centrifugation at 3000 *g* (SS34 rotor, Sorvall, RC5B), the pellet is washed with 5 ml of water and dried for 24 h at 105°C. Tubes are weighed again and the difference permits calculation of cell dry weight (DW). Optical density is measured at 600 nm (OD).

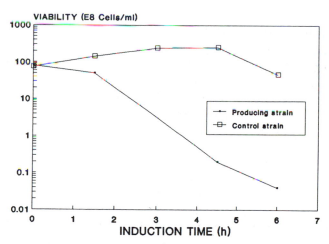

Figure 3. Determination of the viable cell concentration. Samples of the cell suspension are taken from the culture and diluted in 0.9% (w/v) NaCl. Serial dilutions are plated on non-selective medium and incubated at 30°C for 2 days. Colonies are then counted to calculate the concentration of viable cells. The control strain harbours a plasmid from which the expression cassette is removed.

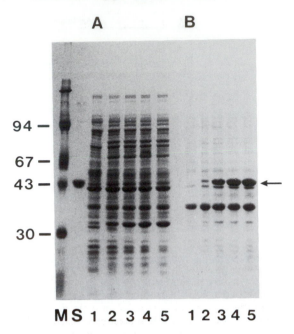

Figure 4. Expression of AAT Ala357, Arg358 after induction at 42°C. Samples corresponding to a 1 ml cell suspension at $OD_{600} = 100$ are taken at 1.5 h intervals and treated as
described in *Figure 2*. The crude extract is then centrifuged in a Sigma centrifuge (2MK)
and the supernatant corresponding to the soluble protein fraction is mixed with sample
buffer (0.6 ml 1 M Tris, pH 6.8; 1 ml glycerol; 2 ml 10% SDS; and 0.5 g DTT). The pellet,
corresponding to the insoluble fraction, is solubilized in 2 ml sample buffer. 2–5 μl of
each preparation are loaded on a 12.5% polyacrylamide gel and electrophoresed at
25 mA for 3–4 h. Lane M, molecular weight markers (kDa); lane S, standard (purified AAT
Ala357, Arg358). (A) Soluble fraction: lane 1, before induction; lane 2, 1.5 h after induction; lane 3, 3.0 h after induction; lane 4, 4.5 h after induction; lane 5, 6 h after induction.
(B) Insoluble fraction: lane 1, before induction; lane 2, 1.5 h after induction; lane 3, 3.0 h
after induction; lane 4, 4.5 h after induction; lane 5, 6 h after induction. AAT is indicated by
the arrow.

- accumulation of AAT in both the soluble fraction and in inclusion bodies
  (*Figure 4*)
- plasmid copy-number before and after induction (*Figure 5*)

From *Figure 2* it can be concluded that AAT accumulates in the fermenter in
proportion to the biomass measured either in OD or DW. *Figure 3* shows that
the concentration of viable cells decreases drastically after induction at 42°C.
This might be due to the overexpression of AAT which accumulates mainly in
inclusion bodies (*Figure 4*). Finally, plasmid copy-number also increases
significantly after induction (*Figure 5*).

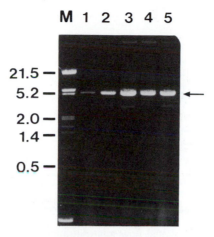

Figure 5. Analysis of plasmid copy-number. Samples corresponding to a 1 ml cell suspension at $OD_{600} = 6$ are taken and centrifuged in an Eppendorf tube (Sigma centrifuge 2MK). Plasmid DNA is extracted by the alkaline lysis method described by Maniatis *et al.* (5). The DNA solution is resuspended in 50 μl TE buffer (10 mM Tris–HCl, pH 7.5; 1 mM EDTA) and 5 μl are digested with the appropriate restriction enzyme to linearize the plasmid. The digest is loaded on an agarose gel (0.8%) and electrophoresed at 100 mA for 1 h. Lanes: M, size marker (kb); 1, before induction; 2, 1.5 h after induction; 3, 3.0 h after induction; 4, 4.5 h after induction; 5, 6 h after induction. Linearized plasmid is indicated by the arrow.

2.3 Cell harvesting

In this example the product of interest is intracellular, thus necessitating harvest of the cells followed by cell disruption. Cells are harvested by centrifugation in a Sorvall H-6000A rotor (Du Pont) at 7000 *g* for 20 min and stored at −20°C. The capacity for a single run is 6 litres.

To harvest cells from larger volumes, tangential flow microfiltration has been employed (see Section 3.1 for a description of the methodology). This technique allows for subsequent scale-up to larger fermentation volumes. In our case, tangential flow microfiltration was performed using a Pellicon filtration system (*Figure 6*).

2.4 Cell disruption

Techniques to release the intracellular content from micro-organisms can be divided into two classes: mechanical (wet milling, high-pressure homogenizer, sonication) and non-mechanical (chemical treatment, enzymatic digestion, osmotic shock) methods. Mechanical disruption is preferred at the pilot scale. High-pressure homogenizers and bead mills are commonly employed for cell disintegration (4). In the present example, pellets of cells are resuspended in extraction buffer (20 mM sodium phosphate, pH 6.8; 300 mM NaCl; 10 mM EDTA) by means of an Ultra Turax T25 (IKA). The volume of the suspension

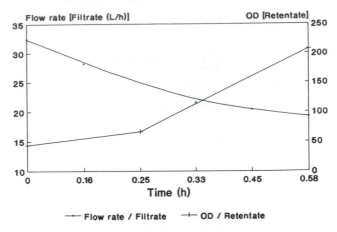

Figure 6. Cell harvest by tangential flow microfiltration on a membrane with 0.45 µM (HVLOP, Millipore) pore diameter. The flow rate of the filtrate decreases throughout the filtration period due to an increase in the viscosity of the retentate. This is reflected by an increasing optical density at 600 nm (OD) caused by an increase in cell concentration.

before breakage corresponds to approximately 10% of the initial fermentation volume. During the process, cells are kept below 5°C to protect the product from proteolytic degradation. Cells are disrupted at 600 bars by two passages through a high-pressure homogenizer (Model 15M 8TA, Gaulin Corp.). The efficiency of disruption can be evaluated by measuring the amount of released protein or by microscopic analysis. This procedure results in approximately 90% cell disruption.

3. Purification of α1-antitrypsin Ala357, Arg358 (AAT Ala357, Arg358)

3.1 Clarification of the crude cell extract by tangential flow microfiltration

Removal of cellular debris from the supernatant after cell breakage is a difficult unit operation in a process for protein isolation. This is due mainly to the heterogeneity in size and density of the insoluble particles that are produced by the generally employed high-pressure homogenizers or bead mills. These particles are of similar dimension to the pores of commonly used microfiltration membranes and show only small density differences as compared to the supernatant. These features may lead to plugging of microfiltration membranes and necessitate high gravitational forces with prolonged centrifugation times to obtain satisfactory separations.

In order to facilitate these operations, especially on a large scale, so called 'processing aids' can be employed, for example submicron latex beads or

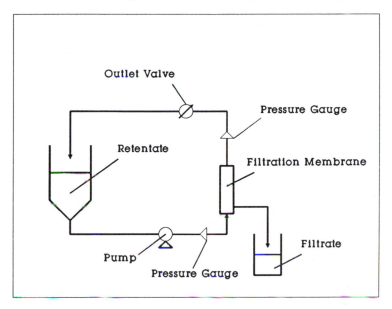

Figure 7. Flow scheme of a tangential flow filtration unit.

soluble macromolecules such as polyethyleneglycols (PEG). In the case of isolating AAT, PEG-4000 (average molecular weight 4000 Da) at a concentration of 5% (w/v) is used to facilitate removal of cellular debris by tangential flow microfiltration or classical centrifugation. A general flow scheme of a typical filtration unit is given in *Figure 7*.

Protocol 4. Tangential flow microfiltration

Special equipment
- Pellicon filtration unit equipped with rotary vane pump (Millipore)
- membrane: 0.2 μm pore diameter, 0.46 m^2 surface area, polyvinyl-difluoridene (PVDF) (GVLP, Millipore)

Method
1. Equilibrate the membrane by passing 5 litres of cell breakage buffer (0.3 M NaCl; 20 mM sodium phosphate, pH 6.8; 10 mM EDTA) through the system.
2. Take the homogenate obtained after cell disruption (3 litres) (Section 2.4) and start the filtration. Do not operate the system above the pressure limit given by the manufacturer, as rupture of the membrane may occur.
3. Concentrate the homogenate to 1.5 litres and add 2.3 litres of extraction buffer.

375

Protocol 4. *Continued*

4. Repeat this operation three times and concentrate to 1.5 litres at the end. Avoid excessive concentration of the retentate, as this may lead to plugging of the fluid channels in the filtration system accompanied by a steep rise in inlet pressure.

The end of the filtration can be estimated from the theoretical curve shown in *Figure 8*, assuming that the protein of interest passes freely through the membrane. In the case of AAT Ala357, Arg358, 3 × 2.3 litres of buffer proved to be sufficient for quantitative recovery of the product in the filtrate, in accordance with the theoretical curve.

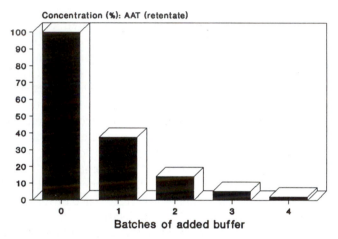

Figure 8. The concentration of a protein in the retentate relative to the initial concentration during tangential flow microfiltration, assuming unrestricted passage through the membrane. The values are calculated based on a starting volume of 1.5 litres and subsequent additions of 2.3 litres of buffer, respectively.

Deviation of the protein concentration from the theoretical curve indicates restricted passage through the membrane, which may arise as follows:

(a) The pores of the membrane may be blocked with cellular debris or adsorbed proteins (fouling). This problem may be overcome when working at a reduced transmembrane pressure, by completely opening the outlet valve (see *Figure 7*) and by reducing the flow rate of the filtrate (clamping of the tube unless a valve is available). Some filtration systems designed for industrial use allow 'backflushing' of the membrane. That is, reversing the direction of the flow across the membrane, which clears the pores of the membrane of particles. However, not all membranes tolerate such a treatment (consult the manufacturer). Inclusion in the feed stream of hydrophilic polymers such as PEG, which have the ability to coat the membrane surface, may also help to reduce fouling. If these measures prove to be unsuccessful, a membrane of different chemical composition and pore structure, or a different method of cell breakage, should be investigated (4). Wet bead milling generally produces larger particles which might be easier to remove.

(b) Soluble components in the homogenate form a secondary filtration layer (gel polarization) at the surface of the membrane. The formation of a secondary filtration layer is very much dependent on the concentration gradient across the membrane. It is thus advantageous to work with a more dilute homogenate. Reduction of the transmembrane pressure, as discussed above, is also advisable.

Cleaning and sterilization of the filtration system after use are crucial for the lifetime of the membranes. Depending on their chemical stability, different agents, such as 50 mM NaOH or basic detergents, may be employed to remove adsorbed material from the membrane surface and thus to re-establish the original flow characteristics (consult the individual manufacturers for detailed information). It is important to verify after each use of the filtration system that the flow rate of the filtrate with water under given conditions of pump speed and transmembrane pressure is identical to the flow that was observed before starting the operation. Differences in flow indicate that the cleaning procedure was not effective (lower flow than before) or that the membrane was modified by too harsh a treatment.

Sterilization procedures of filtration systems depend very much on the chemical nature of the membrane and on the resistance of the materials that are employed by the manufacturer of the hardware. Smaller systems, such as the one described here, have to be sterilized by chemical agents like sodium hypochlorite or formaldehyde. Industrial systems may also be steam-sterilized in conjunction with the production fermenter. The individual manufacturers should be consulted for details.

Tangential flow microfiltration is a technique that can be scaled-up to industrial production levels. An alternative method, which is frequently used at the pilot and production scale, is continuous centrifugation. This technique will not be dealt with here, but an introduction to the methodology can be found in the cited review (6).

3.2 Desalting of the filtrate by tangential flow ultrafiltration

In many cases cell breakage and removal of cellular debris have to be performed in a buffer of elevated ionic strength to avoid significant losses of product due to ionic interactions with the cellular debris. In the case of AAT Ala357, Arg358 0.3 M NaCl is employed. In order to perform the subsequent anion-exchange chromatographic purification step, it is necessary to reduce the ionic strength (measured as electrical conductivity) to a well-defined value. In addition, it is of advantage to remove other low-molecular weight contaminants, such as pigments, in order to increase the binding capacity of the chromatographic matrix.

Tangential flow ultrafiltration is an efficient method by which to achieve these goals, since large volumes can be treated on both the pilot as well as the production scale. In the case of AAT Ala357, Arg358 (44 kDa) tangential flow ultrafiltration on a membrane with an exclusion limit of 10 kDa allows the recovery of the product in the retentate with a concomitant buffer exchange. The general flow scheme shown in *Figure 7* applies also to tangential flow ultrafiltration, although the specific filtration unit may be different.

Protocol 5. Tangential flow ultrafiltration

Special equipment
- ultrafiltration unit DC10LA (Amicon)
- filtration membrane YM10 (regenerated cellulose)/S10 cartridge (Amicon), 0.9 m^2 surface area

Method
1. The filtration membrane is stored in 0.1% (v/v) aqueous formaldehyde to prevent microbial growth. Wash out the storage solution with 2 × 5 litres of water.
2. Equilibrate the membrane with 5 litres of 0.3 M NaCl; 20 mM sodium phosphate, pH 6.8; 10 mM EDTA. Subsequently start the filtration (9.8 litres starting volume).
3. Reduce the volume of the retentate to 4 litres and add 6 litres of 20 mM sodium phosphate, pH 6.8 (conductivity 2–2.5 mS). When the conductivity in the retentate reaches 2.5 ± 0.2 mS (normally 4 × 6 litre additions of buffer) diafiltration is terminated.
4. The retentate can now be loaded on the anion-exchange chromatographic matrix (*Protocol 6*). Clean the filtration unit by rinsing with 3 × 5 litres of water followed by 5 litres of 50 mM NaOH and sufficient water to neutralize the pH. Store the cartridge in 0.1% (v/v) aqueous formaldehyde. Other cleaning solutions can be employed and, alternatively, sodium hypochlorite can be used for sterilization (for additional details consult the manufacturer).

Reduced product recovery after ultrafiltration may be due to the following:

(a) A significant amount of protein either precipitates or remains adsorbed to the filtration membrane. This problem can be addressed by filtering at different pH values (do not filter at a pH value which is close to the isoelectric point of your protein) and salt concentrations, as long as these conditions are compatible with the subsequent chromatographic purification step. Adding a hydrophilic polymer such as PEG to the feed stream may render the membrane more hydrophilic and thus less adsorptive. If these measures do not help, a membrane of different chemical composition should be tried. Screening of membranes from different manufacturers can often be done on a small scale using laboratory equipment. It would be necessary to verify that the membrane of choice can finally be obtained for large-scale use.

(b) A significant amount of product permeates the membrane and is thus found in the filtrate. A membrane with a lower exclusion limit should be tested. If leakage of the product occurs after repeated successful use of a membrane, it is likely that the pore structure has been modified by too harsh a treatment (cleaning procedure) or that mechanical rupture of the membrane has occurred. Storage of membranes that were not correctly cleaned and sterilized may also lead to irreversible deterioration. In all cases replacement of the membrane will be necessary.

3.3 Chromatographic purification

3.3.1 General considerations

Chromatographic purification of proteins depends very much on the properties of the molecule in question and on the characteristics of the contaminants that have to be removed. It is important to note that isolation of a given product from different recombinant organisms represents a specific problem that has to be treated case by case.

The isolation procedure of AAT Ala357, Arg358 from the cytoplasmic fraction of *E. coli* can thus only be regarded as an example. A more complete treatment of the use of chromatographic methods for protein purification can be found in ref. 7.

In order to find the most suitable isolation process for a protein, a number of different chromatographic stationary phases should be screened on a small scale. Stationary phases used at different stages of a purification process have to fulfil different requirements, some of which are outlined below. Development of a process thus necessitates definition of the order in which the chromatographic steps are to be employed, since this has a major influence on the final result. It is advantageous, especially when further scale-up of the isolation procedure is envisaged, to arrange the purification in such a way that handling of pools between the individual steps is minimized.

In practical terms one can classify the individual chromatographic purification steps with respect to their position in the final process:

(a) Initial step: removal of proteinaceous and non-proteinaceous bulk contaminants.

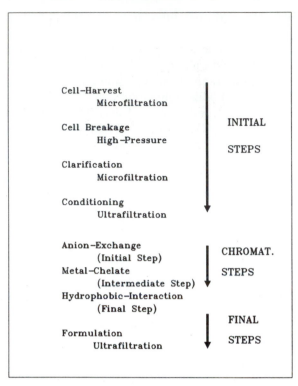

Figure 9. Purification scheme for the isolation of AAT Ala357, Arg358 from the cytoplasmic fraction of *E. coli*, as detailed in the text.

(b) Intermediate steps: purification of the product to near homogeneity.

(c) Final step: removal of impurities that are closely related to the product.

The isolation procedure developed for AAT Ala357, Arg358 from *E. coli* combines these steps, as shown in *Figure 9*.

3.3.2 Isolation of AAT Ala357, Arg358
The purification process for AAT Ala357, Arg358 was designed to meet the following requirements in order to allow a rapid and efficient isolation with the possibility for further scale-up.

(a) no desalting between individual chromatographic purification steps

(b) high product recovery after each step (>80%)

(c) efficient cleaning and sanitation of the stationary phases to allow their repeated use

(d) use of a minimal number of different stationary phases

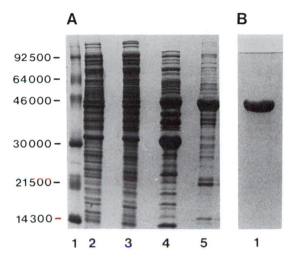

Figure 10. Polyacrylamide gel electrophoresis under denaturing and reducing conditions. (A) Lane 1, molecular weight markers in Da; lane 2, starting material after cell disruption and removal of cellular debris; lane 3, retentate after tangential flow ultrafiltration; lane 4, pool after anion-exchange chromatography; lane 5, pool after metal–chelate chromatography. (B) Final material after formulation, lyophilization, and reconstitution.

(e) use of different types of interactions for the separations (anion-exchange chromatography, electrostatic interactions; metal–chelate chromatography, co-ordinative interactions; hydrophobic-interaction chromatography, salt-induced hydrophobic interactions)

The purity of AAT Ala357, Arg358 after the individual chromatographic steps and in the final formulation, as analysed by polyacrylamide gel electrophoresis (PAGE) under denaturing and reducing conditions, is shown in *Figure 10*, while *Table 1* summarizes the results of a typical purification.

Protocol 6. Chromatographic purification of AAT Ala357, Arg358

Special equipment

- peristaltic pumps type 502SR equipped with a 501R pump head (Watson-Marlow)
- UV absorbance detectors type UVICORD™ (Pharmacia-LKB Biotechnology)
- Chromatography columns type Moduline™ (Amicon)

A. *Anion-exchange chromatography*

1. Equilibrate the anion-exchanger (PAE-300, 50 μm particle diameter, 300 Å pore diameter, 2 litre column volume, column diameter 10 cm) with

Protocol 6. *Continued*

20 mM sodium phosphate, pH 6.8 (conductivity 2–2.5 mS), until both the pH and conductivity of the effluent are identical with those of the equilibration buffer.

2. Load the retentate obtained after diafiltration (Section 3.2) at a flow rate of 50 ml/min.

3. Wash with 10 column volumes of equilibration buffer.

4. Elute AAT Ala357, Arg358 with a gradient from 0 to 0.25 M NaCl in 20 mM sodium phosphate, pH 6.8, at a flow rate of 50 ml/min and a gradient volume of 10 litres. AAT Ala357, Arg358 is recovered between 9 and 20 mS conductivity and can be specifically detected by radial immunodiffusion (RID).

B. *Metal–chelate chromatography*

1. Adjust the pool to 50 mM Tris–HCl, pH 8–8.5 (conductivity ~15 mS) by adding the appropriate volume of 2 M Tris–HCl, pH 8.5.

2. Load it on a chelating Sepharose column charged with Zn^{2+} ions that is equilibrated with 150 mM NaCl, 50 mM Tris–HCl, pH 8.5 (for details concerning the preparation of the stationary phase consult the manufacturer's description) (200 ml column volume, column diameter 4 cm; Pharmacia) at a flow rate of 7 ml/min.

3. Elute AAT Ala357, Arg358 with a gradient from 0 to 0.1 M glycine in the equilibration buffer at 7 ml/min (gradient volume 1.5 litres). AAT Ala357, Arg358 is pooled according to the results of RID and PAGE under denaturing and reducing conditions.

C. *Hydrophobic-interaction chromatography*

1. Adjust the obtained pool to 2 M ammonium sulphate by adding either the appropriate volume of a saturated solution or by slowly adding solid ammonium sulphate under gentle agitation.

2. Remove precipitated proteins by either batch centrifugation at 8000 *g* for 30 min or by placing a 0.2 μm filter at the column inlet.

3. Load the adjusted pool on a 200 ml column (column diameter 4 cm) packed with PAE-300 equilibrated in 2 M ammonium sulphate, 50 mM Tris–HCl, pH 8, and 10 mM EDTA at a flow rate of 14 ml/min.

4. Elute AAT Ala357, Arg358 with a gradient from 2 M to 1 M ammonium sulphate in the equilibration buffer at the same flow rate. Collect AAT Ala357, Arg358 according to the chromatographic profile obtained at 280 nm.

Table 1. Purification chart for the isolation of AAT Ala357, Arg358 from *Escherichia coli*

Purification step	AAT (mg)[a]	Protein (g)[b]	Purification factor	Overall yield (%)	Step yield (%)
After clarification	619	68.5	1.0	100	–
After conditioning	575	62.8	1.0	92	92
After anion-exchange	512	13.9	4.0	82	89
After metal–chelate	435	0.86	55.2	70	85
Final material[c]	288	0.288[a]	109.2	46	66[c]

[a] Quantitated by RID using an antiserum raised against human plasma-derived AAT.
[b] Coomassie blue R-250 binding assay using bovine serum albumin as a standard.
[c] After three further processing steps: hydrophobic-interaction chromatography, formulation by ultrafiltration, and lyophilization.

Protocol 7. Formulation of AAT Ala357, Arg358

1. Desalt and concentrate the isolated AAT Ala357, Arg358 by diafiltration in a type 2000A filtration unit (volume 2 litres; Amicon) using a YM10 membrane with an exclusion limit of 10 kDa (Amicon) into the final formulation buffer of: 150 mM NaCl; 10 mM Tris–HCl, pH 7.5; 0.1% (w/v) polyethyleneglycol (average molecular weight 6000 Da); and 5% (w/v) sucrose to 5 mg/ml AAT Ala357, Arg358.

2. Sterile filter this preparation at 0.2 μm.

3. Lyophilize the filtrate.

The final preparation is stable and can be easily resolubilized upon addition of water, thus allowing the study of the *in vitro* and *in vivo* properties of the protein.

4. Summary

Protein engineering is a promising approach to the development of molecules with novel properties for pharmaceutical applications. At an initial stage of development it is important to obtain sufficient protein to test its efficacy in the envisaged application. These quantities can, in most cases, be provided by laboratory-scale processes. However, industrial development of these molecules necessitates that the production process be performed economically at a much larger scale. This is pivotal, if the protein is to become a viable product in a highly competitive market.

The procedures outlined in this chapter are given with the intention to provide the reader with a view that goes beyond laboratory-scale production.

Techniques that can be scaled-up successfully are described, and priority is given to considerations that are of major importance with regard to developing an industrial production process.

Acknowledgement

The authors would like to thank Jean-Pierre Lecocq for his encouragement and continued interest.

References

1. Remaut, E., Stanssens, P., and Fiers, W. (1981). *Gene,* **15,** 81.
2. Nierman, W. C. and Feldblyum, T. (1985). *Developments in Industrial Microbiology,* **26,** 423.
3. Gherna, R. L. (ed.) (1981). *Manual of methods for general bacteriology,* pp. 208–17. American Society for Microbiology, Washington DC.
4. Kula, M. R. and Schütte, H. (1987). *Biotechnology Progress,* **3,** 31.
5. Maniatis, T., Fritsch, E. F., and Sambrook, J. (ed.) (1982). *Molecular cloning: a laboratory manual*, pp. 368–9. Cold Spring Harbor Press, Cold Spring Harbor, New York
6. Axelsson, H. A. C. (1985). In *Comprehensive biotechnology,* (ed. C. L. Cooney and A. E. Humphrey), Vol. 2, pp. 325–46. Pergamon Press, Oxford.
7. Janson, J. C. and Rydén, L. (ed.) (1989). *Protein purification*. VCH Publishers, New York.

Appendix:
Suppliers of specialist items

Items referred to in the text can be purchased from the following companies. In general, the address of the head office is listed, from which the names and addresses of more local suppliers can be obtained.

Amicon Corp., 17 Cherry Hill Drive, Danvers, MA 01923, USA.

Anglian Biotech. Ltd, Whitehall House, Whitehall Road, Colchester, Essex, UK.

Astrophysics Research Ltd, Vale Road, Windsor, Berkshire SL4 5JP, UK.

Bachem Bioscience Inc., 3700 Market Street, Philadelphia, PA 19104, USA.

Bayer AG, 5090 Leverkusen-Bayerwerk, Germany.

Behringwerke AG., Marburg, FRG.

Bio101 Inc., PO Box 2284, La Jolla, CA 92038-2284, USA.

Bio-Rad Laboratories Ltd, 1414 Harbor Way South, Richmond, CA 94804, USA.

Brighton Systems Ltd, 132 Sherbourne Road, Hove BN3 8BG, UK.

Cambridge Repetition Engineers Ltd, Green's Road, Cambridge CB4 3EQ, UK.

Costar, 205 Broadway, Cambridge, MA, USA.

Crystol Microsystems, K. Harlos, 47 Purcell Road, Oxford OX3 0HB, UK.

Difco Laboratories Ltd, PO Box 14B, Central Avenue, East Molesey, Surrey KT8 0SE, UK.

Douglas Instruments Ltd, 255 Thames House, 140 Battersea Park Road, London SW11 4NB, UK.

Du Pont, Wilmington, DE, USA.

E.I. du Pont de Nemours & Co., Wilmington, DE 19898, USA.

Electrolab Ltd, Oak Lane, Bredon, Tewkesbury GL20 7LR, UK.

Flow Laboratories Inc., McLean, Virginia 22102, USA.

Gaulin Corp., Everett, MA, USA.

Hampton Research, 5225 Canyon Crest Drive, Suite 71-336, Riverside, CA 92507, USA.

IKA Labortechnik, Janke, Staufen, Germany.

Janke & Kunkel GmbH & Co. KG, IKA-Labortechnik, D-7813 Staufen i. Br., FRG.

LSL Biolafitte, 78100 Saint-Germain en Laye, France.

E. Merck, Frankfurter Strasse 250, D-6100 Darmstadt 1, FRG.

Millipore Corporation, 80 Ashby Road, Bedford, MA 01730, USA.

New Brunswick Scientific, Edison, New Jersey, USA.

Appendix

New England Biolabs Inc., 32 Tozer Road, Beverley, MA 01915-5599, USA.

Novo Biolabs, 33 Turner Road, Danbury, CT 06810, USA.

Perkin–Elmer Cetus, 761 Main Avenue, Norwalk, Connecticut 06859, USA.

Pharmacia-LKB Biotechnology AB, Björkgatan 30, S-751 82 Uppsala, Sweden.

Pierce Chemical Co., PO Box 117, Rockford, IL 6115, USA.

Promega Corp., 2800 Woods Hollow Road, Madison, Wisconsin, USA.

Sigma Chemical Company, PO Box 14509, St. Louis, MO 63178, USA.

Wako Chemicals Inc., 12300 Ford Road, Suite 130, Dallas, Texas 75234, USA.

Watson & Marlow Ltd, Falmouth Cornwall TR11 4RU, UK.

386

Glossary of X-ray crystallographic terms used in the text

Absorption: attenuation of X-rays by a combination of photoelectric absorption, incoherent and coherent scattering. Proportional to λ^3.

Accuracy: the lack of error in a measurement.

Amplitude: the magnitude $F(hkl)$ of the structure factor for the reflection (hkl), directly measurable from the observed intensity.

Anisotropic: dependent on direction.

Anomalous scattering: a resonance effect arising when the frequency of the X-rays is close to an absorption edge for the bound electrons in the atom. Causes a **phase** change that breaks some of the symmetry of the diffraction pattern (see **Bijvoet difference**).

Asymmetric unit: minimum unit operated on by the crystallographic symmetry to build up a unit cell and thence a crystal.

Averaging: impose agreement between multiple copies of a protein/subunit in the asymmetric unit.

Beam divergence: the deviation from parallel paths for the X-rays comprising the incident beam.

B (temperature) factor: a factor that models the reduction in the observed scattering factor arising from thermal motion of the atoms and static disorder in the crystal. If atoms are assumed to move isotropically with simple harmonic motion, then B is related to the total mean square displacement of an atom ($<u_t^2>$) by $<u_t^2> = 3B/8\pi^2$.

Bijvoet difference: a difference in magnitude of structure factors, such as $F(hkl)$ and $F(-h-k-l)$, which would normally be identical (see **Freidel's law**), arising from anomalous scattering.

Birefringence: arises from the polarization of light by the anisotropic distribution of enantiomorphic matter in crystals (thus not seen for cubic space groups since they are isotropic); detected by use of crossed polarizers.

Collimation: limitation of diameter of X-ray beam.

Constraint: absolute requirement (cf. **restraint**).

Correlation coefficient: measure of agreement between two sets of data (a fraction, usually 1 = exact agreement, −1 = exact disagreement).

Cross-rotation function: product of Patterson functions (integrated over a certain volume) for structures A and B sampled over a range of orientations of one of the molecules; given structurally similar molecules this will peak when they are in identical orientations.

Crystallographic axes (a, b, and c): the directions of the sides of the unit cell.

Crystallographic symmetry: a set of globally applicable operations which define the crystal.

Derivative: the native molecule with heavy atom(s) bound.

Difference Fourier maps: Fourier syntheses based on the difference between the native and, for example, native plus inhibitor amplitudes calculated using native phases. The difference (e.g. inhibitor) electron density appears at half height.

Difference Patterson: Patterson function based on the squares of the differences in amplitudes between two sets of data. Usually calculated using native and heavy-atom derivative data to yield vectors between heavy-atom sites.

Diffractometry: both crystal and detector are varied in position to individually collect each reflection.

Enantiomorphs: mirror images.

Energy minimization: reduction of the steric strain/poor stereochemistry in a proposed protein structure.

Figure of merit: this reflects the unimodality and sharpness of the phase probability distribution, i.e. how sure you are the phase is right on a scale 0.0–1.0 (the cosine of the likely error in the phase angle).

Flash freezing: very rapid drop to liquid nitrogen temperature to avoid phase transition to ice of water in the crystal.

Fourier synthesis: mathematical operation to move from reciprocal space to real space.

Franks-type optics: two perpendicular bent mirrors providing glancing-angle reflection in horizontal and vertical planes to focus the X-ray beam.

Friedel's law: reflections hkl and $-h-k-l$ have equal amplitudes (only true in the absence of anomalous scattering).

G function: an interference function related to the Fourier transform of the molecular shape.

Harker sections: planes in the Patterson map in which vectors between atoms related by certain crystallographic symmetry operators must appear.

Index (hkl): a unique label of the reflection related to the lattice plane from which the reflection arises.

Integration radius for the rotation function: the radius about the origin of the Patterson map over which the function is calculated.

Intensity: a measurable quantity strictly related to the amplitude $F(hkl)$ by $\sqrt{I} \propto |F|$, where the proportionality is determined by a scale factor, absorption, and certain geometric effects (Lorentz and polarization factors).

Isomorphous: no substantial changes (to the crystal structure).

Isotropic: the same in all directions.

Laue method: the crystal (usually stationary) is irradiated with a broad spectrum of X-ray wavelengths (e.g. 0.5Å–2.5Å).

Local scaling: scale reflections in small blocks of reciprocal space.

Lorentz factor: geometrical factor related to the relative time each reflection spends in the diffracting position during data collection.

Molecular dynamics: simulated motion of the protein atoms under Newton's laws of motion.

Monochromator: means of filtering X-rays to produce a beam of single wavelength, e.g. by diffraction from a graphite crystal.

Mosaicity: degree of deviation from perfect order throughout the entire crystal.

Native data: the data which, when phased, provides the standard electron density map of the protein of interest.

Non-crystallographic symmetry: local symmetry which is only valid for a limited portion of the unit cell.

Normal beam geometry: the crystal rotates about an axis perpendicular to the X-ray beam.

Omit map: map for which part of the proposed protein structure has been deliberately omitted from the phase calculation.

Origin removal: by subtracting $<F^2>$ (i.e. the mean squared structure factor) from each term contributing to a Patterson function the value at the origin of the function becomes zero.

Orthogonalization: convertion to an orthogonal coordinate system from a system based on the crystallographic axes.

Oscillation (rotation) method: the crystal is rotated about a defined direction in the X-ray beam, the detector is held stationary.

Partials: in the oscillation method the finite rocking width of a reflection means that reflections occurring close to the start and end of the oscillation range will be incompletely recorded.

Patterson function: a Fourier synthesis using as amplitudes the observed intensities (with phases of zero). Produces a map of all vectors between atoms in the unit cell.

Phase: the component $\exp[i\alpha(hkl)]$ of the structure factor for reflection (hkl), it cannot be measured directly, hence the phase problem.

Phasing power: the r.m.s. amplitude of the heavy-atom scattering divided by the r.m.s. lack of closure for the vector triangle comprising the derivative, heavy-atom, and native structure factors.

Point group: the space group without the translational symmetry elements.

Polarization factor: the incident X-ray beam may itself be polarized and the diffracted beam will be further polarized such that reflection efficiency varies with the direction of the diffracted beam.

Post-refinement: refinement of various parameters (e.g. missetting angles, cell dimensions) by the analysis of degree of partiality observed for reflections in the oscillation method.

Precession camera: a fiendishly clever device which allows an undistorted image of a plane of the reciprocal lattice to be recorded on photographic film.

Real space: where the electron density is.

Reciprocal space: where the diffraction pattern is.

Resolution: the minimum interplanar spacing in the crystal for which reflections are measured; theoretically the resultant electron density map will distinguish between features separated by more than $0.715 \times$ resolution. *Beware*: in practice this depends on the quality and completeness of the amplitudes and phase angles available. Hence the term 'nominal resolution', translated as 'I'm sure there was a reflection at 2.5Å resolution'.

Restraint: a requirement of variable rigour (may be tight or loose, cf. **constraint**).

R **factor:** the mean fractional disagreement between two sets of observations.

Rigid body refinement: variation of position and orientation of the model protein structure as a whole to minimize disagreement between observed and calculated structure factors.

Rocking width: angle through which the crystal must be rotated to collect the complete reflection, dependent on mosaicity of crystal, beam divergence, and dispersion (spread) in beam wavelength.

Rotating anode: X-rays are emitted at a characteristic wavelength from metal atoms (typically copper) excited by bombardment with high-energy electrons; rotation of the cylindrical metal anode spreads the heat load.

Screw axis: direction in the crystal about which the repeating units are related by rotations and translations to form a helix.

Sealed tube: as in rotating anode but without the rotation, hence less bright (to avoid melting the anode).

Self-rotation function: sum (within a defined volume) of the product of a Patterson function with a rotated copy of itself. This is calculated for a range of rotations; given several copies of a molecule/subunit in the asymmetric unit this function will peak for orientations related by the non-crystallographic symmetry.

Sharpening: application of a negative *B* factor to counter artificially the observed fall off in mean stucture factor amplitude with resolution.

Space group: one of the allowed sets of crystallographic symmetry operations, always refer to the crystallographic tables (see Chapter 1, Section 9).

Standard deviation: a mystical parameter which some philosophers hold to be related to the accuracy (*vide supra*) of an observation.

Structure factors: the vector components of the Fourier transform of the crystal structure, i.e. the diffraction.

Synchrotron: a source of intense X-rays emitted by high-speed electrons when they are accelerated (bent) by magnetic fields. By forming a closed path the electrons may be recycled for many hours.

Systematic absences: reflections of zero intensity for given values of *hkl* arising from space-group-dependent interference effects.

Translation search: a search model of given orientation is moved around the unit cell and the *R*-factor and/or correlation coefficient is used to assess if there really is a molecule there.

Unit cell: usually the smallest unit from which a crystal may be built up by simple unit translations.

Wang solvent flattening: use of the information that the solvent regions in an electron density map must be uniform to improve the original phases.

Wavelength (λ): characteristic parameter of electromagnetic radiation such as light or X-rays.

Weights: various weighting schemes (e.g. unit, Sim, Rayment) are applied on combining data, during refinement, etc. They are important but the source of much dispute and heartache.

Weissenberg camera: the crystal is rotated in the beam as in the oscillation method but the film or imaging plate is also translated in a direction perpendicular to the incident beam, allowing a greater oscillation range per film before reflections overlap.

Glossary

Index

Index

Index